FUNDAMENTALS OF Helicopter Dynamics

FUNDAMENTALS OF Helicopter Dynamics

C. VENKATESAN

CRC Press
Taylor & Francis Group
Boca Raton London New York

CRC Press is an imprint of the
Taylor & Francis Group, an **informa** business

CRC Press
Taylor & Francis Group
6000 Broken Sound Parkway NW, Suite 300
Boca Raton, FL 33487-2742

First issued in paperback 2017

Version Date: 20150709

ISBN 13: 978-1-138-07438-5 (pbk)
ISBN 13: 978-1-4665-6634-7 (hbk)

Library of Congress Cataloging-in-Publication Data

Venkatesan, C.
Fundamentals of helicopter dynamics / C. Venkatesan.
pages cm
Summary: "This is an introductory book on helicopter dynamics. The aim of this book is to introduce the students/engineers to the basic principles of helicopter dynamics. The book focuses on three major topics: (i) rotor blade idealization and blade dynamics in flap, lag and torsion modes; (ii) rotor blade aeroelastic stability {coupled flap-lag and coupled flap-torsion}; (iii) coupled rotor-fuselage dynamics. It covers hover, forward flight and other manoeuvre flights. Starting from basic physics of rotating systems, the equations of motion have been derived in vector form. "-- Provided by publisher.
Includes bibliographical references and index.
ISBN 978-1-4665-6634-7 (hardback)
1. Helicopters--Aerodynamics. 2. Rotors--Dynamics. I. Title.

TL716.V38 2014
629.132'3--dc23 2014009742

Visit the Taylor & Francis Web site at
http://www.taylorandfrancis.com

and the CRC Press Web site at
http://www.crcpress.com

Dedicated to THAT which is the source of everything

Contents

Preface

Due to their unique ability to hover and their ability to land and take off from any terrain, including rooftops, helicopters have established themselves as unique flying vehicles. Their utility has been growing steadily in both civil and military applications. In India, the requirements of high altitude operation and access to remote areas of the country necessitate the use of helicopters as a means of transportation. Helicopters are complex dynamic systems whose design and development require a high level of expertise and technology. Until the early 1990s, in India, very little academic and research activity was focused on helicopter studies. To create trained manpower in the field of helicopter design and development, a short course was organized by the Department of Aerospace Engineering, Indian Institute of Technology (IIT) Kanpur, in 1997 to highlight several important aspects of helicopter technology. The feedback from the participants indicated that the course provided a good insight on helicopter theory. We were very glad to note that this course has created an awareness and interest in helicopters among various organizations. In addition, this course has led to the formation of a strong bond between the academic institution (IIT Kanpur) and the industry (Hindustan Aeronautics Limited [HAL], Bangalore). This book is a culmination of the efforts put in to the preparation of lecture notes for a regular graduate-level introductory course as well as for a series of lectures given to designers, engineers, operators, users, and researchers on the fundamentals of helicopter dynamics and aerodynamics.

The book is written at a basic level to provide a fundamental understanding and an overview of helicopter dynamics and aerodynamics. All the equations are derived from the first principle, and the approximations are clearly explained. The book is divided into 11 chapters, presented in a systematic manner, starting with historical development, hovering and vertical flight, simplified rotor blade model in the flap mode, and forward flight. Two chapters are devoted to the aeroelastic response and stability analysis of an isolated rotor blade in uncoupled and coupled modes. Chapters 8 to 10 address the modeling of coupled rotor–fuselage dynamics and the associated flight dynamic stability. Chapter 11 is devoted to a simplified analysis of the ground resonance aeromechanical stability of a helicopter. Any student with a good knowledge and background in dynamics, vibration, aerodynamics, and undergraduate-level mathematics should be able to follow the contents of the book.

The illustrations given in the introduction chapter are sketches drawn using pictures available in the open literature and various sources from websites to avoid copyright infringment and to comply with Intellectual Property

Rights (IPR) regulations. If any reader is interested in the actual figures, he/she may refer to the literature or use website search options.

Writing this book has been an enlightening and a trying experience for me. Without the support of the following people, it would not have been possible for me to complete this book. My sincere thanks are due to my students Gagandeep Singh, Vadivazhagan, K.R. Prashanth, Puneet Singh, V. Laxman, Rohin Kumar, Sriram Palika, and to project staff Nikita Srivastava and Smita Mishra.

I would like to place on record my sincere thanks to my peers V.T. Nagaraj, P.P. Friedmann, I. Chopra, D. Hodges, D. Peters, G.H. Gaonkar, A.R. Manjunath, and R. Ormiston. As a researcher in the field of helicopter dynamics and aerodynamics, I have immensely benefitted from reading the excellent books written by W. Johnson, G. Padfield, A.R.S. Bramwell, W.Z. Stepniewski, G. Leishman, R.W. Prouty, A. Gessow, and G.C. Myers.

Finally I am indebted to my family and friends for their generosity in their support. The material presented in this book has been developed by the author and any resemblance to the material available in the open literature is not intentional; however, if anyone finds any similarity, it is requested that it may be brought to my attention so that suitable corrective action may be taken.

C. Venkatesan

Author

C. Venkatesan holds a master's degree in physics from the Indian Institute of Technology Madras and a PhD in engineering from the Indian Institute of Science Bangalore. After gaining experience in both industry and academic research, he joined the Department of Aerospace Engineering, Indian Institute of Technology Kanpur as a faculty member. He has initiated strong academic and research activity at IIT Kanpur in the field of helicopter aeroelasticity. With funding from the Department of Science and Technology, Venkatesan established an autonomous helicopter laboratory at IIT Kanpur. He has more than 100 research publications both in journals and conferences. He has received awards from the Aeronautical Society of India. He is a fellow of the Indian National Academy of Engineering and an associate fellow of the American Institute of Aeronautics and Astronautics. Presently he holds HAL Chair Professor position at IIT Kanpur.

List of Symbols

a:	Lift curve slope
$\vec{a}_{\mathrm{P}}$:	Absolute acceleration of any point P on the reference axis of the rotor blade
$\vec{a}_{\mathrm{H}}$:	Absolute acceleration at the hub center due to fuselage motion
A:	Rotor disk area
$[A]$:	System matrix
b:	Blade semi-chord
B:	Tip loss factor
C:	Blade chord
C_{d}, C_{d0}:	Sectional drag coefficient
C_{H}:	Longitudinal in-plane force coefficient at the hub
$\bar{C}_{\mathrm{l}}$:	Mean lift coefficient of the rotor system
$C(k)$:	Theodorsen lift deficiency function
C_{l}:	Sectional lift coefficient
$C_{\mathrm{M}x}$, $C_{\mathrm{M}x\mathrm{H}}$:	Roll moment coefficient at the hub
$C_{\mathrm{M}y}$, $C_{\mathrm{M}y\mathrm{H}}$:	Pitch moment coefficient at the hub
C_{P}:	Power coefficient
C_{Ppd}:	Profile drag power coefficient
C_{Q}:	Torque coefficient (or yaw moment) at the hub
C_{T}:	Thrust coefficient acting at the hub
C_{Y}:	Lateral in-plane force coefficient at the hub
C_x, C_y:	Damping at the base platform in x, y directions
C_ζ:	Damping at blade root in lead–lag motion
$[C]$:	Damping matrix
D:	Sectional aerodynamic drag (or fuselage drag force)
dP:	Differential power
dQ:	Differential torque
dr:	Differential radial length
dT:	Differential thrust
e:	Blade root hinge offset
$\vec{e}_{x1}, \vec{e}_{y1}, \vec{e}_{z1}$:	Unit vectors along the x_1–y_1–z_1 directions
$\vec{e}_{xH}, \vec{e}_{yH}, \vec{e}_{zH}$:	Unit vectors along the x_{H}–y_{H}–z_{H} directions
$\vec{e}_\eta, \vec{e}_\xi$:	Cross-sectional coordinate system of the blade
f:	Equivalent flat plate area of the fuselage
F_x, F_y, F_z:	Sectional forces along the x–y–z directions

F_{x1}, F_{y1}, F_{z1}: Sectional forces along the x_1–y_1–z_1 directions
F_{x2}, F_{y2}, F_{z2}: Sectional forces along x_2–y_2–z_2 directions
H: Longitudinal in-plane force at the rotor hub
$\bar{H}$: Angular momentum
h_x, h_y, h_z: Position vector components of the hub center from the c.g. of the fuselage
h_{xT}, h_{yT}, h_{zT}: Position vector components of the tail rotor hub center from the c.g. of the fuselage
I_b: Mass moment of inertia (second moment of mass) of the blade about the root hinge
I_{ij}: Components of inertia tensor ($i, j, = x, y, z$)
$I_{m\eta\eta}, I_{m\eta\xi}, I_{m\xi\xi}$: Cross-sectional inertia quantities
$I_{\eta\eta}, I_{\eta\xi}, I_{\xi\xi}$: Cross-sectional inertia quantities
$\bar{I}_{\eta\eta}, \bar{I}_{\xi\xi}$: Non-dimensional cross-sectional inertia quantities
k: Reduced frequency
$[K]$: Stiffness matrix
K_c: Control system spring constant in pitch motion
$K_{p\beta}$: Pitch–flap coupling parameter
$K_p\zeta$: Pitch–lag coupling parameter
K_x, K_y: Coefficients for non-uniform inflow variation over the rotor disk
K_β: Root spring constant in flap motion
K_ζ: Root spring constant in lag motion
K_ϕ, K_θ: Root spring constant in torsion motion
l: Length of the blade ($l = R - e$)
L: Sectional lift
L, M, N: Components of the moment acting at the fuselage
M: Figure of merit (or mass of the blade)
$[M]$: Mass matrix in stability analysis
m: Mass per unit length of the blade
$\dot{m}$: Mass flow rate
$\vec{M}_A$: Blade root moment due to the aerodynamic loads acting on the rotor blade
M_b: Mass of the blade
M_{EA}: Aerodynamic moment about the elastic axis
$\vec{M}_{ext}$: Blade root moment due to the external aerodynamic loads acting on the rotor blade
M_F: Mass of the helicopter
$\vec{M}_I$: Blade root moment due to the inertia loads acting on the rotor blade

$MX_{c.g.}$:	First moment of mass of the blade about the root hinge
M_{xH}, M_{yH}, M_{zH}:	Roll, pitch, and yaw moments at the hub center
$M_{\beta A}$:	Aerodynamic flap moment about the root hinge
M_{β}, M_{ζ}:	Flap and lag moments at the blade root due to root springs
N:	Number of blades in the rotor system
P:	Rotor power
p_0:	Atmospheric pressure
p_1:	Pressure just above the rotor disk
p_2:	Pressure just below the rotor disk
$\vec{p}_I$:	Distributed inertia force acting on the blade
P_{Hx}, P_{Hy}:	Hub forces along the x and y directions
P_{pd}:	Profile drag power of rotor blades
p, q, r:	Roll–pitch–yaw angular velocity of the helicopter about the body-fixed coordinate system at the c.g.
$\{q\}$:	State vector
$\vec{q}_I$:	Distributed inertia moment acting on the blade
Q_A:	Aerodynamic moment about the root hinge in lag motion
Q_I:	Inertia moment about the root hinge in lag motion
R:	Rotor radius (or structural flap–lag coupling factor)
r:	Radial location of a typical cross-section of the rotor blade
$\vec{r}_P$:	Position vector of any arbitrary point on the rotor blade
R_x, R_y:	Longitudinal and lateral perturbation motions of the base platform
S_C:	Coupling parameter
T:	Rotor thrust
T_t, T_T:	Tail rotor thrust
$[T_{ij}]$:	Transformation matrix between the jth and the ith coordinate systems
T_∞:	Rotor thrust in out of ground
u, v, w:	Translational velocity components of the helicopter c.g. along the body-fixed system
U:	Resultant relative air velocity at any blade cross-section
U_P:	Flow velocity normal to the rotor disk or at the airfoil cross-section
U_R:	Radial flow velocity along the span of the blade
U_T:	Tangential velocity at the airfoil cross-section
$\vec{v}_P$:	Absolute velocity of any point P on the reference axis of the rotor blade
V, V_C:	Climb velocity of the helicopter

$\vec{V}_{\text{c.g.}}$: Velocity at the c.g. of the helicopter
$\vec{V}_{\text{h}}$: Relative air velocity at the blade cross-section due to helicopter motion
$\vec{V}_{\text{rel}}, \vec{V}_{\text{net}}$: Net-relative air velocity at the blade cross-section due to helicopter and blade motions
w: Induced velocity at far-field downstream
w_{g}: Vertical gust velocity
X_{A}: Distance between the elastic axis and the aerodynamic center of the blade cross-section
$X_{\text{b}}, Y_{\text{b}}, Z_{\text{b}}$: Body-fixed fuselage coordinate system with origin at the c.g. of the helicopter
$X_{\text{ea}}, Y_{\text{ea}}, Z_{\text{ea}}$: Earth-fixed inertial coordinate system
$X_{\text{H}}, Y_{\text{H}}, Z_{\text{H}}$: Hub-fixed non-rotating coordinate system with origin at the hub center
X, Y, Z: Resultant force components acting at the helicopter fuselage
X_1, Y_1, Z_1: Hub-fixed rotating coordinate system
X_2, Y_2, Z_2: Blade-fixed rotating coordinate system
Y: Lateral in-plane force at the rotor hub
Z: Height of the rotor above the ground
α, α_e: Effective angle of attack at any typical cross-section of the blade (or angle between the hub plane and the forward velocity direction)
β_{k}: Flap deflection of kth blade
β_0: Mean value of flap deflection
$\tilde{\beta}$: Time-dependent perturbation motion in the flap
β_{ideal}: Ideal pre-cone angle
$\beta_{\text{nc}}, \beta_{\text{ns}}$: nth harmonic cosine and sine components of flap deflection
β_{p}: Pre-cone angle
β_{1c}, β_{1s}: First harmonic cosine and sine components of flap deflection
γ: Lock number
δ_1: Lag–pitch (or pitch–lag) coupling
δ_3: Flap–pitch (or pitch–flap) coupling
ε: Non-dimensional parameter defined for the ordering scheme
ζ: Lead–lag deformation of the blade
$\tilde{\theta}$: Time-dependent perturbation motion in lead–lag
ζ_k: Lead–lag deformation of the kth blade
ζ_0: Mean value of lead–lag deflection

ζ_{nc}, ζ_{ns}: *n*th harmonic cosine and sine components of lead–lag deflection

η, ξ: Cross-sectional coordinate of any arbitrary point of the blade section

η_m, ξ_m: Cross-sectional coordinate of the mass center

θ: Pitch angle at any cross-section of the blade

$\tilde{\lambda}$: Time-dependent perturbation in pitch input

θ_{con}: Control pitch input given at the blade root

θ_I: Pitch input given at the blade root

θ_{FP}: Flight path angle (angle between the horizon and the velocity direction of the helicopter)

θ_{tip}: Pitch angle at the tip

θ_{tw}: Blade pre-twist

θ_0: Blade root pitch angle (or collective pitch input)

θ_{1c}, θ_{1s}: First harmonic cosine and sine components of the blade pitch input (cyclic pitch input)

$\theta_{0.75}$: Sectional pitch angle at a radial station 0.75*R*

Θ: Pitch attitude of the helicopter

κ: Empirical correction factor for induced power

λ: Non-dimensional total inflow (or inflow ratio)

λ_0: Mean value of inflow

$\tilde{\lambda}$: Time-dependent perturbation in inflow

λ_c: Non-dimensional inflow due to the climb speed of the helicopter

λ_h: Non-dimensional inflow during hover

λ_i: Non-dimensional induced flow

λ_{NFP}: Non-dimensional induced flow defined in a no-feathering plane

μ: Advance ratio (non-dimensional forward speed)

ν: Induced flow or inflow velocity at the rotor disk

ν_0: Mean inflow in forward flight

ν_h: Induced flow during hover

ρ: Density of air (or mass per unit length of the blade)

ρ_b: Mass per unit volume of the blade

σ: Solidity ratio

σ_j: Real part of the eigenvalue representing damping

ϕ: Induced angle of attack (or blade elastic torsion deformation)

ϕ_k: Torsional deformation of the *k*th blade

ϕ_0: Mean value of torsional deformation

$\tilde{\phi}$: Time-dependent perturbation of torsional motion

ϕ_{nc}, ϕ_{ns}: *n*th harmonic cosine and sine components of torsional deformation

Φ: Roll angle of the helicopter

χ: Wake skew angle

ψ: Azimuth location of the rotor blade or non-dimensional time

ψ_k: Azimuth location of the *k*th blade

$\vec{\omega}$: Angular velocity vector of the rotor blade

ω_j: Imaginary part of the eigenvalue representing frequency

$\bar{\omega}_{NRF}$: Non-dimensional non-rotating natural frequency in flap motion

$\bar{\omega}_{NRL}$: Non-dimensional non-rotating natural frequency in lead–lag motion

$\bar{\omega}_{NRT}$: Non-dimensional non-rotating natural frequency in the torsion mode

$\bar{\omega}_{RF}$: Non-dimensional rotating natural frequency in flap motion

$\bar{\omega}_{RL}$: Non-dimensional rotating natural frequency in lead–lag motion

$\bar{\omega}_{RT}$: Non-dimensional rotating natural frequency in the torsion mode

$\bar{\omega}_{T}$: Non-dimensional non-rotating natural frequency in the torsion mode

Ω: Main rotor angular velocity

$(\ )_e$: Equilibrium quantities

$(\ )_T$: Quantities related to the tail rotor

$\tilde{(\ )}$: Perturbation quantities

1

Historical Development of Helicopters and Overview

Nature has played a significant role in creating a desire in human beings to develop flying machines. It is well known that bird (of course with flapping wings) flying had a strong influence in the development of gliders and aircraft. Even in plants, the concept of flying over reasonably long distance to spread the seeds exists. For the sake of illustrating the beauty of nature, a few examples of flying fixed wing (Figure 1.1) and rotating wing (Figures 1.2–1.5) seeds are shown. The rotating wing seeds can be further divided into single or multiple wing seeds.

The fixed wing seed (Figure 1.1), known as *Alsomitra macrocarpa,* has a wing span of about 13 cm. It can glide through air in wide circles in the rain forest and can travel a substantial distance.

There are several types of rotating winged seeds. Some of them have only one wing attached to a seed at the root (Figures 1.2 and 1.3). The wing is a membrane with a slight twist. It spins as it falls, and it is similar to the autorotation of a helicopter when it descends after a power loss. Depending on height and wind speed above the ground, these seeds can be swept far away from the parent tree. Some of the seeds have two, three, or more wings (Figures 1.4 and 1.5).

Unlike a fixed wing aircraft, helicopters use rotating wings to provide lift, propulsion, and control. In addition, helicopters are capable of hovering, landing, and takeoff from any terrain. Efficient accomplishment of vertical flight (with minimum power) is the fundamental characteristic of the helicopter. In the following section, a brief history of the development of helicopter is provided.

Historical Development

During the initial development of helicopters, three fundamental problems had to be overcome by the designers.

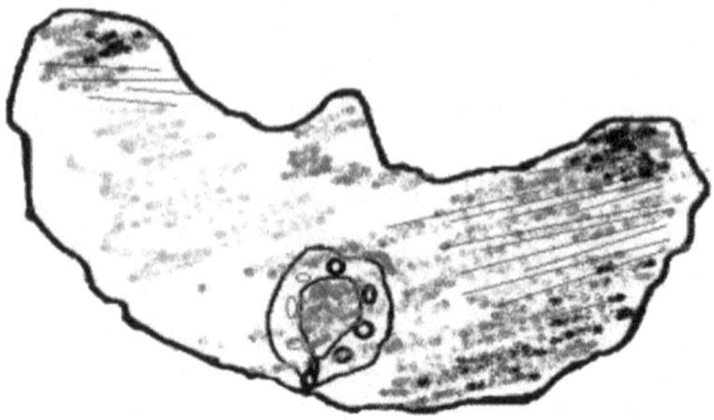

FIGURE 1.1
Fixed wing seed (*Alsomitra macrocarpa*).

FIGURE 1.2
One-seeded fruit with one wing (South American Tipuana tipu tree).

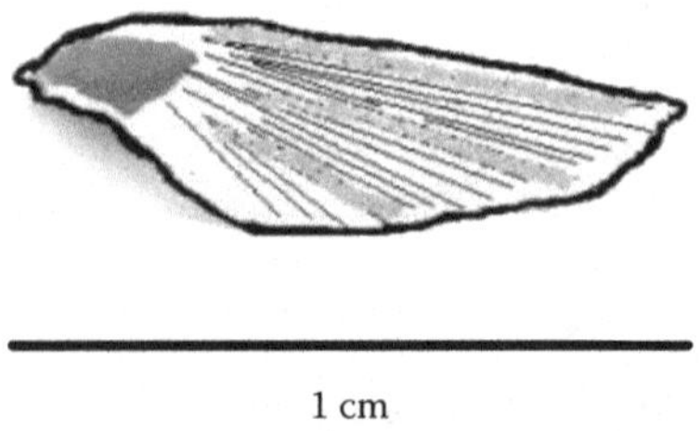

FIGURE 1.3
Pine tree winged seed from woody cone rather than from flowers.

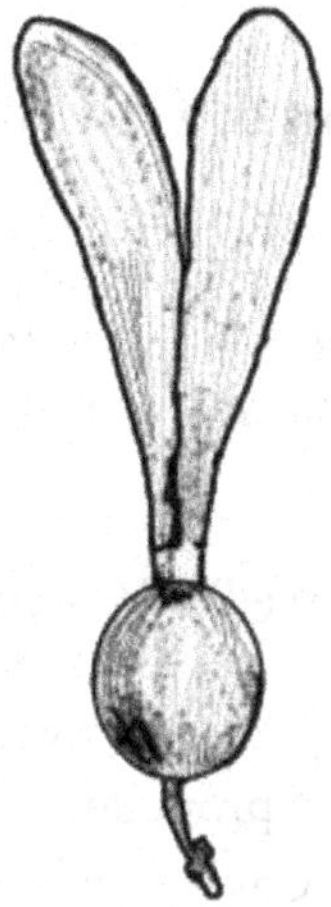

FIGURE 1.4
One-seeded two-winged fruit (*Gyrocarpus*, Hawaiian Islands).

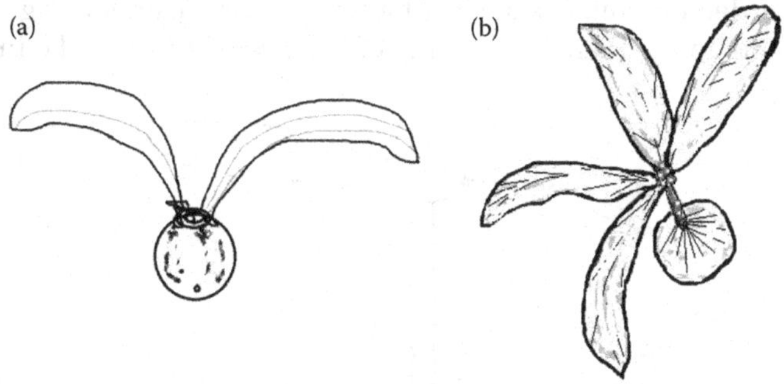

FIGURE 1.5
Multiwinged spinning fruit/seed (Thailand). (a) *Diptocarpus obtusifolios.* (b) Gluta *Melanorrhoea usitata.*

- Keeping the structural weight and the engine weight at a minimum so that the engine develops enough power to lift itself and some useful load
- Developing a light, strong structure for rotor hub and blades
- Understanding and developing schemes for controlling the helicopter, including the torque balance

These problems were essentially similar to those faced by the developers of the airplane, which were solved by the Wright brothers. The development of the helicopter took a longer duration possibly due to the difficulty in vertical flight, which required a further development in aeronautical technology.

Chronological Development of Helicopters

The material presented in the following is based on the information available in Bramwell (1976); Johnson (1980); Stepniewski (1984); Prouty (1990); Padfield (1996); Leishman (2000); Seddon (1990); Bielawa (1992) and the articles appeared in *Vertiflite*, a publication of AHS International, The Vertical Flight Technical Society.

- B.C.: The Chinese flying top (400 B.C.), shown in Figure 1.6, is a stick with a propeller on top, which was spun by the hands and released.
- 15th century: Leonardo da Vinci sketched a machine for vertical flight utilizing a screw-type propeller (Figure 1.7).
- 18th century: Sir George Cayley constructed models (Figure 1.8) powered by elastic elements and made some sketches. Mikhail V. Lomonosov of Russia demonstrated a spring-powered model to the Russian Academy of Sciences. Similarly, Launoy and Bienvenu of France demonstrated a model having turkey feathers (Figure 1.9) for rotors to the French Academy of Sciences. This model climbed a

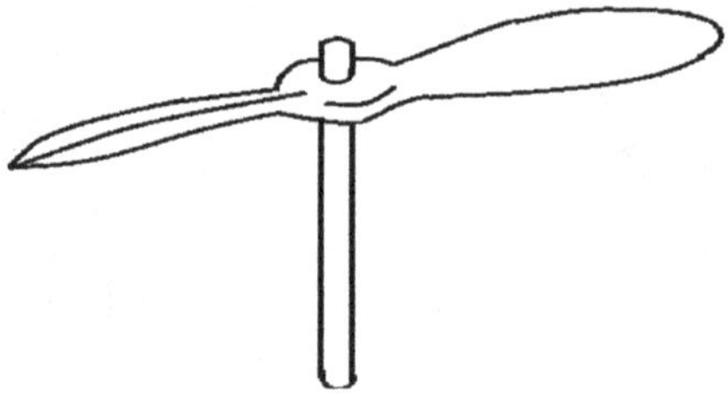

FIGURE 1.6
Chinese top, B.C.

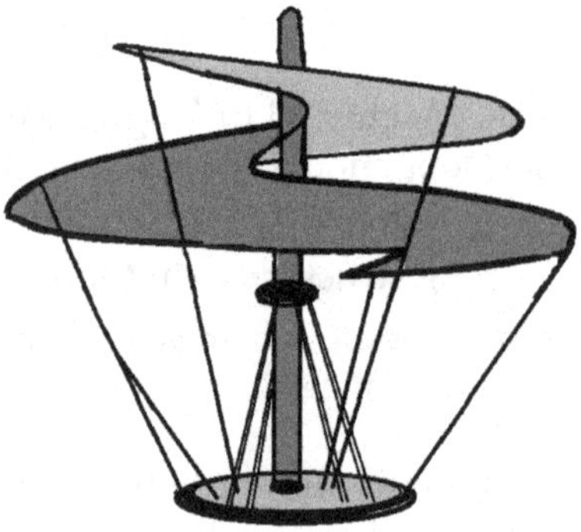

FIGURE 1.7
Sketch of Leonardo da Vinci's "Helicopter," 15th century.

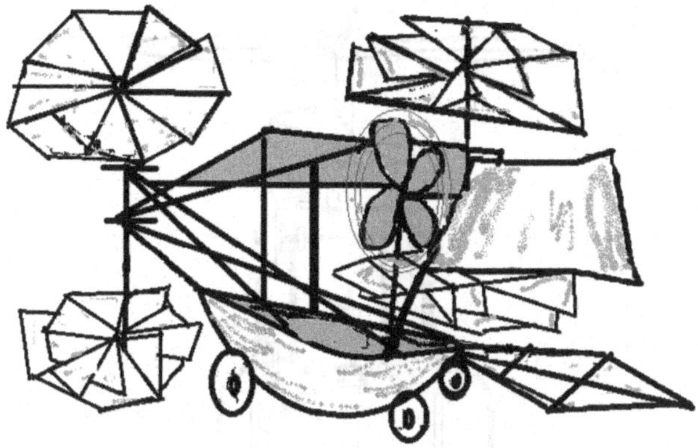

FIGURE 1.8
Sir George Cayley's helicopter, 1796.

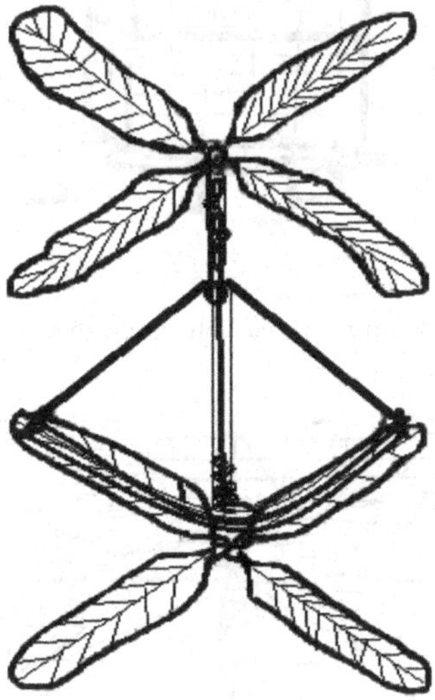

FIGURE 1.9
Launoy and Bienvenu: turkey-feather rotors, 18th century.

height of 20 m, creating an enormous interest in flying. These models had little impact on full-scale helicopter development.

- 19th century: The last half of the 19th century saw some progress, but there was no successful vehicle. The problem was due to the lack of a cheap, reliable, and light engine. Attempts were made to use steam engine, and W. H. Phillips (England, 1842) constructed a 10-kg steam-powered model. Similarly, Ponton d'Amecourt (France, 1862) built a 4-kg steam-powered model (Figure 1.10) having coaxial contra-rotating rotors. At full power, this model reportedly tried to bob lightly on its

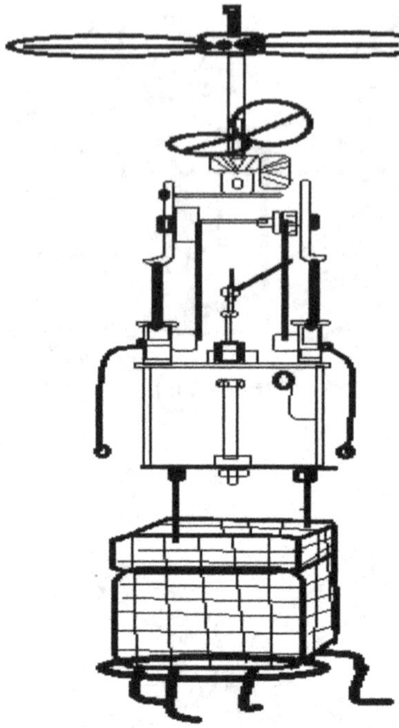

FIGURE 1.10
Gustave Ponton d'Amecourt's steam-powered helicopter, 1865.

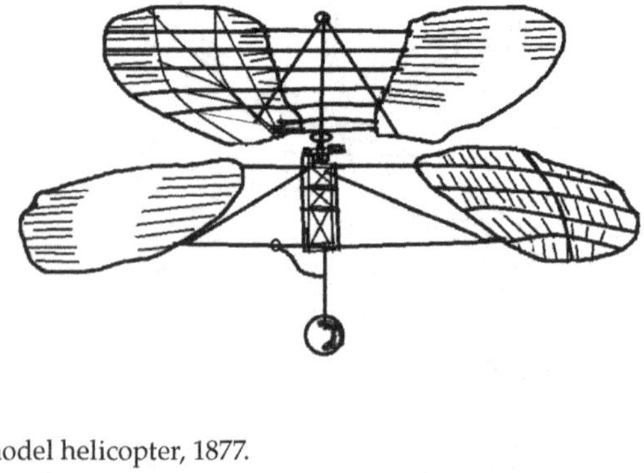

FIGURE 1.11
Forlanini's steam model helicopter, 1877.

base. Ponton d'Amecourt was responsible for inventing the word *helicopter* derived from coining the two words *helix* (which means "screw" or "spiral") and *petron* (which means "wing"). Enrico Forlanini (Italy, 1878) built a 3.5-kg flying steam-driven model (Figure 1.11). This model climbed a height of about 12 m and stayed aloft for about 20 min.

Thomas Edison's experiments with models led to the important conclusion that no helicopter would fly until engines with a weight-to-power ratio below 1 or 2 kg/hp were available. (It may be noted that present-day turbo shaft engines have a weight-to-power ratio of the order of 0.5–0.15 kg/hp or, in other words, 2–5 hp/kg.) Around 1900, internal combustion reciprocating engines became available, which made airplane and helicopter flight possible.

- 20th century
  - Renard (France, 1904) built a helicopter with two side-by-side rotors using a two-cylinder engine. He introduced the flapping hinge.
  - Paul Cornu (France, 1907) constructed the first man-carrying helicopter with two contra-rotating rotors of 6-m diameter, in tandem configuration (Figure 1.12). The total weight of the vehicle was 260 kg and was powered by a 24-hp engine. This helicopter achieved an altitude of 0.3 m for about 20 s. It had problems with stability. However, Cornu became the first person to succeed in actual helicopter flight.
  - Louis Charles Brequet (France, 1907) built a machine that he called "helicoplane" (Figure 1.13). It had four rotors (8-m diameter rotors), with a gross weight of 580 kg, and an engine with 45 hp. This vehicle made a tethered flight at an altitude of 1 m for about 1 min. The vehicle had no control mechanism. Breguet was one of the foremost pioneers to lift a full-scale helicopter with a pilot.
  - Henry Berliner (U.S.A., 1920–1922), son of Emile Berliner, an inventor, built a two-engine co-axial vehicle (Figure 1.14), which lifted a pilot. He also built a side-by-side rotor helicopter (Figure 1.15), in which forward flight control was achieved by titling the rotor shaft. These vehicles were highly unstable.

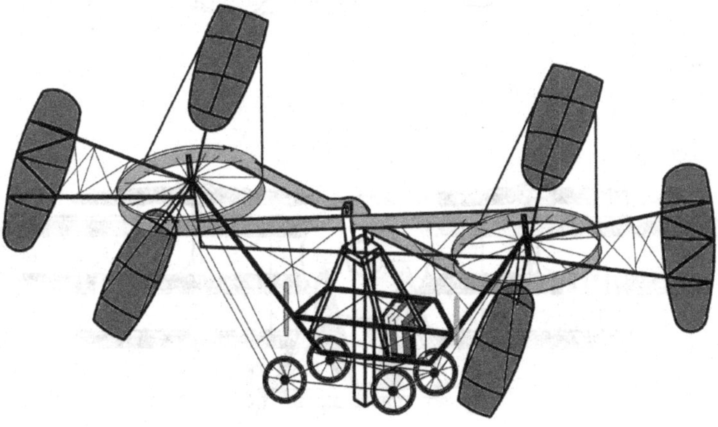

FIGURE 1.12
Paul Cornu's first man-carrying helicopter, 1907.

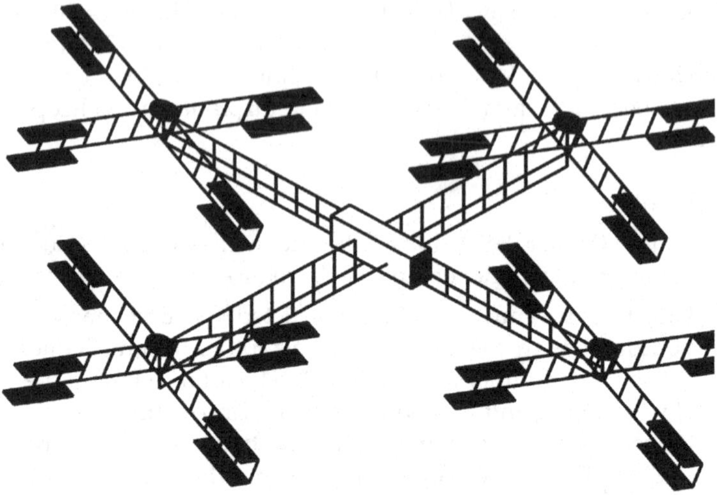

FIGURE 1.13
Louis Breguet's large four-rotor helicopter, 1907.

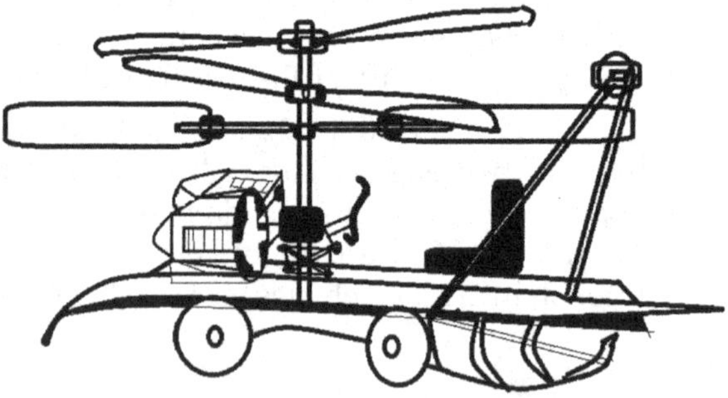

FIGURE 1.14
Berliner coaxial vehicle, 1920–1922.

FIGURE 1.15
Berliner helicopter with side-by-side rotors, 1924.

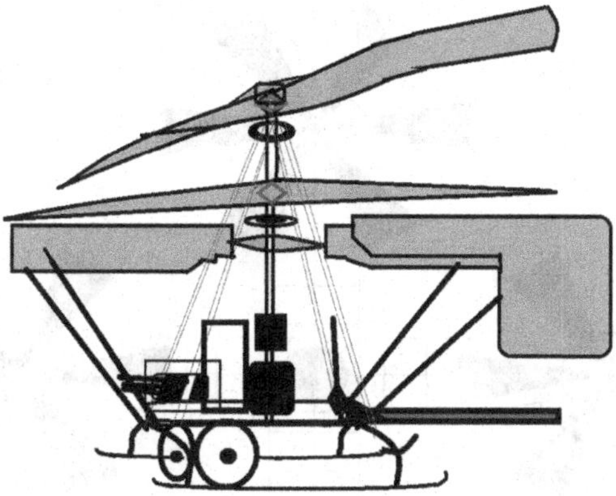

FIGURE 1.16
Sikorsky's coaxial rotor vehicle, 1910.

- Sikorsky (Russia, 1910) built a vehicle with coaxial rotors (Figure 1.16) having a rotor diameter of 5.8 m and a 25-hp engine. This vehicle could lift its own weight of 180 kg, but not with a pilot.
- B. N. Yuriev (Russia, 1912) built a two-bladed main rotor and a small anti-torque tail rotor (main rotor: 8-m diameter, 200-kg weight, 25-hp engine). This helicopter did not make any successful flight.
- Lts. Petroczy and von Karman (Austria, 1916) constructed a tethered contra-rotating, coaxial helicopter with three engines of 40 hp each (Figure 1.17). The rotor diameter was 6 m. This vehicle was designed as an observation platform. Although it made several flights, it had control problems.

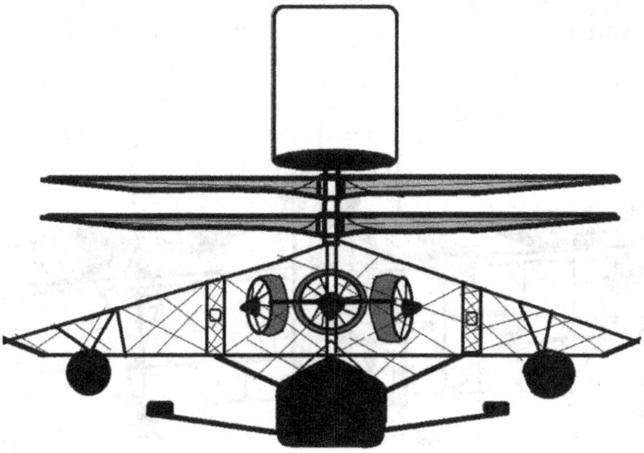

FIGURE 1.17
Lts. Petroezy and von Karman's tethered contra-rotating coaxial helicopter, 1916.

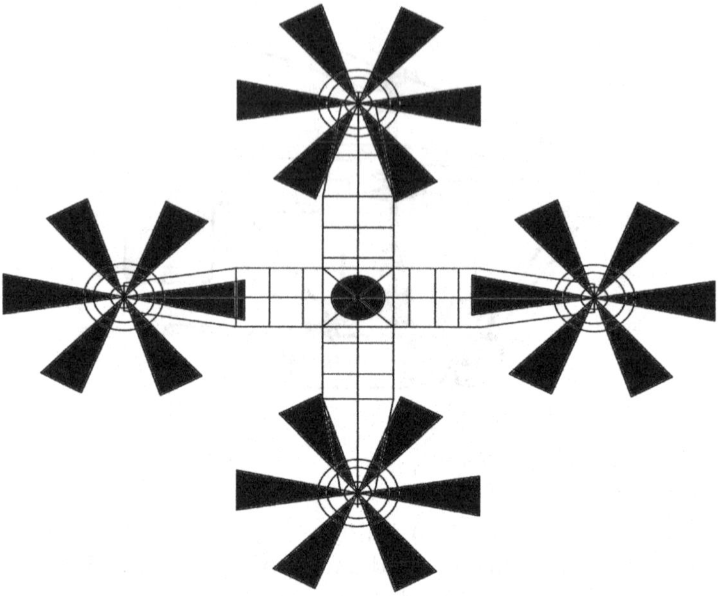

FIGURE 1.18
George de Bothezat's large 6-bladed rotor helicopter, 1922.

- George de Bothezat (U.S.A., 1922) built a large helicopter (Figure 1.18) with four six-bladed rotors at the ends of intersecting beams having a weight of 1600 kg and a 180-hp engine. Good control behavior was obtained by utilizing differential collective. This vehicle made many flights with passengers at an altitude of 4 to 6 m. This was the first rotorcraft ordered by the U.S. Army. Later, it was abandoned due to the mechanical complexity of the vehicle.
- Etienne Oehmichen (France, 1924) built a machine (Figure 1.19) with four two-bladed rotors (7.6- and 6.4-m diameter) to provide

FIGURE 1.19
Etienne Oehmichen's four two-bladed rotor vehicle, 1924.

lift and five horizontal propellers for attitude control, two propellers for propulsion, and one propeller for yaw control, all powered by a 120-hp engine. It had 13 separate transmission systems. This vehicle made several flights and set a distance record of 360 m.

- M. Raoul P. Pescara (Spain, 1920–1926) built a helicopter (Figure 1.20) with two coaxial rotors of four blades each having a diameter of 6 m. This was powered by a 120-hp engine. For forward flight control, he warped the biplane blades to change their pitch angle over one cycle. Pescara was the first to demonstrate cyclic pitch control for helicopters. This vehicle set a record distance of 736 m but had stability problems.
- von Baumhaver (Holland, 1924–1929) developed a single main rotor and a tail rotor vehicle. This had a two-bladed main rotor of 15-m diameter, a total weight of 1300 kg, and a 200-hp engine. A separate engine of 80 hp was used for the tail rotor. Blade control was achieved by cyclic pitch using a swash plate, similar to modern helicopters. Several flights were made around 1-m altitude. There were difficulties in control because of separate engines for the main and tail rotors. The project was abandoned after a bad crash in 1929.
- Corradino d'Ascanio (Italy, 1930) built a helicopter with two coaxial rotors having a diameter of 13 m and powered by a 95-hp engine. The blades had flap hinges and free feathering (pitching of blades) hinges. Collective (constant pitch angle over the

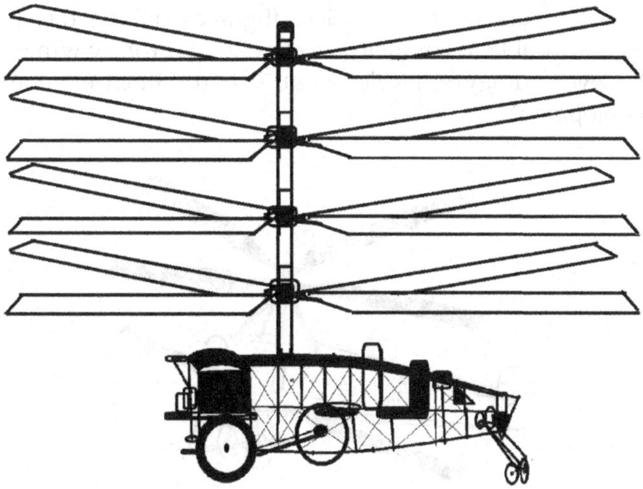

FIGURE 1.20
Pescara coaxial helicopter, 1925.

azimuth) and cyclic (variation of pitch angle once in a revolution) pitch changes were achieved by controlling the servo tabs on the blade. For many years, this vehicle held records for altitude (18 m), endurance (8 min, 45 s), and distance (1078 m).

- Although the development of helicopter was fairly well advanced, it had severe stability and control problems. It was in the 1920s to the 1930s that autogyro was developed. An autogyro is essentially an airplane with wings replaced by rotors. A propeller is used for propulsive forces. It was developed by Juan de la Cierva (Spain, 1920–1930). In this vehicle, the rotor acts as a windmill and generates the lift. The initial design even used the conventional airplane-type control surfaces with no power to the rotor (Figure 1.21). In autogyro, hover and vertical flight are not possible, but a very slow forward flight can be possible. The reason for this development was that Cierva's airplane crashed due to stall in 1919. He then became interested in designing an aircraft with a low takeoff and landing speed. In 1922, Cierva built the C-3 autogyro with a five-bladed rigid rotor, and the vehicle had a tendency to fall over sideways. Cierva incorporated flapping hinges in his design, which eliminated the rolling moment on the aircraft in forward flight. He was the first to use the flap hinges successfully in a rotary wing vehicle. In 1925, Cierva founded the Cierva Autogyro Company in England. In the next decade, about 500 autogyros were produced. In 1927, a crash led to the understanding of high inplane loads due to flapping (Coriolis effect), which led to the incorporation of lag hinges. In 1932, Cierva added rotor control to replace the airplane control surfaces. Lacking true vertical flight capability, the autogyro was never able to compete effectively with rotary wing aircraft. However, autogyro development had an influence on helicopter development.

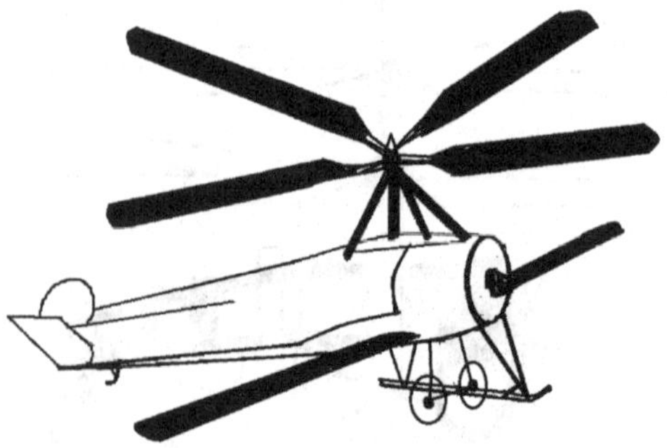

FIGURE 1.21
Cierva autogyro, 1920–1930.

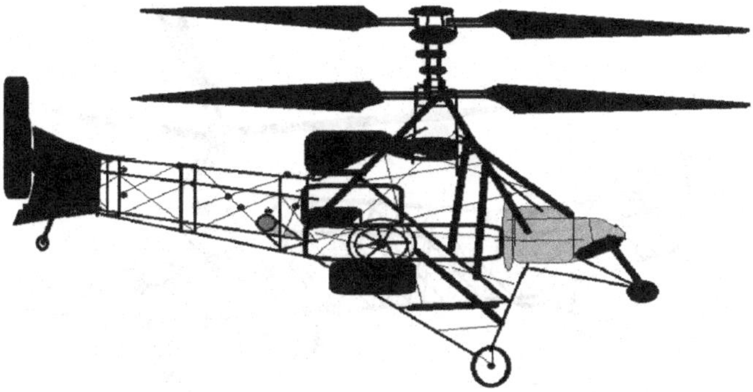

FIGURE 1.22
Louis Charles Brequet's coaxial rotor helicopter, 1935.

- Louis Charles Brequet (France, 1935) built a coaxial rotor helicopter having an 18-m diameter rotor powered by a 350-hp engine. This vehicle (Figure 1.22) exhibited promising control and stability characteristics. Even though it was damaged before the completion of the test, it is thought to be the first successful helicopter ever built.
- E. H. Henrich Focke (Germany, 1936) constructed a helicopter with two three-bladed rotors mounted on side-by-side configuration (7-m diameter, 950-kg weight, 160-hp engine). The rotors were contra-rotating, and the small propeller in the nose was used to cool the engine. The rotors had an articulated hub. Directional and longitudinal control was achieved by cyclic pitch and roll by differential collective. Vertical and horizontal tail surfaces were used for stability and trim in forward flight. The vehicle set records for speed (122.5 kmph), altitude (2440 m), endurance (1 h, 21 min), and distance (224 km).
- Igor Sikorsky (U.S.A., 1930–1941) pursued helicopter development in America, after leaving Russia. He built the VS-300 in 1941, which had a single three-bladed main rotor having a 9-m diameter and a tail rotor with an all-up weight of 520 kg, and a 100-hp engine. Lateral and longitudinal control was by the main rotor cyclic, and directional control, by the tail rotor. The tail rotor was driven by a shaft from the main rotor. Initially, this vehicle had three tail rotors (one vertical, two horizontal). Later, it was reduced to two and, finally, to one vertical tail rotor. In 1942, the R-4, a derivative of VS-300 (Figure 1.23), was constructed. This helicopter had a single main rotor and one tail rotor (main rotor: diameter, 11.6 m; weight, 1100 kg; 185-hp engine). This model went into production, and several hundreds were built during World War II. Sikorsky's vehicle is considered to be the first practical, truly operational, and mechanically simple and controllable vehicle.

FIGURE 1.23
Igor Sikorsky's helicopter, 1941.

The success of Sikorsky's vehicle led to the development of other successful helicopters in America. Lawrence Bell (two-bladed main rotor helicopters), Frank N. Piasecki (tandem rotor helicopters), Stanley Hiller, Charles Kaman, and McDonnell built different types of helicopters. These names later became helicopter manufacturing companies in America. Piasecki's company became the Boeing–Vertol company. In Russia, Mikhail Mil (Mil helicopters), Nikolai Kamov (coaxial rotor helicopters), and Alexander Yakolev (Yak-24 helicopter) built helicopters in 1949 to 1955. The invention of helicopters can be considered to be completed by 1950s. However, new developments are still being pursued, which are possible due to the overall technological development in several associated fields. Modern developments include

(a) Notar (No-Tail Rotor) McDonnell–Douglas
(b) Tilt rotor (V-22 Osprey) Bell–Boeing
(c) ABC (Advancing Blade Concept, Sikorsky) (Coaxial, Contra-Rotating Rotors)
(d) X-Wing (NASA) (Circulation Control Rotors)
(e) Tilt Wing
(f) Compound helicopter

Helicopter Configurations

Helicopter configurations may be broadly classified into five types. However, with newly emerging configurations such as NOTAR, X-Wing, and compound helicopter, additional classifications may come into the picture.

However, the basic principle of helicopter flight may not differ in these configurations. In the following, a brief description of the conventional classifications is provided.

1. Single main rotor: This configuration has a single main rotor and one tail rotor for torque control (Figure 1.24). It has the advantage of being simple. The disadvantage is the danger of the vertical tail rotor to ground personnel. One French design (AS 365 N2 Dauphin 2) has a fan-tail configuration. This is known as the Fenestron or the Fan-in-Fin type.
2. Jet rotor: This configuration provides the simplest solution to the torque balance. Fuselage directional control is achieved by rudder or vane, which uses the rotor downwash in hover (Figure 1.25). In this configuration, the presence of a jet engine at the tip of the blade leads to complex blade dynamics.
3. Coaxial rotor: In this configuration, the fuselage torque is balanced by two main rotors rotating in opposite directions. This design has the advantage of having its overall dimension defined by the main rotor diameter and also saving the power required for the tail rotor. However, the rotor hub and controls design is more complex (Figure 1.26). Kamov Ka-32A and Ka-226 are examples of coaxial contra-rotating rotor configuration.
4. Side-by-side rotors: The basic advantage of side-by-side rotor configuration is that it requires less power to produce lift in forward flight, which is similar to a large aspect ratio airplane wing (Figure 1.27). The disadvantages are increased parasite drag and high structural weight. In addition, compared to single rotor configuration, side-by-side configuration has complex gearing and transmission systems (e.g., Bell–Boeing V-22 Osprey).

FIGURE 1.24
Conventional single main rotor and one tail rotor helicopter configuration.

FIGURE 1.25
Tip jet rotor configuration.

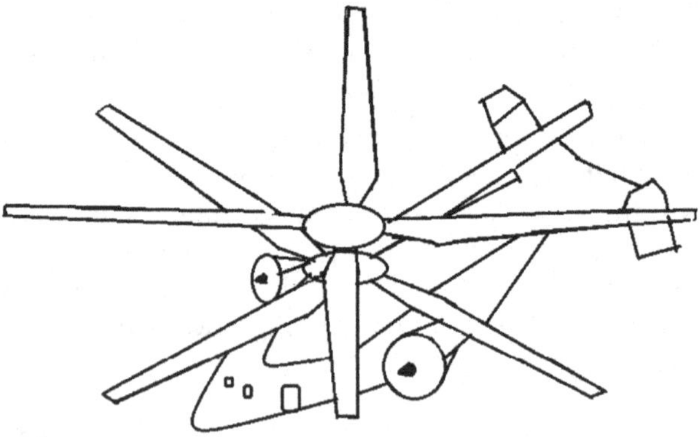

FIGURE 1.26
Coaxial contra-rotating rotor helicopter configuration.

5. Tandem rotors: The main advantage of the tandem configuration lies in its clean fuselage, together with a large available centre of gravity range (Figure 1.28). Total load may be distributed between the two rotors. Disadvantages are due to complexities in transmission and gear systems, which are similar to side-by-side configuration. Another disadvantage is the loss in efficiency of the rear rotor since it operates in the wake of the front rotor. The loss in efficiency may be minimized by placing the rear rotor above the front rotor (e.g., Boeing CH-47 series).

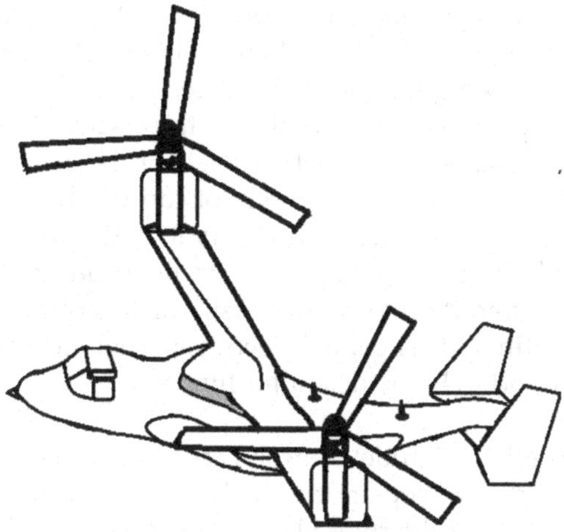

FIGURE 1.27
Side-by-side rotor configuration.

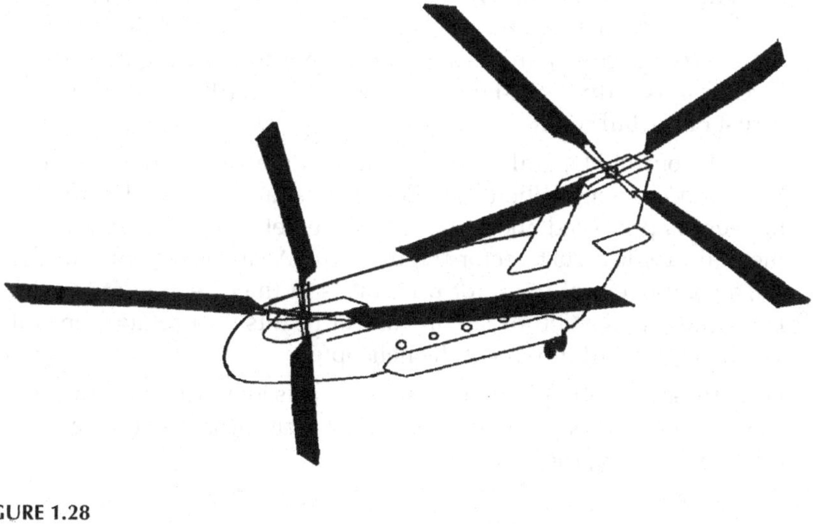

FIGURE 1.28
Tandem rotor configuration.

Control Requirements

The position and attitude of a rigid body (or vehicle) in space can be controlled by forces and moments applied along the respective direction. For a six-degrees-of-freedom control, one requires six control inputs. In general, it would be difficult for a human being to control a machine having six independent controls. However, it is possible to reduce the number of controls by coupling together a few independent controls. But, such couplings involve some sacrifice on complete freedom of control both in the position and the attitude of the vehicle simultaneously in space.

For example, the pilot requires an ability to produce moments about all the three orthogonal axes to correct the vehicle in a maneuver or when the vehicle is disturbed in a gust. If the pilot can produce a moment that is accompanied by force, he sacrifices the ability to maintain force equilibrium. Therefore, by coupling pitching moment with longitudinal force and rolling moment with side (lateral) force, the necessity of two of the six independent controls can be eliminated. Thus, in helicopters, there are four independent controls that are found to be sufficient for flying.

1. Vertical (or height or altitude) control: This control is required to change the altitude of the helicopter. It is achieved by increasing or decreasing the pitch angle of the rotor blades to increase or reduce the thrust, respectively. (Note that, in helicopter terminology, "thrust" implies the total lift force generated by all the rotor blades in the rotor system.)
2. Directional control: Directional control denotes the control of the heading of the helicopter by rotating about a vertical axis. In single main rotor helicopter, this is achieved by producing a moment about a vertical axis through the c.g. of the helicopter by modifying the thrust of the tail rotor.
3. Lateral control: Lateral control involves the application of both force and moment. When the pilot applies lateral control, a rolling moment is produced about the c.g. of the helicopter due to a tilt of the main rotor thrust vector (Figure 1.29). As a consequence of the tilt, a component of the rotor thrust acts in the direction of the tilt. Hence, an application of lateral control results in a rotation in roll and in a sideward motion of the helicopter.
4. Longitudinal control: Longitudinal control is identical to lateral control. Pitching moment is coupled with longitudinal force when the pilot applies longitudinal control.

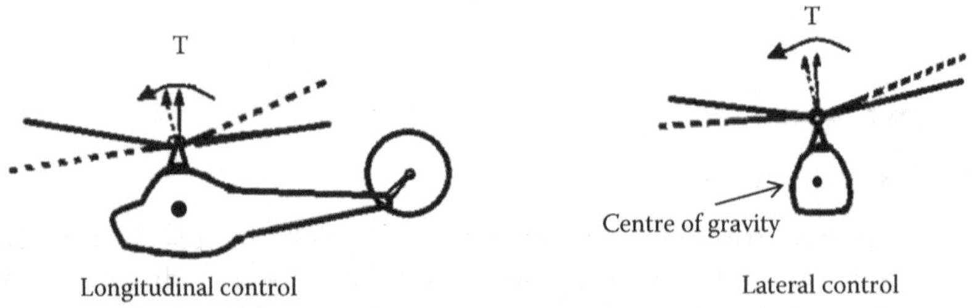

FIGURE 1.29
Helicopter control by tilt of the rotor thrust for longitudinal and lateral controls.

TABLE 1.1

Rotor Control Input for Various Helicopter Configurations

	Height	Longitudinal	Lateral	Directional	
Helicopter Configuration	**Vertical Force**	**Pitch Moment**	**Roll Moment**	**Yaw Moment**	**Torque Balance**
Single main rotor and tail rotor	Main rotor Collective	Main rotor Cyclic	Main rotor Cyclic	Tail rotor Collective	Tail rotor Thrust
Coaxial[a]	Main rotor Collective	Main rotor Cyclic	Main rotor Cyclic	Main rotor Differential Collective	Main rotor Differential Torque
Tandem[a]	Main rotor Collective	Main rotor Differential Collective	Main rotor Cyclic	Main rotor Differential Cyclic	Main rotor Differential Torque
Side by side[a]	Main rotor Collective	Main rotor Cyclic	Main rotor Differential Collective	Main rotor Differential Cyclic	Main rotor Differential Torque

[a] Combined pitch differential control.

In general, cross coupling between various degrees of freedom of the helicopter is undesirable. For example, in a single main rotor machine, an increase in vertical force results in an increase in rotor torque (due to drag force acting on the blade), so a correction is required in the directional control to maintain the fuselage heading. Therefore, any control application to produce a required moment or force needs some compensating control inputs on other axes as well. Moreover, without an automatic stability augmentation system, the helicopter is not stable, particularly in hover. Consequently, the pilot is required to provide the feedback control to stabilize the vehicle continuously; therefore, flying a helicopter demands constant attention. The use of an automatic control system to augment the stability and control characteristics is desirable, but such systems increase the cost and complexity of the helicopter.

Table 1.1 provides a comparison of the control inputs required for various types of helicopter configurations. Except for the conventional single main rotor and one-tail rotor configuration, all other configurations use differential collective or cyclic inputs, that is, when the input to one rotor increases, the input to the other rotor decreases by the same amount.

Rotor Systems

The classification of rotor systems is essentially based on the type of the mechanical arrangement of the rotor hub and the blade attachment to accommodate the flap, lag motion, and the ability to change the pitch angle of the blade.

1. Articulated rotor: A schematic of the articulated rotor is shown in Figure 1.30. The rotor blades are attached to the rotor hub with flap (out-of-plane motion) and lag (in-plane deformation) hinges. A pitch bearing is employed to provide changes in the pitch angle of the rotor blade. The location of the hinges (offset distances) from the center of the hub plays a significant role in the performance, stability, control, and ground resonance characteristics of the vehicle. The advantage of an articulated rotor system is that the blade root bending moment is zero. The disadvantage is that there are many moving parts due to the hinges, which require frequent maintenance.
2. Teetering rotor: Two blades forming a continuous structure are attached to the rotor shaft with a single flap hinge in a teetering or seesaw arrangement, as shown in Figure 1.31. There is no lag hinge, but a pitch bearing is provided to change the pitch angle of the blade. These rotor systems are simple but can be used for small helicopters. As the weight of the helicopter becomes large, more blades may be required for the distribution of loads. Since there are only two blades, these rotor systems provide high vibratory loads.
3. Hingeless rotor (or rigid rotor): In hingeless rotor systems, the rotor blades are attached to the hub without flap and lag hinges, but it has a feathering (pitch) bearing. The blades are attached to the hub with cantilever root restraint (Figure 1.32) so that blade flap and lag motions occur through elastic bending near the root. The flexible section of the blade near its root acts like virtual hinges in the flap and the lag. This rotor system is also called "rigid rotor." However, the limit of a fully rigid blade is applicable only to propellers. The

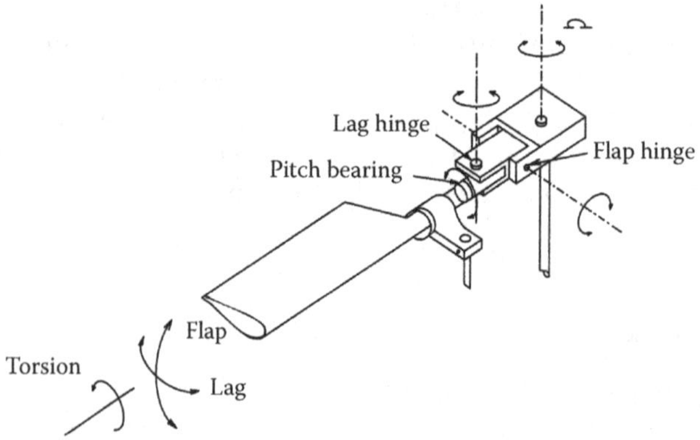

FIGURE 1.30
Schematic of articulated rotor hub/blade configuration.

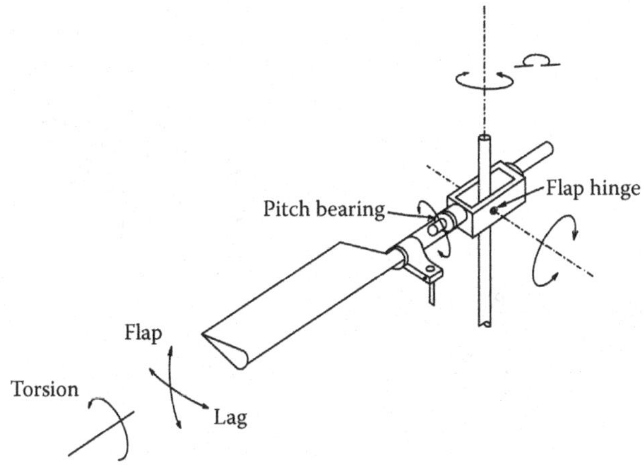

FIGURE 1.31
Schematic of a teetering (or seesaw)-type rotor configuration.

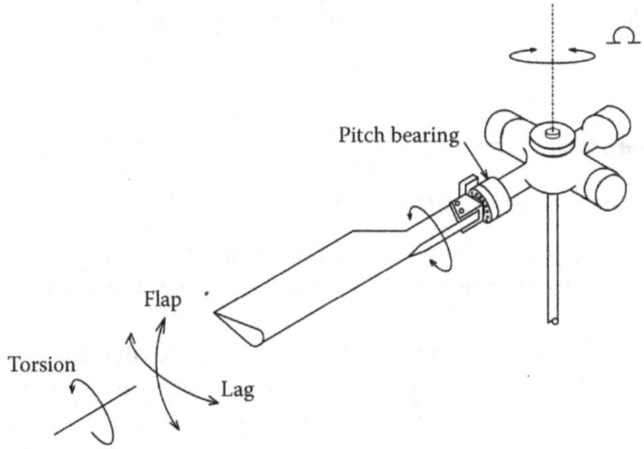

FIGURE 1.32
Schematic of a hingeless rotor configuration.

advantage of a hingeless rotor system is that it has fewer moving parts. The disadvantage is that these rotor systems will give rise to high vibratory loads at the hub.

4. Bearingless rotor: In this rotor system, not only are the flap and lag hinges eliminated, but also the pitch bearing. The rotor blades are attached to the rotor hub through a composite beam (Figure 1.33) called the "flex beam." This composite beam is not only designed to provide the required stiffness in the flap and lag deformations of the blade, but also acts as a pitch change mechanism. The pitch change in the blade is achieved by twisting a composite beam, thereby

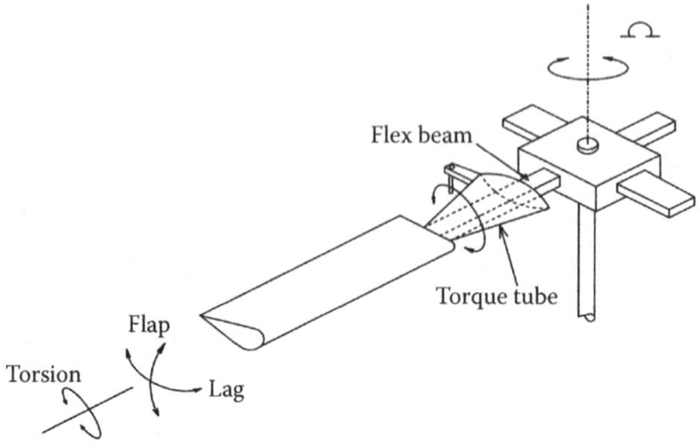

FIGURE 1.33
Schematic of a bearingless rotor configuration.

eliminating the pitch bearing. The main advantage of this rotor system is that it has fewer moving parts, but the simplicity in configuration hides the complexity in the design aspects.

Performance: Power Requirement

To fly any vehicle, power is required. In the case of helicopters, engine power must be supplied to the rotor for the following three main reasons:

- Power is required to produce lift, which is generated by pushing the air down. The power required for generating lift is referred to as "induced power." In hover, this power is about 60%.
- Power is required to drag the blades through air. This is known as "profile drag power," and in hover, it is about 30%.
- During forward flight, power must be supplied to drag the fuselage through the air in addition to the induced and profile drag losses. This power is known as "parasite drag power."

The variation of power loss with forward speed is shown in Figure 1.34. The performance capabilities of the helicopter are determined by the power available to power required curves.

From Figure 1.34, it is clear that, while parasite power increases rapidly with airspeed, the power required for producing lift, that is, the induced power, decreases with increasing speed. As the rotor moves forward, the rotor encounters a large mass of air per second. Therefore, to produce the same thrust, the rotor needs to impact less velocity to the mass of air, and

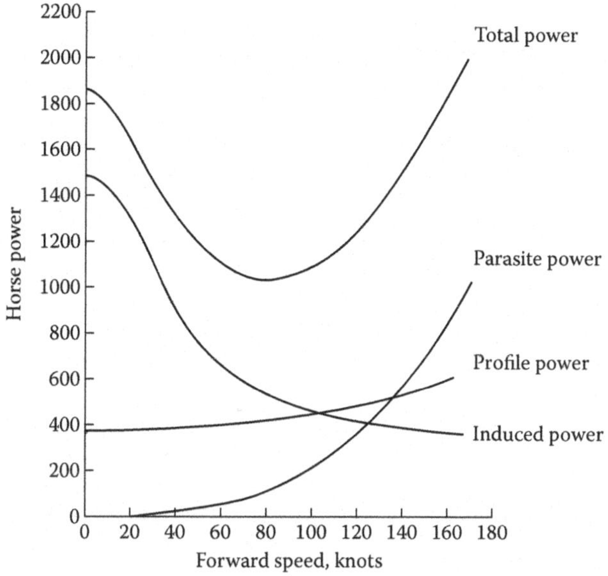

FIGURE 1.34
Power variation with forward speed.

hence, the energy imparted to air is reduced. Profile drag power increases slightly initially and increases at a higher rate at high speeds. The sum of these power losses is also plotted in the figure. It is clear that, if power available is just equal to the power required to hover, the performance of the machine is marginal. The vehicle can barely hover and will be unable to climb. It may be noted that, near the ground, a helicopter will be able to hover even when it has insufficient power to hover away from the ground. This is due to a phenomenon known as "ground effect." The ground stops the rotor downwash (or induced velocity), thus decreasing the induced power required to hover. It is also noted that the reduction in power required with forward speed enables an overloaded helicopter to take off in wind or to make a ground run to attain a small forward speed.

Apart from these three power losses, there are additional losses due to nonuniform flow, swirl in the wake, tip losses, loss in transmission systems, rotor-fuselage aerodynamic interference, tail rotor, etc.

- Nonuniform inflow: 5–7%
- Swirl loss: 1%
- Tip loss: 2–4%
- Engine transmission: 4–5% (turbine)
- Tail rotor: 7–9%
- Rotor-fuselage aerodynamic interference: ~2%

2

Introduction to Hovering and Vertical Flight Theory

The lifting rotor is assumed to be in hovering condition when both the rotor and the air outside the slip stream are stationary, that is, there is no relative velocity between the rotor and the air outside the slip stream. The airflow developed due to the rotor is confined inside a well-defined imaginary slip stream, as shown in Figure 2.1. There is an axial symmetry in the airflow inside the slip stream.

The hovering theory (or momentum theory) was formulated for marine propellers by W. J. Rankine in 1865 and was later developed by R. E. Froude in 1885. Subsequently, Betz (1920) extended the theory to include rotational effect.

Momentum Theory

Momentum theory is based on the basic conservation laws of fluid mechanics (i.e., conservation of mass, momentum, and energy). The rotor is continuously pushing the air down. As a result, the air in the rotor wake (inside the slip stream) acquires a velocity increment or a momentum change. Hence, as per Newton's third law, an equal and opposite reaction force, denoted as rotor thrust, is acting on the rotor due to air. It may be noted that the velocity increment of the air is directed opposite to the thrust direction.

Assumptions of Momentum Theory

The rotor is assumed to consist of an infinite number of blades and may therefore be considered as an "actuator disk." The actuator disk is infinitely thin so that there is no discontinuity in the velocity of air as it flows through the disk. The rotor is uniformly accelerating the air through the disk with no loss at the tips. The axial kinetic energy imparted to the air in the slip stream is equal to the power required to produce the thrust. In addition, air

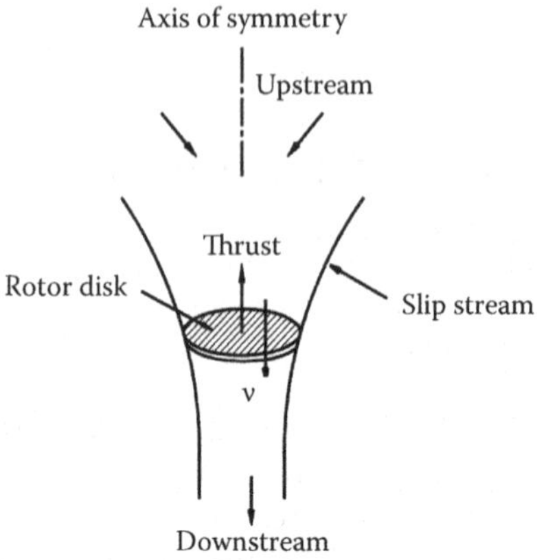

FIGURE 2.1
Rotor disk and slip stream.

is assumed to be incompressible and frictionless. There is no profile drag loss in the rotor disk, and the rotational energy (swirling motion of air) imparted to the fluid is ignored.

NOTE: The actuator disk model is only an approximation to the actual rotor. The momentum theory is not concerned with the details of the rotor blades or the flow, and hence, this theory by itself is not sufficient for designing the rotor system. However, it provides an estimate of the induced power requirements of the rotor and also of the ideal performance limit. The slip stream of the actuator disk in hovering condition is shown in Figure 2.2.

The rotor disk is represented by a thin disk of area A (= πr^2); the far field upstream is denoted as station 1, and the far field downstream is denoted as station 4. The pressure of air at stations 1 and 4 is atmospheric pressure P_0.

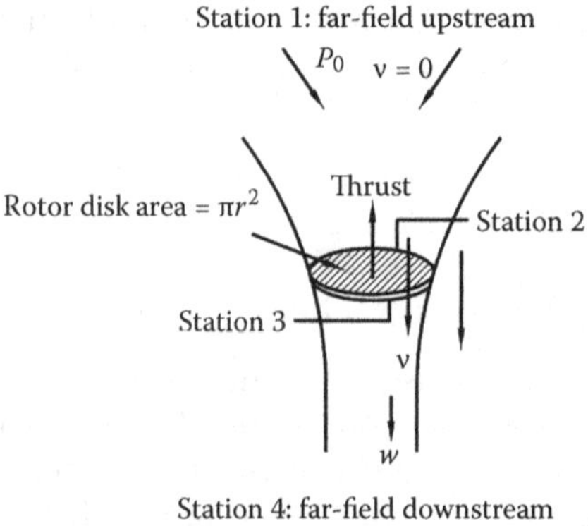

FIGURE 2.2
Flow condition in the slip stream.

Stations 2 and 3 represent the locations just above and below the rotor disk, respectively. It is assumed that the loading is uniformly distributed over the disk area. The induced velocity or inflow velocity is ν at the rotor disk, and w is the far field wake–induced velocity. The fluid is assumed to be incompressible, having a density ρ. The conservation laws are as follows:

Mass flow rate is given as

$$\dot{m} = \rho A \nu \tag{2.1}$$

Momentum conservation is obtained by relating the force to the rate of momentum change, which is given as

$$T = \dot{m}(w - 0) = \rho A \nu w \tag{2.2}$$

Energy conservation relates the rate of work done on the air to its change in kinetic energy per second, which is given as

$$T\nu = \frac{1}{2}\dot{m}(w^2 - 0) = \frac{1}{2}\dot{m}w^2 \tag{2.3}$$

Substituting for $\dot{m}$ from Equation 2.1 and using Equation 2.2, Equation 2.3 can be written as

$$\rho A \nu w \nu = \frac{1}{2}\rho A \nu \cdot w^2$$

Cancelling the terms results in

$$\nu = \frac{1}{2}w \quad \Rightarrow \quad w = 2\nu \tag{2.4}$$

This shows that the far field–induced velocity is twice the induced velocity at the rotor disk.

Substituting for w in Equation 2.2, the expression for rotor thrust in terms of induced velocity at the rotor disk is given by

Rotor thrust:

$$T = \rho A \nu 2\nu \tag{2.5}$$

or the induced velocity is given as

$$\nu = \sqrt{\frac{T}{2\rho A}} \tag{2.6}$$

The induced power loss or the power required to develop the rotor thrust T is given as

$$P = Tv = T\sqrt{\frac{T}{2\rho A}} \tag{2.7}$$

The induced power per unit thrust for a hovering rotor can be written as

$$\frac{P}{T} = v = \sqrt{\frac{T}{2\rho A}} \tag{2.8}$$

The above expression indicates that, for a low inflow velocity, the efficiency is higher. This is possible if the rotor has a low disk loading (T/A). In general, the disk loading of helicopters is of the order of 100–500 N/m^2, which is the lowest disk loading for any vertical take-off and landing (VTOL) vehicle. Therefore, the helicopters have the best hover performance. Note that the parameter determining the induced power is essentially $T/(\rho A)$. Therefore, the effective disk loading increases with an increase in altitude and temperature.

The pressure variation along the slip stream can be determined from the steady-state Bernoulli equation. The pressure between stations 1 and 2 and between stations 3 and 4 are related, respectively, as

$$p_0 = p_1 + \frac{1}{2}\rho v^2 \quad \text{between stations 1 and 2} \tag{2.9}$$

and

$$p_2 + \frac{1}{2}\rho v^2 = p_0 + \frac{1}{2}\rho w^2 \quad \text{between stations 3 and 4} \tag{2.10}$$

The pressure variation along the slip stream is shown in Figure 2.3. There is a pressure jump across the rotor disk even though the velocity is continuous.

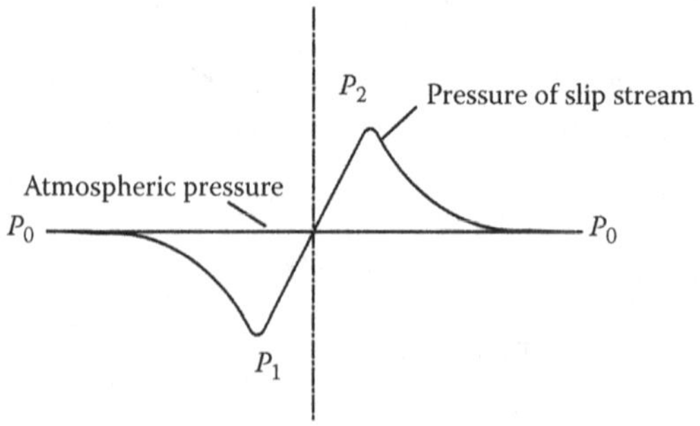

FIGURE 2.3
Pressure variation along the slip stream.

The jump in pressure is caused by the power given to the rotor to push the air downstream.

From Equations 2.9 and 2.10, the rotor thrust can be evaluated.

$$T = (p_2 - p_1)\ A = \frac{1}{2}\rho w^2\ A = \frac{1}{2}\rho(2v)^2\ A = 2\rho A v^2 \tag{2.11}$$

The various quantities can be written in nondimensional form. Using the tip speed ΩR of the rotor blade as reference, the rotor inflow is represented in nondimensional form as

$$\text{Inflow ratio: } \lambda = \frac{v}{\Omega R} = \sqrt{\frac{C_T}{2}} \tag{2.12}$$

Thrust coefficient C_T and power coefficient C_P are defined, respectively, as

$$\text{Thrust coefficient: } C_T = \frac{T}{\rho A(\Omega R)^2} \tag{2.13}$$

$$\text{Power coefficient: } C_P = \frac{P}{\rho A(\Omega R)^3} = \frac{T}{\rho A(\Omega R)^2}\frac{v}{\Omega R} = \frac{C_T^{3/2}}{\sqrt{2}} \tag{2.14}$$

The hovering efficiency of the rotor is defined as

$$M = \frac{\text{minimum power required to hover (ideal power)}}{\text{Actual power required to hover}} = \frac{Tv}{P_{\text{actual}}} \tag{2.15}$$

M is called the figure of merit (FM). The ideal value of M is equal to 1. However, for practical rotors, the value will be less than 1. For good rotors, M is in the range of 0.75 to 0.8. For inefficient rotors, FM will have a value around 0.5. FM can be used for comparing the efficiency of different rotor systems. It should be noted that FM is defined only for the hovering condition of the rotor.

Blade Element Theory

Blade element theory (BET) is the foundation for all analyses of helicopter dynamics and aerodynamics because it deals with the details of the rotor system. This theory relates the rotor performance and the dynamic and aeroelastic characteristics of the rotor blade to the detailed design parameters. In contrast, although momentum theory is useful to predict the rotor-induced velocity for a given rotor thrust, it cannot be used to design the rotor system and the rotor blades.

The basic assumption of BET is that the cross section of each rotor blade acts as a two-dimensional airfoil to produce aerodynamic loads, that is, sectional lift, drag, and pitching moment. The effect of the rotor wake is entirely represented by an induced angle of attack at each cross section of the rotor blade. Therefore, this theory requires an estimate of the wake-induced velocity at the rotor disc. This quantity can be obtained either from the simple momentum theory (as given in the previous section) or from more complex theories, such as the prescribed wake or the free wake vortex theories, or nonuniform inflow calculations using acceleration potential.

History of the Development of BET

The origin of BET can be attributed to the work of Willium Froude (1878). However, the first major treatment was by Stefan Drzewiecki, during 1892 to 1920. Drzewiecki considered different blade sections to act independently but was not certain about the aerodynamic characteristics to be used for the airfoils. The two velocity components considered in the theory are (1) tangential velocity Ωr due to rotation and (2) axial velocity V of the propeller. Note that the inflow component at the rotor disk was not included. In Drzewiecki's calculations, the estimated performance exhibited a significant error, which was attributed to the airfoil characteristics. Since the aspect ratio affects the aerodynamic characteristics (in fixed wings), Drzewiecki proposed that the three-dimensional wing characteristics (with appropriate aspect ratio) be used in the BET. The results of this theory had the correct general behavior but were found to be quantitatively inaccurate.

Several attempts (Betz [1915] and Bothezat [1918]) were made to include the increased axial velocity from the momentum theory into the BET. However, Prandtl's finite wing theory provided a proper framework for the correct treatment and inclusion of the influence of the propeller wake on the aerodynamic environment at the blade section. In fixed wing, the lifting line theory is used for the calculation of induced velocity from the properties of the vortex wake. Thus, following the same approach, the vortex wake was used to define the induced velocities at each cross section of the rotor blade. This theory is called the "vortex theory." It was through this approach rather than through the momentum theory that induced velocity was finally incorporated correctly in the BET. Therefore, during the initial stages of development, the vortex theory dominated the momentum theory in evaluating the inflow at the rotor disk. The vortex theory is also regarded as a reliable approach in both fixed and rotary wing analyses.

BET for Vertical Flight

In this section, as a general case, BET is applied to a rotor that has a vertical velocity. If the vertical velocity term is set equal to zero, it represents the hovering condition of the rotor. While developing the BET, several assumptions are made. The important assumptions are as follows:

1. In the preliminary highly simplified case, the rotor blade is assumed to be a rigid beam with no deformation. The blade can have a pretwist. (The effects of blade deformation will be considered in later chapters.)
2. The rotor system is rotating with a constant angular velocity Ω.
3. The plane of rotation of the rotor blades is perpendicular to the shaft.
4. The rotor is operating at low disk loading, i.e., a small value of inflow velocity.
5. Compressibility and stall effects are neglected.

Figure 2.4 shows a rotor system along with the nonrotating hub fixed axes system (X_H–Y_H–Z_H) and the rotating blade fixed axes system (X_1–Y_1–Z_1). Figure 2.5 shows a typical cross section of the rotor blade at a radial distance r from the center of rotation (or the hub center), various velocity components, and the resultant forces acting on the airfoil section. The blade is set at a pitch angle θ measured from the plane of rotation. U_T and U_P are, respectively, the tangential and the perpendicular relative air velocity components, as viewed by an observer on the blade section.

In vertical flight condition, U_P consists of the climb velocity V_C of the helicopter (or rotor) and the induced velocity v. Note that $V_C = 0$ for hovering condition. The tangential component of velocity U_T is due to rotor rotation.

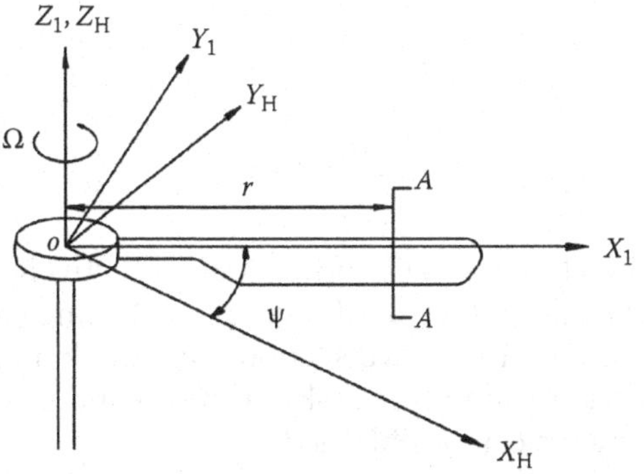

FIGURE 2.4
Nonrotating hub fixed and rotating blade fixed coordinate systems.

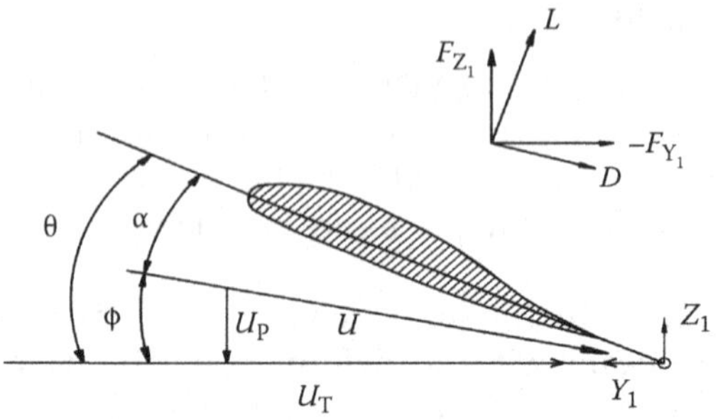

FIGURE 2.5
Typical cross section of a rotor blade at radial location *r* and the velocity components.

The relative air velocity components U_T and U_P for vertical flight can be written as

$$U_T = \Omega r \quad \text{and} \quad U_P = V_C + \nu \tag{2.16}$$

The resultant air velocity U and the inflow angle are given by

$$U = \sqrt{U_T^2 + U_P^2}$$
$$\tan\phi = \frac{U_P}{U_T} \tag{2.17}$$

It can be seen that the vertical component of velocity U_P modifies the angle of attack. Thus, the effective angle of attack at the blade section is given as

$$\alpha = \theta - \phi \tag{2.18}$$

The sectional lift and drag forces can be written as

$$L = \frac{1}{2}\rho U^2 C C_l$$
$$D = \frac{1}{2}\rho U^2 C C_d \tag{2.19}$$

where C is the blade chord, and C_l and C_d are, respectively, the lift and drag aerodynamic coefficients, which are functions of the angle of attack and the Mach number. Resolving these two sectional forces along parallel and perpendicular directions to the rotor disk, the vertical and horizontal (or in-plane) force components can be obtained as

$$F_Z = L\cos\phi - D\sin\phi$$
$$-F_{Y1} = L\sin\phi + D\cos\phi \tag{2.20}$$

Combining the forces due to all the blades in the rotor system, the elemental thrust, torque, and power due to all the blades in the rotor system can be obtained. The following expressions are obtained by noting that the sectional forces acting on the blade cross section at a radial distance r are the same on all the blades.

$$\begin{aligned} dT &= NF_Z dr \\ dQ &= -NF_{Y1} r dr \\ dP &= \Omega |dQ| = N |F_{Y1}| \Omega r dr \end{aligned} \tag{2.21}$$

where N is the total number of blades in the rotor system. The negative sign in the torque expression indicates that aerodynamic drag force on the blade provides a clockwise torque when the rotor is rotating in a counterclockwise direction, as shown in Figure 2.1. The positive quantity of the torque essentially represents the torque applied to the rotor by the engine to keep the rotor rotating at the prescribed angular velocity.

Assuming that $U_P \ll U_T$, one can make a small angle assumption for the inflow angle. Note that this assumption is not valid near the blade root. However, since the dynamic pressure near the root is very small, the aerodynamic loads are also of small magnitude. Hence, the error due to the small angle assumption can be considered to be negligible. In addition, due to root cutout, the cross sections near the root (~20% of the rotor radius) do not produce any significant aerodynamic lift.

Making a small angle assumption, one can write

$$\begin{aligned} &\cos\phi \approx 1 \\ &\sin\phi \approx \phi \approx \frac{U_P}{U_T} \\ &U \approx U_T \\ &\text{and } C_l = a\alpha = a(\theta - \phi) \end{aligned} \tag{2.22}$$

where a is the lift curve slope.

Substituting the above approximations, the expressions for lift and drag per unit length, from Equation 2.19, can be written as

$$\begin{aligned} L &\cong \frac{1}{2}\rho U_T^2 Ca\left(\theta - \frac{U_P}{U_T}\right) = \frac{1}{2}\rho Ca\left(U_T^2\theta - U_P U_T\right) \\ D &\approx \frac{1}{2}\rho U_T^2 C C_d \end{aligned} \tag{2.23}$$

Also, the force components can be approximated as

$$F_Z \approx L$$
$$-F_{Y1} \approx L\phi + D \tag{2.24}$$

Using this approximation, the elemental thrust, torque (applied torque), and power (Equation 2.21), can be expressed as

$$dT \approx NLdr$$
$$dQ \approx N(L\phi + D)rdr \tag{2.25}$$
$$dP \approx N(L\phi + D)\Omega rdr$$

The elemental thrust, torque, and power quantities can be nondimensionalized by using appropriate reference quantities, as

$$dC_T = \frac{dT}{\rho A(\Omega R)^2}$$
$$dC_Q = \frac{dQ}{\rho A(\Omega R)^2 R} \tag{2.26}$$
$$dC_P = \frac{dP}{\rho A(\Omega R)^3}$$

Thrust Coefficient

Using Equations 2.16, and 2.23 to 2.26, the elemental thrust coefficient can be written as

$$dC_T = \frac{N\frac{1}{2}\rho(\Omega r)Ca[\Omega r\theta - (V_C + v)]dr}{\rho\pi R^2(\Omega R)^2}$$
$$dC_T = \frac{1}{2}\frac{NC}{\pi R}a[\theta\bar{r}^2 - \lambda\bar{r}]d\bar{r} \tag{2.27}$$
$$dC_T = \frac{\sigma a}{2}[\theta\bar{r}^2 - \lambda\bar{r}]d\bar{r}$$

Solidity ratio is defined as σ = blade area/rotor disk area, where $\sigma = \frac{NCR}{\pi R^2}$ is the solidity ratio for the constant chord blade.

$$\lambda = \frac{V_C + v}{\Omega R} \text{ is the total inflow ratio}$$

$$\bar{r} = \frac{r}{R} \tag{2.28}$$

The differential expressions can be integrated over the length of the blade by assuming a constant chord, and a uniform inflow.

The thrust coefficient is obtained as

$$C_T = \int_0^1 \frac{\sigma a}{2}[\theta \bar{r}^2 - \lambda \bar{r}]d\bar{r} \tag{2.29}$$

Assuming a linear twist variation for the blade pitch angle along the span of the blade as

$$\theta = \theta_0 + \bar{r}\theta_{tw}(\text{or } \theta = \theta_{0.75} + (\bar{r} - 0.75)\theta_{tw}) \tag{2.30}$$

where θ_{tw}, θ_0, and $\theta_{0.75}$ are the blade twist rate due to the pretwist of the blade, the pitch angles at the blade root, and at a radius 0.75 R, respectively. Substituting Equation 2.30 in Equation 2.29 and integrating over the length of the blade, the thrust coefficient becomes

$$C_T = \frac{\sigma a}{2}\left[\frac{\theta_{0.75}}{3} - \frac{\lambda}{2}\right] \tag{2.31}$$

On the other hand, a twist distribution of $\theta = \frac{\theta_{tip}}{\bar{r}}$ (known as the "ideal twist distribution") results in a thrust coefficient expression given as

$$C_T = \frac{\sigma a}{4}[\theta_{tip} - \lambda] \tag{2.32}$$

This ideal twist distribution, while physically not possible to achieve, is of interest because it gives a uniform inflow over the rotor disk for constant chord blades. This twist distribution is denoted as ideal twist distribution because the momentum theory shows that uniform inflow provides a minimum induced power loss.

Now, it is shown that BET provides a relationship between rotor thrust coefficient, pitch angle, and inflow ratio. On the other hand, momentum theory gives a relationship between thrust coefficient and inflow ratio.

For hovering condition, in the case of constant chord and a linearly twisted blade, thrust coefficient is written as (note: $V_C = 0$)

$$C_T = \frac{\sigma a}{2}\left[\frac{\theta_{0.75}}{3} - \frac{\lambda}{2}\right] \tag{2.33}$$

From the momentum theory, the inflow ratio in hover is given as (Equation 2.12)

$$\lambda = \sqrt{\frac{C_T}{2}}$$

Combining the above two expressions, the relationship between $\theta_{0.75}$ and C_T can be written as

$$\theta_{0.75} = \frac{6C_T}{\sigma a} + \frac{3}{2}\sqrt{\frac{C_T}{2}} \tag{2.34}$$

The first term corresponds to the mean angle of attack of the rotor blade, while the second term is the additional pitch angle required due to the induced inflow. The above three relationships can be used to obtain an estimate of the thrust coefficient, the pitch angle at 0.75 R, and the inflow ratio under hovering condition.

Using Equations 2.12 and 2.33, a relationship between the inflow ratio and the blade pitch angle $\theta_{0.75}$ can be obtained as

$$\lambda = \frac{\sigma a}{16}\left\{\sqrt{1 + \frac{64}{3\sigma a}\theta_{0.75}} - 1\right\} \tag{2.35}$$

Torque/Power Coefficient

The elemental torque and power coefficients can be obtained from Equations 2.16, 2.23, 2.25, and 2.26. They are shown to be equal, and it is given as

$$dC_P = dC_Q = \left[\frac{\sigma a}{2}\left(\bar{U}_P\bar{U}_T\theta - \bar{U}_P^2\right) + \frac{\sigma C_d}{2}\bar{U}_T^2\right]\bar{r}d\bar{r}$$
$$dC_P = dC_Q = \left[\frac{\sigma a}{2}(\theta\bar{r}\lambda - \lambda^2) + \frac{\sigma C_d}{2}\bar{r}^2\right]\bar{r}d\bar{r} \tag{2.36}$$

The velocity quantities with overbar represent nondimensional quantities obtained by dividing with reference velocity ΩR, which is the tip speed of the rotor blade.

The elemental power coefficient can be written as

$$dC_p = \left[\lambda\frac{\sigma a}{2}\left[\theta\bar{r}^2 - \lambda\bar{r}\right] + \frac{\sigma C_d}{2}\bar{r}^3\right]d\bar{r} \tag{2.37}$$

Noting from Equation 2.27 that the first term contains an elemental thrust coefficient, Equation 2.37 can be written as

$$dC_p = \lambda dC_T + \frac{\sigma C_d}{2}\bar{r}^3 d\bar{r} \tag{2.38}$$

Integrating both sides over the rotor system, the power coefficient can be obtained as

$$C_P = \int \lambda dC_T + \int \frac{\sigma C_d}{2}\bar{r}^3\, d\bar{r}$$
$$C_p = C_{pi} + C_{Ppd} \tag{2.39}$$

The power coefficient C_{pi} represents the power loss due to total induced flow, and C_{Ppd} is the power loss due to profile drag.

For uniform inflow, the power due to total induced flow is $C_{pi} = \lambda C_T$.

Since total inflow $\lambda = \frac{V_C + \nu}{\Omega R}$, during climb, C_{pi} includes the power required for climbing as well as the induced power loss. Therefore, one can write the induced power in terms of two quantities, namely climb power and induced power, that is,

$$C_{pi} = C_{pc} + C_{piv}$$

In other words, this power term can be expressed as

$$P_i = P_c + P_{iv} = V_C T + \nu T = (V_C + \nu)T \tag{2.40}$$

During hover $\lambda = \sqrt{\frac{C_T}{2}}$, therefore, the induced power loss becomes

$$C_{P_i} = C_{P_{iv}} = \frac{C_T^{3/2}}{\sqrt{2}} \tag{2.41}$$

This is the induced power loss in hover under ideal condition. However, for a real rotor with a practical twist, a planform, and a finite number of blades, the induced power loss will be higher than the ideal value $\frac{C_T^{3/2}}{\sqrt{2}}$. One way to compute the induced power loss is to integrate $\int \lambda dC_T$ under real conditions, taking into account the nonuniform inflow over the rotor disk. On the other hand, one can use the same expression as the momentum theory expression for power loss but with an empirical factor κ to account for the additional losses due to the real situation, that is,

$$C_{pi} = \frac{\kappa C_T^{3/2}}{\sqrt{2}} \tag{2.42}$$

The factor κ accounts for nonideal conditions, and its value is usually taken as 1.15, which implies an additional power of about 15% more than the ideal power.

Next, consider the profile power term from Equation 2.39, which is given as

$$C_{Ppd} = \int \frac{\sigma C_d}{2} \bar{r}^3 \, d\bar{r} \tag{2.43}$$

Assuming the constant chord for the blade with the drag coefficient $C_d = C_{do}$ (a constant), the profile power loss can be obtained by integrating over the rotor radius. It is given as

$$C_{Ppd} = \frac{\sigma C_{do}}{8} \tag{2.44}$$

Combining Equations 2.42 and 2.44, the total power loss in hover can be expressed as

$$C_p = \kappa \frac{C_T^{3/2}}{\sqrt{2}} + \frac{\sigma C_{do}}{8}$$

The efficiency of the rotor is expressed as the ratio of the ideal power over the actual power. This ratio is denoted as figure of merit, which is given by

$$M = \frac{C_P \text{ ideal}}{C_{Pi} + C_{Ppd}} = \frac{\dfrac{C_T^{3/2}}{\sqrt{2}}}{\kappa \dfrac{C_T^{3/2}}{\sqrt{2}} + \dfrac{\sigma C_{do}}{8}} \quad (2.45)$$

A plot of figure of merit M as a function of thrust C_T (for a set of given values of (σ, C_{do}, κ)) is shown in Figure 2.6. It can be noted from the figure that, for low values of thrust coefficient, due to a comparatively high value of profile drag, the figure of merit is low. Thus, the rotor is not operating efficiently. As C_T increases, the figure of merit increases. For high values of C_T ($\approx$0.006–0.01), figure of merit does not show a large variation. This functional form indicates that figure of merit asymptotically approaches unity as the C_T is increased. In practical situations, for a large value of C_T, the angle of attack has to be large, which can lead to blade stall and an associated increase in drag. Hence, there will be a reduction in figure of merit.

Figure of merit is a useful parameter for comparing the efficiency of different rotors having the same disk loading. For a given value of C_T, figure of merit M will have a high value for a low value of solidity ratio σ and drag coefficient C_{d0}. If the rotor solidity is too low, a high value of angle of attack will be required to achieve the given thrust (Equation 2.34). Therefore, the rotor should have as low a value of solidity as possible with an adequate stall margin for the airfoil. Blade twist and variable chord also influence the induced and profile power losses. A study of these effects requires a more detailed analysis, which can be taken as an exercise.

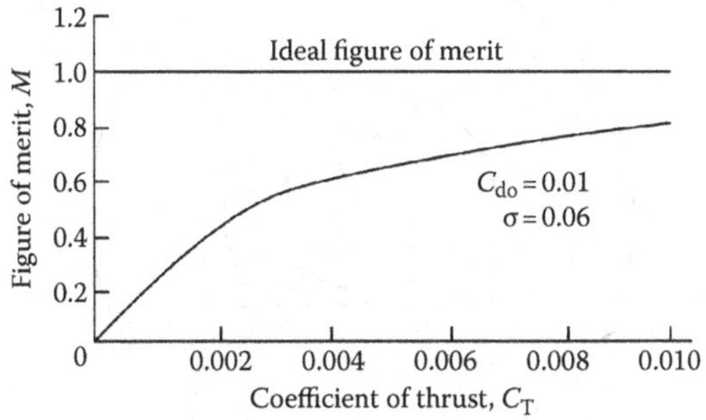

FIGURE 2.6
Variation of figure of merit with thrust coefficient.

Combined Differential Blade Element and Momentum Theory: Nonuniform Inflow Calculation

The nonuniform inflow distribution as a function of radial distance from the center of the rotor disk can be obtained by comparing the differential thrust expressions obtained from both the momentum theory and the BET. Consider a circular strip of differential area $dA = 2\pi r dr$ at a radial distance r from the center of the rotor disk, as shown in Figure 2.7. From BET (Equation 2.27), the differential thrust expression over this differential area is given as

$$dC_T = \frac{\sigma a}{2}(\theta \bar{r}^2 - \lambda \bar{r})d\bar{r} \quad (2.46)$$

The differential thrust expression from the momentum theory can be written as

$$dT = \rho(V + \nu)2\nu(2\pi r dr) \quad (2.47)$$

In nondimensional form, Equation 2.47 can be written as

$$dC_T = 4\lambda\lambda_i \bar{r} d\bar{r} \quad (2.48)$$

where total inflow $\lambda = \frac{V+\nu}{\Omega R}$ and induced inflow $\lambda_i = \frac{\nu}{\Omega R}$ and $\lambda_c = \frac{V}{\Omega R}$. Hence, $\lambda_i = \lambda - \lambda_c$.

Equating the two differential thrust expressions given in Equations 2.46 to 2.48, an equation for λ_i can be formed:

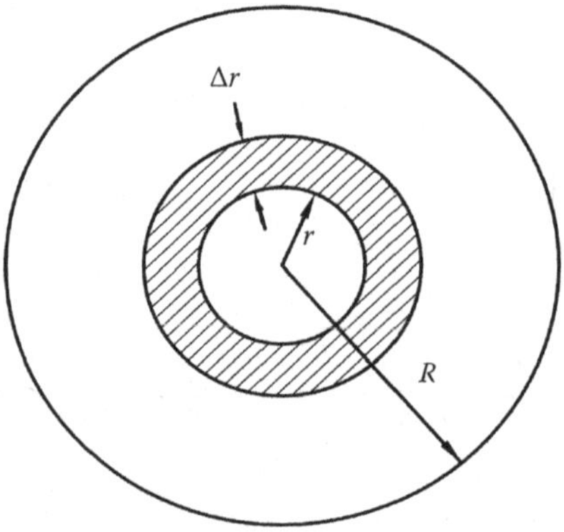

FIGURE 2.7
Annular area for nonuniform inflow calculation.

$$4(\lambda_c + \lambda_i)\lambda_i = \frac{\sigma a}{2}(\theta \bar{r} - \lambda_c - \lambda_i)$$
$$4\lambda_i^2 + \lambda_i\left[4\lambda_c + \frac{\sigma a}{2}\right] - \frac{\sigma a}{2}(\theta \bar{r} - \lambda_c) = 0 \tag{2.49}$$

Solving for λ_i and taking the positive root,

$$\lambda_i = \frac{-\left(4\lambda_c + \frac{\sigma a}{2}\right) + \sqrt{\left(4\lambda_c + \frac{\sigma a}{2}\right)^2 + 8\sigma a(\theta \bar{r} - \lambda_c)}}{8}$$

or

$$\lambda_i = -\left(\frac{\lambda_c}{2} + \frac{\sigma a}{16}\right) + \sqrt{\left(\frac{\lambda_c}{2} + \frac{\sigma a}{16}\right)^2 + \frac{\sigma a}{8}(\theta \bar{r} - \lambda_c)} \tag{2.50}$$

For the hovering condition, $\lambda_c = 0$, and the inflow ratio becomes

$$\lambda = \lambda_i = \frac{\sigma a}{16}\left[\sqrt{1 + \frac{32}{\sigma a}\theta \bar{r}} - 1\right] \tag{2.51}$$

Equation 2.51 provides an expression for the nonuniform inflow distribution in hover. It can be seen that the inflow ratio is a function of radial distance. (Compare this expression with the uniform inflow expression derived earlier and given in Equation 2.35.) Using the above equation, for given values of pitch, twist, and chord, the inflow λ_i can be calculated as a function of radius. It can be observed from Equation 2.51 that a uniform inflow distribution requires that $\theta \bar{r} = \text{a constant}$. In other words, the ideal twist distribution is given by the expression

$$\theta = \frac{\theta_{\text{tip}}}{\bar{r}} \tag{2.52}$$

It can be shown that uniform inflow results in uniform disk loading (i.e., equal areas of rotor disk support equal thrust, that is, $\frac{dT}{dA}$ = a constant) and also provides minimum induced power loss. (The proof can be taken as an exercise.)

Although the rotor performance evaluated using nonuniform inflow may be more accurate than that evaluated using uniform inflow, the differential momentum theory is still only an approximate theory. A more refined inflow distribution can be obtained using the vortex theory or from computational fluid dynamics (CFD) calculations.

Tip Loss Factor

When the chord at the blade tip is finite, BET gives a nonzero lift. However, the lift drops to zero at the tip due to three dimensional flows. Since the dynamic pressure is proportional to r^2, the blade loading is concentrated near the tip and drops steeply to zero at the tip, as shown in Figure 2.8. The loss of lift at the tip is very important in calculating rotor performance.

A rigorous treatment of tip loading requires a lifting surface analysis. An approximate method to account for tip losses is to assume that the blade element outboard of the radial station $r = BR(B < 1.0)$ has only profile drag but produces no lift. The parameter B is called "tip loss factor." A number of methods are available for evaluating B. Prandtl gave an expression for tip loss factor in terms of thrust coefficient and number of blades, which can be approximately given as (Bramwell, 1976):

$$B = 1 - \frac{\sqrt{C_T}}{N}$$

Generally, tip loss factor is taken as $B = 0.97$, and it provides a good correlation with the experimental results.

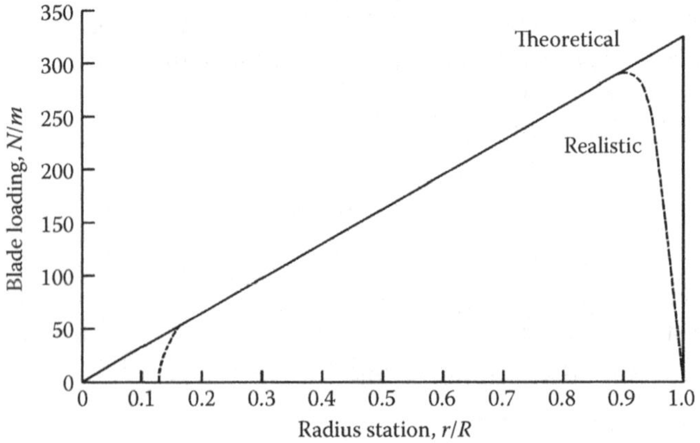

FIGURE 2.8
Spanwise lift distribution on the blade, showing the tip and the root offset effects.

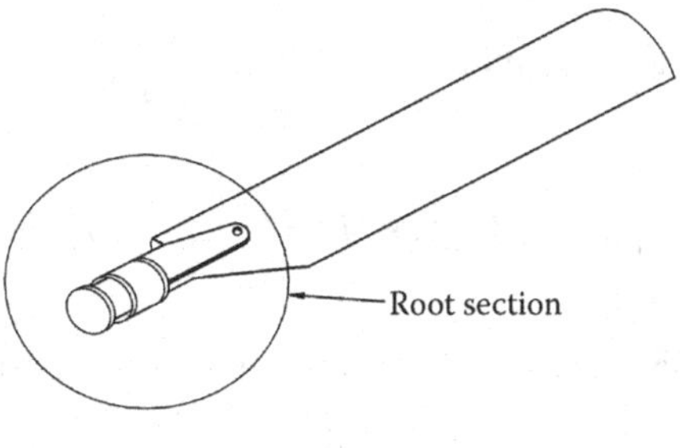

FIGURE 2.9
Rotor blade root section/root cutout.

Blade Root Cutout

Performance losses can also occur due to root cutout (Figure 2.9). The lifting portion of the blade generally starts at a radial station of about 10% to 30% of the rotor radius. The root cutout portion is required for installing flap-lag hinges, pitch bearings, blade root attachment to the hub, etc.

Considering both root cutout and tip losses, in BET, the integration for thrust is performed from $\bar{r} = \bar{r}_{\text{root}}$ to $\bar{r} = B$, that is,

$$C_T = \int_{\bar{R}_{\text{root}}}^{B} dC_T \tag{2.53}$$

Since the dynamic pressure at the root is very low, the correction to the thrust calculation is also very small. Therefore, in the following, these effects have been neglected. However, in design calculations, one must take into account these effects.

Mean Lift Coefficient

The operating condition of the rotor can be represented by defining the mean aerodynamic lift coefficient $\bar{C}_l$, which is obtained as follows. From Equation 2.46, one can write the thrust expression in terms of the lift coefficient C_l.

$$dC_T = \frac{\sigma a}{2}(\theta\bar{r}^2 - \lambda\bar{r})d\bar{r} = \frac{\sigma}{2}C_l\bar{r}^2 d\bar{r} \tag{2.54}$$

where the lift coefficient is given as $C_l = a\left(\theta - \frac{\lambda}{\bar{r}}\right)$

$$C_T = \int_0^1 dC_T = \int_0^1 \frac{\sigma}{2} C_l \bar{r}^2 \, d\bar{r} = \bar{C}_l \int_0^1 \frac{\sigma}{2} \bar{r}^2 \, d\bar{r} = \bar{C}_l \frac{\sigma}{6} \tag{2.55}$$

From Equation 2.55, the mean lift coefficient can be expressed as

$$\bar{C}_l = \frac{6C_T}{\sigma} \tag{2.56}$$

$\frac{C_T}{\sigma}$ is the ratio of the thrust coefficient to the solidity ratio. This quantity is a measure of the mean lift coefficient of a blade. Correspondingly, $\frac{6C_T}{\sigma a}$ may be interpreted as the mean angle of attack of the blade. It can be shown that $\frac{C_T}{\sigma}$ represents dimensionless blade loading, whereas $\frac{T}{\pi R^2}$ is the disk loading.

$$\frac{C_T}{\sigma} = \frac{T/(\rho \pi R^2 (\Omega R)^2)}{(\text{Total blade area}/\pi R^2)} = \frac{T}{\rho(\text{Total blade area})(\Omega R)^2} \tag{2.57}$$

$\frac{C_T}{\sigma}$ plays an important role in rotor aerodynamics since many characteristics of the rotor and the helicopter depend on the mean lift coefficient of the blade. The rotor figure of merit in hover, from Equation 2.45, can be expressed as

$$M = \frac{\lambda_h C_T}{\kappa \lambda_h C_T + \frac{\sigma C_{d0}}{8}}$$

where $\lambda_h = \sqrt{\frac{C_T}{2}}$ is the inflow ratio during hover. This expression can also be written as

$$M = \frac{1}{\kappa + \frac{\sigma C_{d0}}{8 \lambda_h C_T}}$$

Substituting for C_T in terms of $\bar{C}_l$ from Equation 2.56, as $C_T = \frac{\sigma \bar{C}_l}{6}$,

$$M = \frac{1}{\kappa + \frac{3}{4} \frac{C_{do}}{\bar{C}_l} \frac{1}{\lambda_h}} \tag{2.58}$$

This expression indicates that, for a high value of figure of merit, the blade must have an airfoil that has a high value of lift-to-drag ratio.

Ideal Rotor versus Optimum Rotor

Ideal rotor is one where the induced power loss is a minimum. Hence, it requires a uniform inflow, with the rotor blade having an ideal twist distribution. Optimum rotor is one which is optimized to have both induced and profile power losses at a minimum. Minimum induced power requires uniform inflow. Minimum profile power requires that each blade section operates at its optimum condition with a maximum value of $\frac{C_l}{C_d}$. These two criteria determine the twist and taper for the optimum rotor, which has the best hover performance. Due to manufacturing considerations, almost all rotor blades have a constant chord over a major portion of the rotor blade.

Momentum Theory for Vertical Flight

Vertical flight of the helicopter at a speed V includes climb ($V > 0$), hover ($V = 0$), descent ($V < 0$), and also the special case of autorotation (i.e., power-off descent). Between hover and autorotation descend speed, the helicopter is descending at a reduced power. At autorotation descend speed, the helicopter rotor does not require any power to keep the rotor rotating. Beyond autorotation descend speed, the rotor is actually generating power. An interpretation of induced power losses requires a discussion of the flow states of the rotor in axial flight. Consider two cases of the actuator disk theory, namely vertical climb and vertical descent. Assume that the flow is uniform in the slip stream. Figure 2.10 shows the velocity profile in both cases. The arrows represent the positive direction of flow velocity in the slip stream. It may be noted that, for climb, velocity V is positive, and for descent, it is

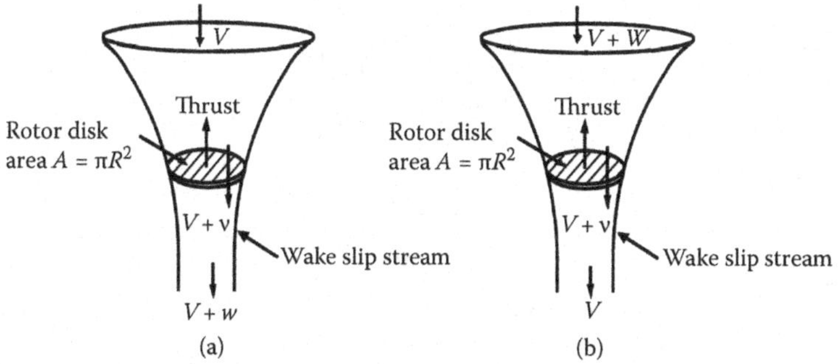

FIGURE 2.10
Flow velocities in climb and descent. (a) Climb ($V > 0$). (b) Descent ($V < 0$).

negative. In the following, the derivation of the induced velocity expression for both climb and descent is provided side by side.

Climb ($V > 0$)	Descent ($V < 0$)
Mass flow through the rotor disk	
$\dot{m} = \rho A(V + \nu)$	$\dot{m} = \rho A(V + \nu)$
Momentum conservation	
$T = \dot{m}(V + w) - \dot{m}V = \dot{m}w$	$T = \dot{m}V - \dot{m}(V + w) = -\dot{m}w$
Energy conservation	
$P = \frac{1}{2}\dot{m}(V + w)^2 - \frac{1}{2}\dot{m}V^2$	$P = \frac{1}{2}\dot{m}V^2 - \frac{1}{2}\dot{m}(V + w)^2$
Simplifying,	
$P = \frac{1}{2}\dot{m}(2Vw + w^2)$	$P = -\frac{1}{2}\dot{m}(2Vw + w^2)$
Since $P = T(V + \nu)$, we have	
$\dot{m}w(V + \nu) = \frac{1}{2}\dot{m}(2Vw + w^2)$	$-\dot{m}w(V + \nu) = -\frac{1}{2}\dot{m}(2Vw + w^2)$
$w = 2\nu$	$w = 2\nu$
Hence, thrust	
$T = \rho A(V + \nu)2\nu$	$T = -\rho A(V + \nu)2\nu$

Since the induced velocity in hover is given as $\nu_h^2 = \frac{T}{2\rho A}$, and assuming that the actuator disk in steady vertical flight is supporting the same weight as in hover (i.e., $T = T_h = \rho A 2\nu_h^2$). Equating the thrust expressions in hover and vertical flight, the equation for the induced flow ν can be obtained for both climb and descent as

Climb ($V > 0$)

$$\frac{\nu}{\nu_h}\left(\frac{V}{\nu_h} + \frac{\nu}{\nu_h}\right) = 1$$

Descent ($V < 0$)

$$\frac{\nu}{\nu_h}\left(\frac{V}{\nu_h} + \frac{\nu}{\nu_h}\right) = -1$$

Solving these equations, the induced velocity $\frac{\nu}{\nu_h}$ as a function of V can be obtained for climb and descent flight conditions.

Note that induced velocity $\frac{v}{v_h}$ is always positive. Hence, it may be noted that the negative root of the radical is not valid for climb, but for descent, both the positive and negative roots of the radical will provide a positive induced velocity. It will be shown later that the positive root of the radical is not a valid root due to the physical condition of the flow.

Climb ($V > 0$)

$$\frac{v}{v_h} = -\frac{V}{2v_h} + \sqrt{\left(\frac{V}{2v_h}\right)^2 + 1} \tag{2.59a}$$

Descent ($V < 0$)

$$\frac{v}{v_h} = -\frac{V}{2v_h} \pm \sqrt{\left(\frac{V}{2v_h}\right)^2 - 1} \tag{2.59b}$$

The net flow velocity at the rotor disk is

Climb ($V > 0$)

$$\frac{V+v}{v_h} = \frac{V}{2v_h} + \sqrt{\left(\frac{V}{2v_h}\right)^2 + 1} \tag{2.60a}$$

Descent ($V < 0$)

$$\frac{V+v}{v_h} = \frac{V}{2v_h} \pm \sqrt{\left(\frac{V}{2v_h}\right)^2 - 1} \tag{2.60b}$$

The net velocity at the far wake is

Climb ($V > 0$)

$$\frac{V+2v}{v_h} = +\sqrt{\left(\frac{V}{v_h}\right)^2 + 4}$$

Descent ($V < 0$)

$$\frac{V+2v}{v_h} = \pm\sqrt{\left(\frac{V}{v_h}\right)^2 - 4}$$

Using Equations 2.59a and 2.59b, the variation of induced velocity $\frac{v}{v_h}$ versus climb (or descent) velocity $\frac{V}{v_h}$ is plotted, and it is shown in Figure 2.11.

The dashed portions of the curve are branches of the solution, which are extrapolated beyond the assumed conditions. These dashed lines do not

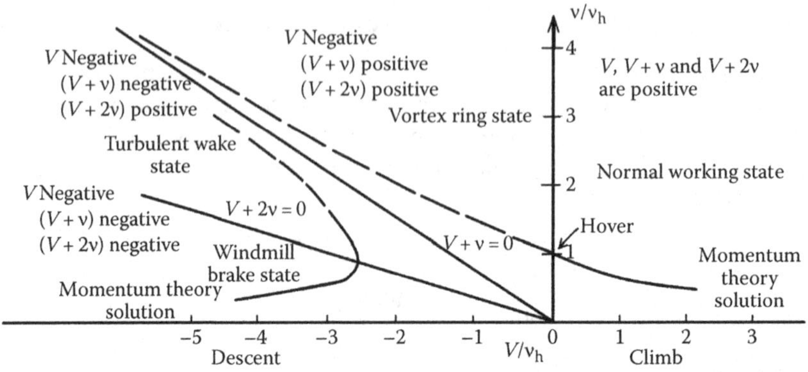

FIGURE 2.11
Variation of induced flow as a function of climb and descent speed.

correspond to the assumed flow state. The line $V + \nu = 0$ is where the direction of flow through the rotor disk and the total induced power $P = T(V + \nu)$ change sign. At the line, $V + 2\nu = 0$, the flow in the far wake changes sign. The three lines $V = 0$, $V + \nu = 0$, and $V + 2\nu = 0$ divide the graph into four regions. These regions are denoted as the normal working state (hover and climb), the vortex ring state, the turbulent wake state, and the windmill brake state. The flow characteristics in each of the states are described below.

Normal Working State

The normal working state includes climb and hover. During climb, the velocity in the slip stream throughout the flow field is downward, with both V and ν positive. For mass conservation, the wake contracts downstream of the rotor. The momentum theory gives a good estimate of the performance. Hover ($V = 0$) is the limit of the normal working state. Even in hover, the momentum theory gives a good estimate of the performance. The flow pattern in the normal working state is shown in Figure 2.12.

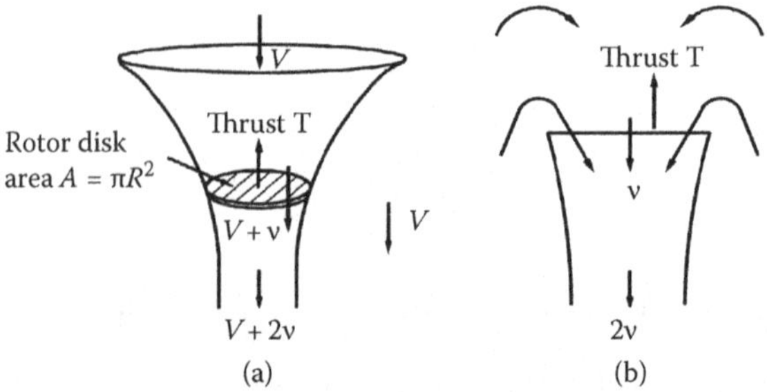

FIGURE 2.12
Flow pattern in the normal working state. (a) Climb ($V > 0$). (b) Hover ($V = 0$).

Vortex Ring State

When the rotor starts to descend, definite slip stream ceases to exist because the flow inside the slip stream changes its direction as we move from far upstream to far wake downstream. Therefore, there will be a large recirculation and turbulence. In the vortex ring state, the induced power ($P = T(V + \nu) > 0$) is positive in the sense that the engine has to supply power to the rotor to keep it rotating. The flow pattern and the directions of the flow are shown in Figure 2.13.

The flow pattern in the vortex ring state is like that of a vortex ring in the plane of the rotor disk. The upward velocity in the free stream keeps the tip vortices piled up as a ring. As the strength builds up, it breaks away from the disk plane, leading to a sudden breakdown of the flow. The flow is highly unsteady and produces a highly disturbing low-frequency vibration. The momentum theory is not valid since the flows inside the slip stream are in opposite directions.

The limiting case of the vortex ring state is when the flow through the disk is zero, that is, $V + \nu = 0$. It may be noted that, during descent, $V < 0$ and induced velocity $\nu > 0$. Hence, the power required for the induced flow ν is exactly equal to the gain in power due to descent. This state corresponds to ideal autorotation (in the absence of profile power loss).

Turbulent Wake State

The turbulent wake state corresponds to the region in Figure 2.11, where $V + \nu = 0$ to $V + 2\nu = 0$. Under the condition $V + \nu = 0$, there is no flow though the rotor disk. In reality, there is a considerable recirculation and turbulence. The flow pattern in the turbulent wake state is shown in Figure 2.14. The flow state is somewhat similar to the flow past a circular plate of the same area of the rotor disk, with no flow through the disk and a turbulent wake behind it.

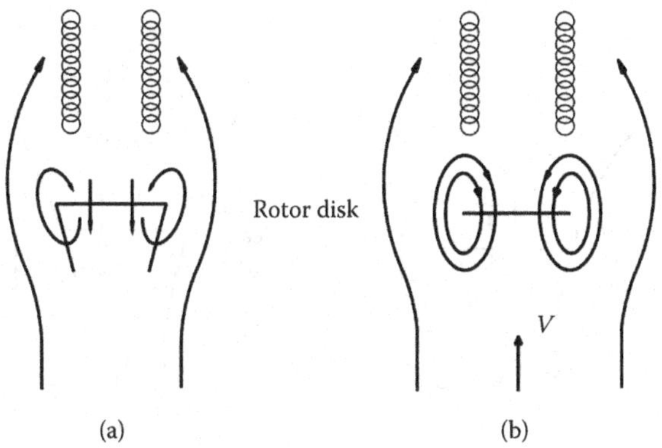

FIGURE 2.13
Flow pattern and velocities in the vortex ring state. (a) Low descent rates. (b) High descent rates.

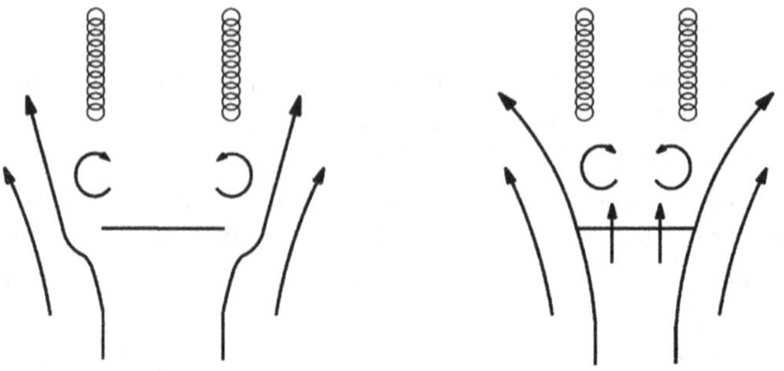

FIGURE 2.14
Flow pattern in the turbulent wake state. (a) Ideal autorotation ($V + \nu = 0$). (b) Tubulent wake state.

When the descent speed increases, $V + \nu < 0$, that is, the flow at the rotor disk is upward with less recirculation through the rotor. The flow above the rotor is highly turbulent. The rotor in this state experiences some roughness due to turbulence, but not like the high vibration in the vortex ring state.

Windmill Brake State

At the high rate of descent ($V < -2\nu$), the flow once again becomes smooth, with a definite slip stream. The flow is upward throughout the slip stream, and the momentum theory is valid in this condition. In this state, the power $P = T(V + \nu)$ is less than 0, implying that the rotor is producing power (or power is extracted from the flow). The flow pattern in the windmill brake state is shown in Figure 2.15.

It is important to note that induced velocity is almost impossible to measure in flight. Therefore, the induced velocity curve is drawn by calculating

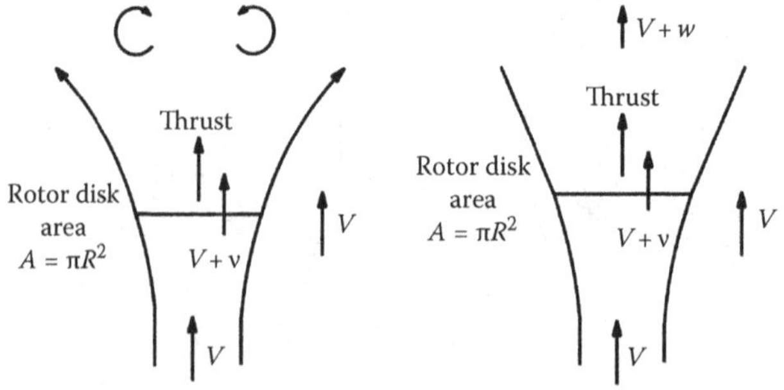

FIGURE 2.15
Flow pattern in the windmill brake state. (a) Boundary ($V + 2V = 0$). (b) Windmill brake state.

the induced velocity from the power measurements. The power supplied to the rotor can be expressed as a sum of three components, given as

Shaft power = climb power + induced power + profile power

$$P = TV + T\nu + \frac{\sigma C_{d0}}{8}\rho\pi R^2(\Omega R)^3 \tag{2.61}$$

For given values of shaft power, gross weight ($T = W$), rotor angular velocity, rate of descent, and blade drag coefficient, the mean effective induced velocity can be determined from the above equation.

Another way of presenting the induced velocity variation was developed by Lock (1947). In this representation (Figure 2.16), the variation of total induced velocity $\left(\frac{V+\nu}{\nu_h}\right)$ is plotted as a function of $\frac{V}{\nu_h}$. This curve is also known as the "universal inflow curve."

In the vortex ring and the turbulent wake regions, the inflow curve is represented by a band, corresponding to practical situations. The universal inflow curve crosses the ideal autorotation line at about $\left(\frac{V}{\nu_h}\right) = -1.71$ (i.e., in the range of –1.6 to –1.8). In practical situations, because of fuselage drag effects, autorotation occurs at a higher rate of descent, which is in the turbulent wake state. In the turbulent wake region, the inflow curve can be approximated by

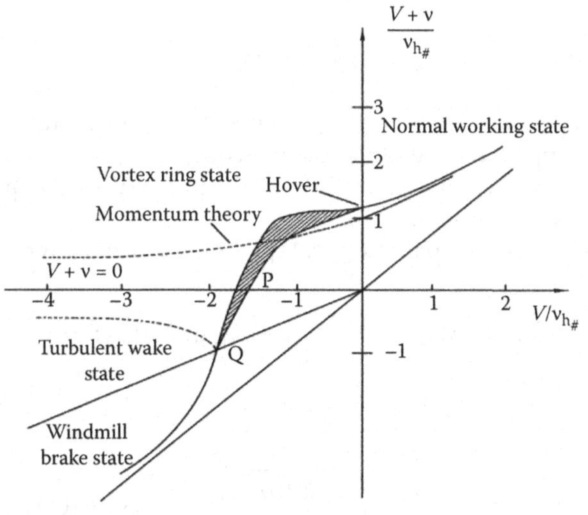

FIGURE 2.16
Total inflow as a function of climb and descent speed.

a straight line PQ. The coordinates of P and Q are, respectively, ($\bar{X}$,0) and Q is (–2,–1). The equation of line PQ can be written as

$$\frac{V+\nu}{\nu_h} = \frac{-\bar{X}}{2+\bar{X}} + \frac{1}{2+\bar{X}}\frac{V}{\nu_h} \tag{2.62}$$

where $\bar{X}$ is the intercept at the $\frac{V}{\nu_h}$ axis.

When $\bar{X} = -1.71$, the inflow equation in the turbulent wake state becomes

$$\frac{V+\nu}{\nu_h} = 5.9 + 3.4\frac{V}{\nu_h} \tag{2.63}$$

Autorotation in Vertical Descent

Autorotation is the state of rotor operation where there is no net power requirement from the power plant. The source of power is due to the decrease in gravitational potential energy. The descent velocity supplies the power to the rotor, and the helicopter is capable of power-off autorotation in vertical flight. It may be noted that the lowest descent rate is achieved in forward flight, which will become obvious when we deal with forward flight power requirements in the following chapter.

The net power to the rotor during autorotation is zero. Hence, from Equation 2.61,

$$P = T(V + \nu) + P_{pd} = 0 \tag{2.64}$$

In Equation 2.64, the first term represents the power required to generate the thrust to support the weight of the helicopter through total inflow at the rotor disk, and the second term represents the profile power required to drag the rotor blades through air.

Using nondimensional parameters, Equation 2.64 can be written as

$$C_T\rho\pi R^2(\Omega R)^2\left(\frac{V+\nu}{\nu_h}\right)\nu_h + C_{Ppd}\rho\pi R^2(\Omega R)^3 = 0 \tag{2.65}$$

or

$$C_T\rho\pi R^2(\Omega R)^2\left(\frac{V+\nu}{\nu_h}\right)\sqrt{\frac{C_T\rho\pi R^2(\Omega R)^2}{2\rho\pi R^2}} + C_{Ppd}\rho\pi R^2(\Omega R)^3 = 0 \tag{2.66}$$

Using Equations 2.56 and 2.61, Equation 2.66 can be simplified as

$$\frac{V+v}{v_h} = -\frac{C_{Ppd}}{C_T^{3/2}/\sqrt{2}} \propto \frac{C_{d0}}{\sigma^{3/2}\bar{C}_l^{3/2}} \tag{2.67}$$

This expression indicates that a low value of $\frac{C_{d0}}{\bar{C}_l^{3/2}}$ provides a low descent velocity in autorotation in vertical descent. Knowing the thrust and profile drag coefficients, the descent rate can be obtained from the universal inflow curve (Figure 2.16). The value of $\frac{V+v}{v_h}$ is typically about −0.3, which is in the turbulent wake state. Since the slope of the curve (the slope is 3.5 from Equation 2.63) is very large in this region, the increase in the descent rate required to overcome profile drag is very small. Tail rotor and other aerodynamic interference losses must also be included in evaluating the autorotation descent rate. Such losses are usually about 15% to 20% of the profile power. The limits of the descent rate in vertical autorotation are essentially the limits of the turbulent wake state (i.e., $\frac{V}{v_h}$ in the range of −1.71 to −2). For practical purposes, one can assume that autorotation occurs for $\frac{V}{v_h} = -1.81$. For typical values of inflow v_h, the autorotation descent velocity is in the range V = 15 to 25 m/s. (For the values of helicopter disk loading T/A = 100–500 N/m², v_h is in the range of 6.4–14 m/s, with density ρ = 1.225 kg/m³.) The autorotation performance may also be evaluated in terms of rotor drag coefficient. During steady autorotation descent, the drag coefficient of the rotor can be defined as

$$C_D = \frac{T}{1/2\rho V^2 A} = \frac{T/(2\rho A)}{V^2/4} = \left(\frac{2}{V/v_h}\right)^2 \tag{2.68}$$

For typical values of $\frac{V}{v_h} = -1.71 \text{ to } -1.81$, the drag coefficient has a value in the range of 1.22 to 1.38. For real helicopters, C_D is in the range of 1.1 to 1.3. For comparison, a circular plate of area A has a drag coefficient of about C_D = 1.28 and a parachute of frontal area A has C_D = 1.40. This shows that a helicopter rotor in power-off descent is quite efficient in producing the thrust to support the helicopter. The rotor is almost as good as a parachute of the same diameter.

Forces on a Blade Element during Autorotation

The physics of autorotation can be best understood by studying the forces acting on the typical cross section of the rotor blade during descent. The velocity at the blade element is made up of rotational speed $U_T = \Omega r$ and normal velocity

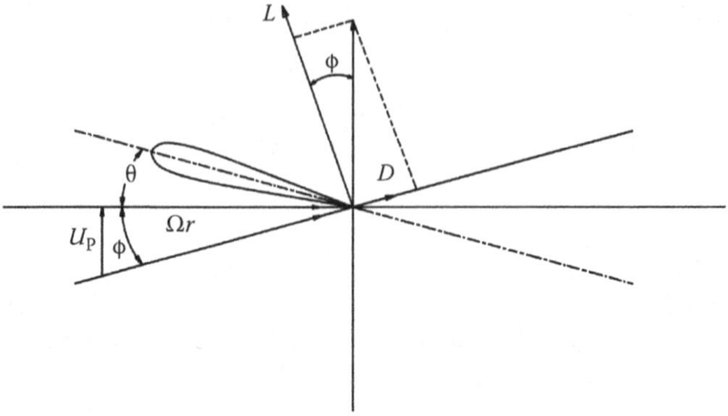

FIGURE 2.17
Components of relative airflow and aerodynamic forces acting on a blade section during descent.

$U_P = V + \nu$ (note that V is negative and ν is positive, but U_P is negative; hence, it is shown with an upward arrow) to the rotor plane, as shown in Figure 2.17.

The inflow angle

$$\tan\phi = \frac{U_P}{\Omega r} = \frac{D}{L} = \frac{C_{d0}}{C_l} \tag{2.69}$$

where C_{d0} and C_l are sectional drag and lift coefficients, respectively. In autorotative equilibrium, there is no net force component in the plane of rotation. The resultant force is acting only along the rotor axis. Hence, from Figure 2.17, we have

$$D\cos\phi - L\sin\phi = 0 \tag{2.70}$$

when $D\cos\phi - L\sin\phi > 0$ or $\tan\phi < \frac{D}{L}$, there is a net decelerating force acting on the airfoil in the plane of rotation. Similarly, when $D\cos\phi - L\sin\phi < 0$ or $\tan\phi > \frac{D}{L}$, there is a net accelerating force acting on the airfoil in the plane of rotation.

An interesting and useful diagram for studying the autorotation characteristics of a particular airfoil is shown in Figure 2.18. The figure consists of the airfoil section characteristics $\tan^{-1}\frac{C_{d0}}{C_l}$ plotted against the section angle of attack α. Note that both X and Y scales are the same.

Let us assume that the diagram is applied to a particular cross section of the rotor blade by first marking off the pitch angle (θ) of the section from the origin along the α axis (i.e., X-axis) and then constructing a 45° line from the pitch angle, as shown in Figure 2.18. A perpendicular from any point on the line to the horizontal axis will define the inflow angle ϕ. Point A corresponds

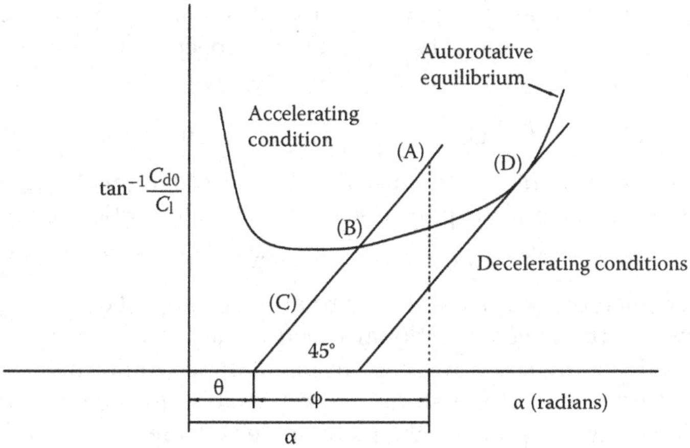

FIGURE 2.18
Effect of operating pitch angle at autorotation condition.

to the case, when $\tan\phi > \frac{C_{d0}}{C_l}$. This implies that there is a resultant accelerating force acting on the blade element, in the direction of rotation. As Ωr increases, the inflow angle ϕ will decrease (Equation 2.69). The element continues to accelerate until its rotational speed has increased to such a value that the autorotative equilibrium condition is established, as indicated by point B. In this condition, $\tan\phi = \frac{C_{d0}}{C_l}$.

There are several important facts that can be drawn from this curve. For a given pitch angle θ, and the 45° line (looking at both Figures 2.17 and 2.18),

1. Any point above the curve (point A) represents an accelerating condition, where the resultant aerodynamic force vector falls ahead of the rotor axis;
2. Any point on the curve (point B) represents autorotative equilibrium, where the resultant force vector is along the rotor axis; and
3. Any point below the curve (point C) represents a decelerating condition, where the resultant aerodynamic vector falls behind the rotor axis.

The highest possible value of pitch (θ_m) at which autorotation is possible corresponds to the value of θ for which the 45° line is tangent to the curve (i.e., point D). It is important to note that autorotation is a stable phenomenon, so long as the pitch angle is less than the maximum θ_m. Therefore, one can conclude that any disturbance slowing down the rotor will increase the angle ϕ and will accelerate the rotor to an autorotative equilibrium. Similarly, any disturbance that speeds up the rotor decreases ϕ and decelerates the rotor.

The autorotation diagram also demonstrates some important concepts concerning the relationship between rotational speed and blade pitch angle. If changes in inflow velocity due to θ are neglected, then inflow angle ϕ varies inversely with $\Omega r \left(\tan\phi = \dfrac{U_P}{\Omega r} \right)$. Therefore, the highest value of rotation speed corresponds to the lowest value of ϕ. The pitch angle (θ) for maximum rotor speed is the pitch defined by the intersection of the 45° line through the minimum value of the $\dfrac{C_{d0}}{C_l}$ curve and the X-axis. Operating at a higher pitch angle implies lower rotational speed. Point D, therefore, corresponds to the lowest rotational speed for autorotation. This point is a discontinuity—any slight increase in pitch angle would make the blade decelerate, and the resulting stopping and rotation in the opposite direction would be catastrophic. As the pitch angle is decreased from the maximum value θ_m, the rotational speed increases until the minimum value of $\dfrac{C_{d0}}{C_l}$ is reached. Then, a further reduction in pitch angle will decrease the rotational speed.

Efficient Angle of Attack for Autorotation

The aim in the design of an autorotating rotor is to obtain a minimum rate of descent for a given helicopter weight and horizontal speed. The optimum angle of attack corresponds to angle α, which provides the lowest value of $\dfrac{C_{d0}}{C_l^{3/2}}$.

Figure 2.19 shows the region of the optimal operating angle attack (i.e., in the range of the angle of attack, which is 10–14°) for a typical airfoil. So long as the optimum part of the curve is flat, the operating angle could be anywhere in a reasonably high range of angles but below the stalling angle. It must be remembered, however, that the foregoing description pertains to a

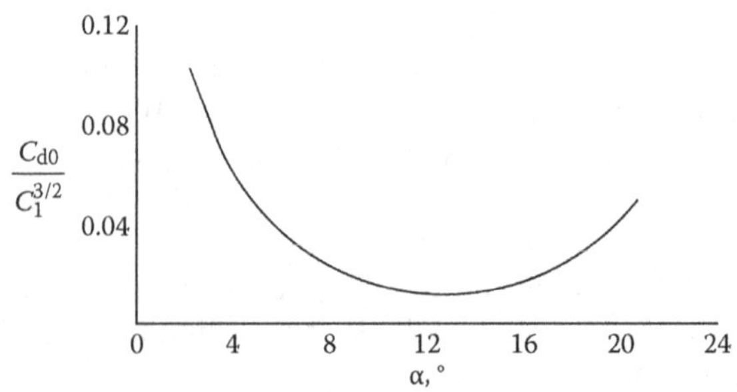

FIGURE 2.19
Variation of $\dfrac{C_{d0}}{C_l^{3/2}}$ with an angle of attack for a typical airfoil.

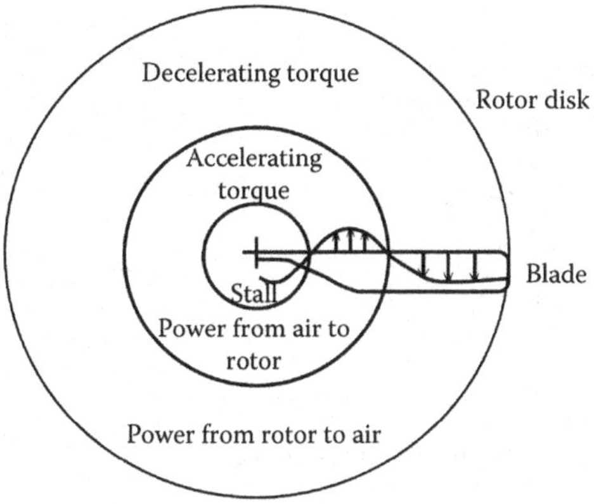

FIGURE 2.20
Aerodynamic condition in terms of the in-plane sectional loads acting along the span of the blade during autorotation.

single blade section. In reality, each cross-sectional element of the rotor blade operates at a different velocity and angle of attack. Therefore, some sections may have accelerating forces, while other elements may have decelerating forces. It is necessary to integrate the forces to obtain the overall effect on the blade. Figure 2.20 shows the different regions of the rotor disk during autorotation: (1) near root: there is stall; (2) mid region: there is an accelerating force due to moderate blade speed; and (3) outer region: there is a decelerating force due to high blade speed.

Ground Effect

The proximity of the ground to a hovering rotor increases the rotor thrust for a given power. In other words, the influence of the ground reduces the induced velocity at the rotor disk. Because of this phenomenon, a helicopter can hover in ground effect at a higher gross weight than is possible with out-of-ground effect. Analytically, ground effect is studied by a method of images (Figure 2.21), where a mirror image rotor is placed beneath the ground plane so that the boundary condition of no flow through the ground is automatically satisfied. The effect of the image vortex is to reduce the inflow velocity at the rotor disk above the ground. Useful information, however, is obtained from measurements.

The increase in thrust $\frac{T}{T_\infty}$ at a constant power can be plotted as a function of the height above the ground $\left(\frac{Z}{R}\right)$. It may be noted that T_∞ is the thrust

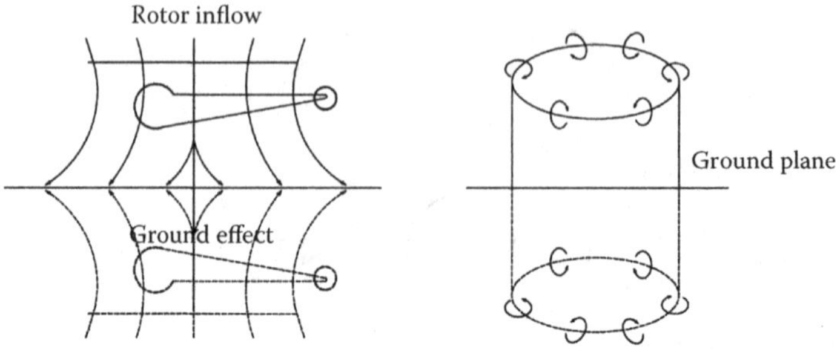

FIGURE 2.21
Rotor hovering near the ground and method of images.

when the rotor is far away from the ground and T is the thrust when it is near the ground. Considering only induced power, if it is a constant, then

$$\lambda C_T = \lambda_\infty C_{T_\infty} \tag{2.71}$$

or

$$\frac{C_T}{C_{T_\infty}} = \frac{T}{T_\infty} = \frac{\lambda_\infty}{\lambda} = \frac{v_\infty}{v} \tag{2.72}$$

Cheeseman and Bennett (1955) (Cheeseman and Bennett, 1955) derived a simple expression for the ground effect of a hovering rotor, which is given as

$$\frac{T}{T_\infty} = \frac{1}{1-(R/4Z)^2} \tag{2.73}$$

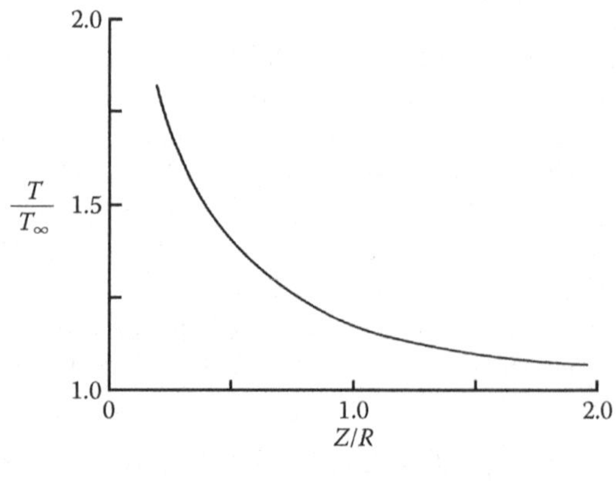

FIGURE 2.22
Thrust variation in hover with rotor height from ground.

The thrust variation with ground effect, given in Equation 2.73, is plotted on Figure 2.22. It can be seen that ground effect is generally negligible when the rotor is one diameter above the ground ($Z/(2R) > 1.0$). Ground effect decreases rapidly with forward speed of the helicopter, since the rotor wake is swept backward. Therefore, it is evident that ground effect is also sensitive to crosswinds, which will displace the wake from the rotor. It is observed that ground effect also depends on rotor blade loading $\frac{C_T}{\sigma}$ (Bramwell, 1976; Johnson, 1980).

3

Introduction to Forward Flight Theory

During forward flight, as the rotor blades go around the azimuth, they experience a time-varying periodic oncoming airspeed. It is assumed that the rotor blade is rotating in a counterclockwise direction when viewed from above the rotor. This condition has been followed throughout the book. The rotor disk can be divided into two half regions: one denoted as the "advancing side" (Figure 3.1), where the relative air velocity of the blade is higher than the rotational speed, and the other denoted as the "retreating side," where the relative air velocity is lower than the rotational speed of the blade. This asymmetry in the aerodynamic environment leads to time-varying periodic loads. Unlike the hovering condition, the time-varying nature of the blade loading creates a dynamic response of the blade, which must be included in the formulation in forward flight.

Momentum Theory in Forward Flight

First, let us formulate the momentum theory in forward flight to evaluate the induced flow through the rotor disk. Consider a rotor operating at a forward speed V, with an angle of attack α between the free-stream velocity and the rotor disk, as shown in Figure 3.2.

Following the momentum theory in hover, let us assume that the induced velocity at the disk is ν, and in the far wake, $w = 2\nu$. These induced velocities are assumed to be in the direction that is parallel to the rotor thrust vector T, but in the opposite direction.

From momentum conservation, one can write

$$T = \dot{m}2\nu \tag{3.1}$$

The mass flow rate is given by

$$\dot{m} = \rho A U \tag{3.2}$$

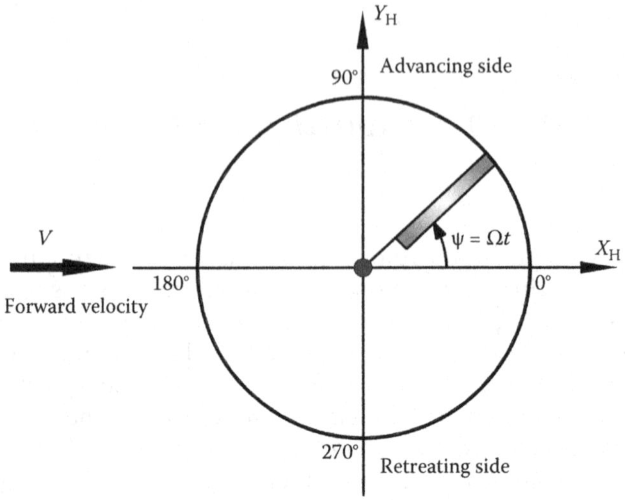

FIGURE 3.1
Rotor disk in forward flight.

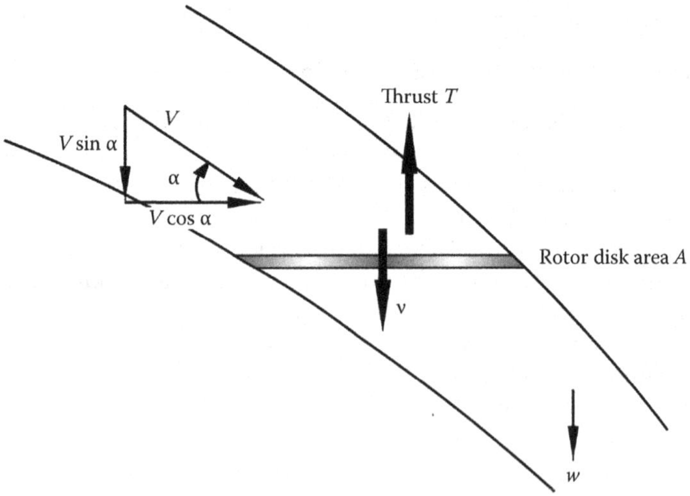

FIGURE 3.2
Direction of flow velocities in the forward-flight momentum theory.

Following Glauert (Bramwell, 1976), the resultant flow velocity at the rotor disk is given by

$$U^2 = (V\cos\alpha)^2 + (V\sin\alpha + \nu)^2$$
$$U^2 = V^2 + 2V\nu\sin\alpha + \nu^2 \tag{3.3}$$

Using Equations 3.1 and 3.2, the expression for thrust in forward flight can be expressed as

$$T = 2\rho A\nu\sqrt{(V\cos\alpha)^2 + (V\sin\alpha + \nu)^2} \tag{3.4}$$

Using energy conservation, the relationship between rotor power and change in kinetic energy can be written as

$$P = \dot{m}\{(1/2)[(V\cos\alpha)^2 + (V\sin\alpha + 2\nu)^2 - V^2]\} = T(V\sin\alpha + \nu) \tag{3.5}$$

For high forward speeds ($V \gg \nu$), one can approximate the thrust expression as

$$T \cong \rho AV\, 2\nu \tag{3.6}$$

In hover ($V = 0$), the thrust expression (Equation 3.4) reduces to

$$T = 2\rho A\nu^2 \tag{3.7}$$

It may be noted that the resultant velocity expression (Equation 3.3), used in the definition of mass flow rate through the rotor disk, provides the thrust expressions in both limits of hover and high forward speed. In the following, as a note, a brief derivation is provided using the fixed wing theory to relate the expression for thrust at high forward speeds.

BRIEF NOTE: The finite wing theory provides the expression for the induced drag coefficient in terms of the lift coefficient and the aspect ratio as (Ref. 4)

$$C_{Di} = \frac{C_L^2}{\pi AR} \tag{3.8}$$

Expressing in dimensional form, the induced drag for a planar circular wing of span $b = 2R$ and area $A = \pi R^2$ (aspect ratio $AR = b^2/A$), operating at a speed V is

$$D_i = \frac{T^2}{2\rho AV^2} \tag{3.9}$$

(NOTE: Lift is denoted by the symbol T, for ease of comparison.)

Induced velocity ν at the wing is related to the ratio of induced power to thrust, which can be written as (using Equation 3.9)

$$\nu = \frac{P_i}{T} = \frac{VD_i}{T} = \frac{T}{2\rho AV} \tag{3.10}$$

This expression matches with the induced velocity expression derived for a rotor operating at a high forward speed, given in Equation 3.6. At high forward speeds, the rotor wake is swept behind the plane of the disk, as in the fixed wing. Hence, at high forward speeds, the rotor behaves like a circular wing. The lifting line theory interprets ν as the induced velocity at the wing. For a circular wing, the aspect ratio becomes $AR = (2R)^2/(\pi R^2)$. This value of aspect ratio ($AR = 4/\pi = 1.27$) is very low; hence, considerable variation in the induced velocity over the rotor disk can be expected.

Rewriting Equation 3.4, for a given thrust, the induced velocity at any forward speed is obtained as a solution of the following equation:

$$T = 2\rho A\nu\sqrt{(V\cos\alpha)^2 + (V\sin\alpha + \nu)^2} \tag{3.11}$$

Since the thrust $T = 2\rho A\nu_h^2$ in hover, one can write the induced velocity in forward flight in terms of the induced velocity in hover by equating the thrust expressions in hover and forward flight as (since thrust is basically equal to the weight of the helicopter)

$$\nu = \frac{\nu_h^2}{\sqrt{(V\cos\alpha)^2 + (V\sin\alpha + \nu)^2}} \tag{3.12}$$

Defining the following nondimensional quantities:

$$\text{Advance ratio: } \mu = \frac{V\cos\alpha}{\Omega R} \tag{3.13}$$

$$\text{Total induced velocity (or total inflow): } \lambda = \frac{V\sin\alpha + \nu}{\Omega R} = \mu\tan\alpha + \lambda_i \tag{3.14}$$

Using Equations 2.12 and 3.12, the induced velocity [$\lambda_i = \nu/(\Omega R)$] expression in forward flight can be written as

$$\lambda_i = \frac{C_T}{2\sqrt{\mu^2 + (\mu\tan\alpha + \lambda_i)^2}} \tag{3.15}$$

or, using Equation 3.14, one can write

$$\lambda = \mu \tan\alpha + \frac{C_T}{2\sqrt{\mu^2 + \lambda^2}} \tag{3.16}$$

For a given value of thrust coefficient C_T, advance ratio μ, and disk angle α, the induced velocity λ_i can be obtained by numerically solving Equation 3.15. When $\alpha = 0$, at high forward speeds, $\mu \gg \lambda_i$, induced velocity can be approximately expressed as (from Equation 3.15)

$$\lambda_i \approx \frac{C_T}{2\mu}\left\{\text{or } v = \frac{T}{2\rho A V \cos\alpha}\right\} \tag{3.17}$$

Figure 3.3 shows the variation of rotor-induced velocity with forward speed, when $\alpha = 0$. The approximation $\lambda_i = \dfrac{C_T}{2\mu}$ is quite good for $\dfrac{\mu}{\lambda_h} > 1.5$. For typical helicopters, when $\dfrac{\mu}{\lambda_h} > 1.5$, it corresponds to an advance ratio $\mu > 0.1$. Hence, except at very low speeds the rotor wake is similar to the wake of a circular wing. The speed range $0 < \mu < 0.1$ is called the "transition zone," where the rotor wake has both vertical and horizontal velocity components. The transition region has a number of special characteristics, such as a high level of blade loads and vibration due to blade–vortex interaction. A schematic of the rotor wake in forward flight is shown in Figure 3.4.

The helical vortices trailing from the blade are swept downward by the normal velocity component λ at the disk. The wake skew angle $\chi = \tan^{-1}\left(\dfrac{\mu}{\lambda}\right)$

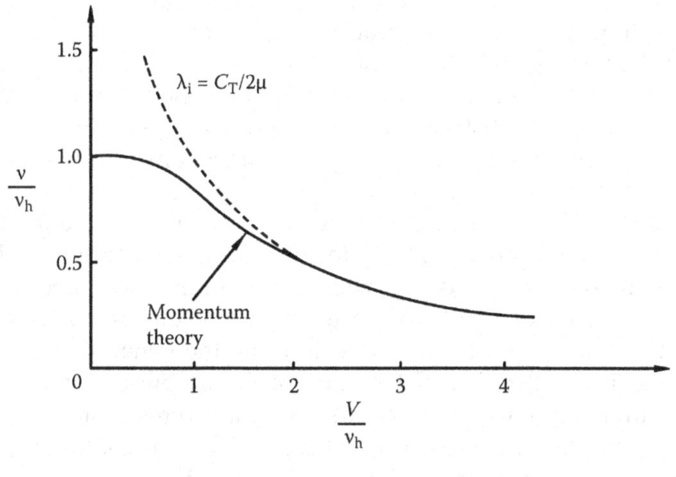

FIGURE 3.3
Variation of induced velocity with forward speed.

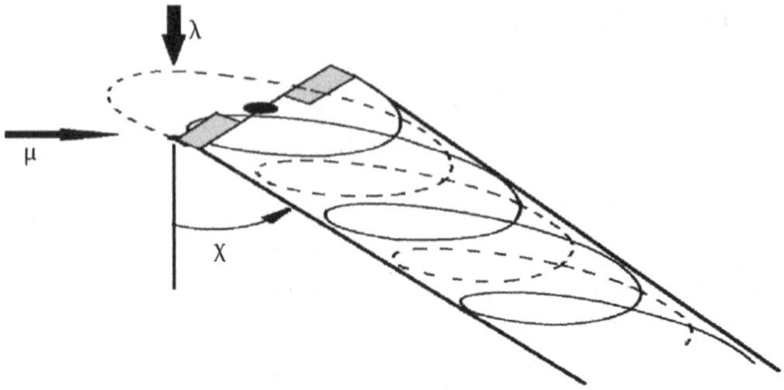

FIGURE 3.4
Rotor wake structure at a moderate forward speed.

can be estimated fairly accurately using the momentum theory. The transition region $0 < \mu/\lambda_h < 1.5$ corresponds approximately to the wake skew angle $\chi = 0°$ to $60°$. The relative position of the rotor blade and the individual wake vortices vary periodically as the blade rotates. This periodicity produces a strong variation in the wake-induced velocity encountered by the blade and also in the blade loading. The induced velocity is highly nonuniform in forward flight. Realizing the variation in induced velocity, Glauert (1926) proposed an inflow model, which is of the form

$$\nu = \nu_0\left(1+\frac{r}{R}K_x\cos\psi\right) \tag{3.18}$$

where ν_0 is the mean induced velocity at the rotor disk and K_x is assumed to be equal to 1.2. This model simulated a longitudinal variation of induced velocity, with an upwash at the leading edge and an increase in induced velocity at the trailing edge of the rotor disk.

The classical vortex theory for forward flight is based on the actuator disk model. The vorticity is distributed throughout the wake rather than being treated as concentrated discrete lines. Oftentimes, uniform loading is also assumed so that the vorticity is only on the surface of the wake cylinder and in a root vortex. These assumptions yield a simple wake model, but still, the problem is complicated. With uniform loading, the results are the same as from the momentum theory, particularly at high speed. Because of the limitations of the wake model, the vortex theory results based on the actuator disk model are presently useful to indicate the general features of the induced velocity. A detailed distribution of the inflow has to be obtained from nonuniform inflow computations, using a discrete vorticity model of the wake. A simple wake structure of the shed (γ_s) and trailing (γ_t) vortices along with a strong tip vortex is shown in Figure 3.5.

Coleman, Feingold, and Stempin (1945) performed a vortex theory analysis for the induced velocity variation along the fore–aft diameter of the rotor.

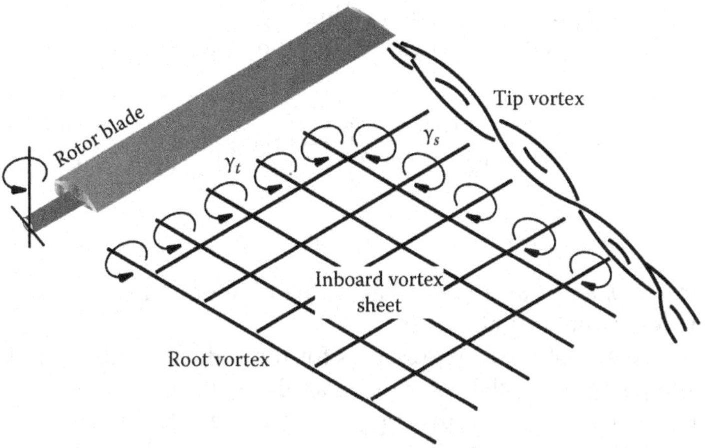

FIGURE 3.5
Sample wake structure of a rotor blade in forward flight.

They considered a uniform loading rotor. Drees (1949) calculated the rotor-induced velocity using the vortex theory with a bound circulation having an azimuthal variation given by $\Gamma = \Gamma_0 - \Gamma_1 \sin\psi$. Hence, he treated both trailing and shed vorticity. Mangler and Squire (1950) treated the rotor as a lifting surface and formulated the problem of pressure jump across the surface in elliptic coordinates. Using pressure–shape functions, they obtained the induced velocity distribution over the rotor disk.

All these theories provide an expression for the induced flow at the rotor disk, which can be written in a general form as

$$\nu = \nu_0\left(1 + K_x \frac{r}{R}\cos\psi + K_y \frac{r}{R}\sin\psi\right) \tag{3.19}$$

where ν_0 is the mean value of the induced velocity obtained from the momentum theory. Equation 3.19 represents the variation of induced velocity as a function of both radial and azimuthal locations on the rotor disk. Different researchers have derived expressions for K_x and K_y. It must be noted that Equation 3.19 represents the steady inflow in forward flight. Several research studies have been undertaken to formulate time varying unsteady inflow. This is beyond the scope of this book.

Coleman, Feingold, and Stempin (1945) provided the expression for the constants as

$$K_x = \tan(\chi/2) = \sqrt{1 + \left(\frac{\lambda}{\mu}\right)^2} - \left|\frac{\lambda}{\mu}\right| \tag{3.20}$$

$$K_y = 0 \tag{3.21}$$

In this model, K_x approaches unity at a high speed.

Drees (1949) obtained the expressions for the constants as

$$K_x = \frac{4}{3}\frac{1-\cos\chi - 1.8\mu^2}{\sin\chi} = \frac{4}{3}\left[(1-1.8\mu^2)\sqrt{1+\left(\frac{\lambda}{\mu}\right)^2} - \frac{\lambda}{\mu}\right] \tag{3.22}$$

$$K_y = -2\mu \tag{3.23}$$

$K_x = 0$ for $\mu = 0$, and has a maximum value of about 1.1 for $\mu = 0.16$ and is approximately equal to 1 at $\mu \approx 0.3$.

One can use any of these expressions for the induced velocity in forward flight while performing the rotor blade analysis. It must be borne in mind that all these theories are only approximate. There has been a continued research effort to improve the inflow model in forward flight. Currently, the focus is directed toward using the computational fluid dynamics approach to estimate the induced velocity and blade loads. However, several basic studies on helicopter dynamics and rotor blade aeroelasticity use the momentum theory because of its simplicity and ease of implementation.

Blade Element Theory in Forward Flight

Before developing the mathematical expressions of the aerodynamic loads in forward flight, it is important to discuss the physics of the blade motion in forward flight. During forward flight, the rotor disk is moving almost horizontally in the air. Thus, the rotor blade, as it goes around the azimuth, experiences a periodically varying relative air velocity due to the forward speed as well as its angular motion (Figure 3.6). Note that out-of-plane deformation of the rotor blade due to time-varying aerodynamic lift is denoted as the flapping motion of the blade.

On the advancing side, the velocity of relative airflow is more than that on the retreating side. If a constant angle of attack is assumed, then forward flight will produce different lift forces on the two halves of the rotor disk, with the advancing side producing more lift than the retreating side. (This differential lift results in a rolling moment on the rotor.) Thus, the rotor blades experience once per revolution (1/rev) variation in aerodynamic loads due to the asymmetry in the airflow. In response to this variation in loads, there is a 1/rev dynamic response of the blade in out-of-plane bending, which is the flapping motion of the blade. The inertia load associated with the flapping motion results in a net reduction in the blade root loads. To alleviate the root bending moment in flap, earlier rotor designs incorporated a root hinge (articulated rotor). In the last three decades, rotors without

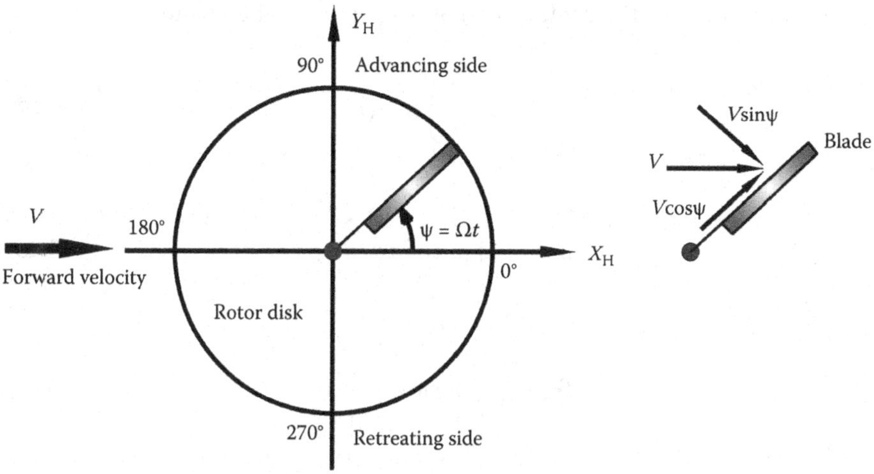

FIGURE 3.6
Description of relative air velocity in forward flight.

hinges (hingeless rotors) have been successfully designed. In these rotors, the blade root is strong enough to withstand the load and flexible enough to provide the flapping motion necessary to reduce the root loads. Increased hub moments in hingeless rotors have a significant effect on the handling qualities of the helicopter, which will be discussed in later chapters. In summary, one can state that the flapping motion of the blade reduces the asymmetry of the rotor moment during forward flight. Thus, flap motion is of principal concern in the analysis and/or design of the helicopter for good performance in forward flight.

During steady forward flight, the blades experience a 1/rev load variation due to asymmetry in the flow, that is, the loads and the blade response undergo a periodic variation with period 2π (one revolution of the rotor around the azimuth). The periodic variation of the flapping motion of the rotor blade can be represented as a Fourier series.

$$\begin{aligned}\beta_k(\psi) &= \beta_0 + \beta_{1c}\cos\psi_k + \beta_{1s}\sin\psi_k + \beta_{2c}\cos 2\psi_k + \beta_{2s}\sin 2\psi_k + \ldots\ldots \\ &= \beta_0 + \sum_{n=1}^{\infty}\beta_{nc}\cos n\psi_k + \beta_{ns}\sin n\psi_k\end{aligned} \tag{3.24}$$

where $\psi_k = \Omega t + 2\pi\frac{k-1}{N}$. The subscript "$k$" represents the kth blade in the rotor system, and N is the total number of blades in the rotor system. Ω is the rotor angular speed in radians per second.

The harmonics (or the coefficients) of the series can be written as

$$\beta_0 = \frac{1}{2\pi}\int_0^{2\pi} \beta_k \, d\psi \tag{3.25}$$

$$\beta_{nc} = \frac{1}{\pi}\int_0^{2\pi} \beta_k \cos n\psi_k \, d\psi_k \tag{3.26}$$

$$\beta_{ns} = \frac{1}{\pi}\int_0^{2\pi} \beta_k \sin n\psi_k \, d\psi_k \tag{3.27}$$

In the analysis of rotor blade flapping motion, only a finite number of harmonics is considered to describe the periodic motion of the rotor blade.

Let us examine the physical meaning of these harmonics of the flap motion. β_0 represents the mean flap angle, which is called the "coning angle." β_{1c} and β_{1s} generate a 1/rev variation of the flap angle, as shown in Figure 3.7.

β_{1c} represents a longitudinal tilt of the tip-path plane (TPP) with respect to the reference plane. β_{1s} represents the lateral tilt. The higher harmonic components represent the warping of the TPP (Figure 3.8). Higher harmonic components become important at high speeds, and they must be included while performing vibration analysis. Beyond second harmonics, the harmonic contents decrease with an increase in harmonic number. For ease of understanding of the physics of the rotor response problem, only first harmonic contents are included in the analysis, and all the other harmonics are neglected.

Similar to flap motion, one can express harmonic motion for both the lead–lag and the elastic torsional deformation of the rotor blade.

The blade lead–lag motion can be expressed as

$$\varsigma_k = \varsigma_0 + \varsigma_{1c} \cos \psi_k + \varsigma_{1s} \sin \psi_k + \ldots\ldots. \tag{3.28}$$

The elastic torsional deformation of the blade can also be expressed as

$$\varphi_k = \varphi_0 + \varphi_{1c} \cos \psi_k + \varphi_{1s} \sin \psi_k + \ldots\ldots.. \tag{3.29}$$

The main source of blade pitch is due to the pilot command or pilot input. These control inputs are specified by a mean and a first harmonic variation as

$$\theta_k = \theta_0 + \theta_{1c} \cos \psi_k + \theta_{1s} \sin \psi_k \tag{3.30}$$

θ_0 is the collective pitch input. θ_{1c} and θ_{1s} are known as cyclic pitch inputs. Collective pitch angle θ_0 controls the average blade force or, hence, the

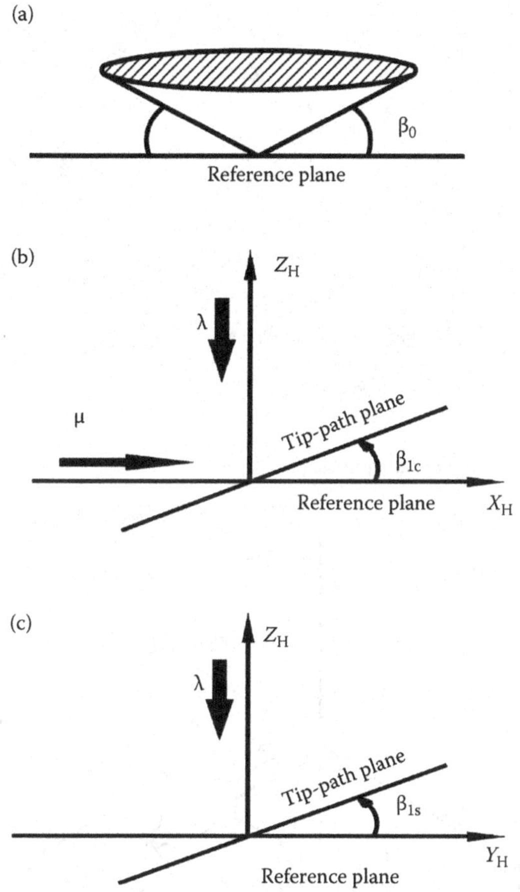

FIGURE 3.7
Physical meaning of rotor flapping in zero and first harmonics. (a) Coning. (b) Longitudinal tilt (view from port side). (c) Lateral tilt (view from aft).

magnitude of the rotor thrust. The cyclic pitch control inputs influence the TPP tilt (1/rev flapping), that is, the orientation of the thrust vector. θ_{1c} controls the lateral (β_{1s}) tilt and θ_{1s} controls the longitudinal (β_{1c}) tilt of the TPP. This aspect will become clear while relating pilot input to flap response.

The pitch input variation of the blade is achieved by a swash plate mechanism. There are also other means of achieving the pitch control of the blade, such as the Kaman servo-flap or the pitch–flap kinematics coupling. A schematic of a swash plate mechanism is shown Figure 3.9.

The swash plate mechanism has rotating and nonrotating parts. The nonrotating part is connected to the control mechanism from the pilot. The pitch control link rod from each rotor blade is connected to the rotating part of the swash plate. During collective input, the pilot moves the nonrotating part of the swash

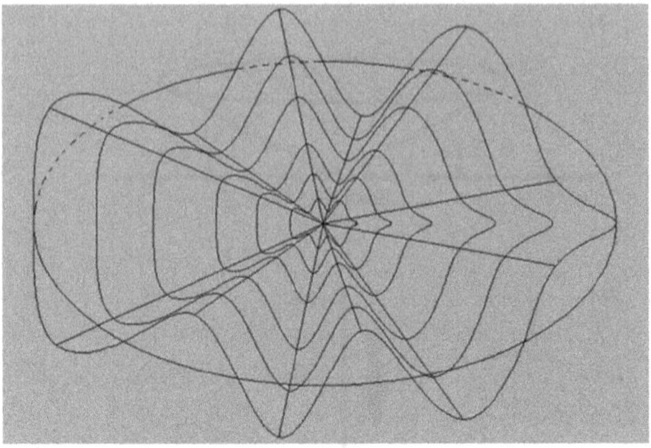

FIGURE 3.8
Rotor blade flapping in higher harmonics.

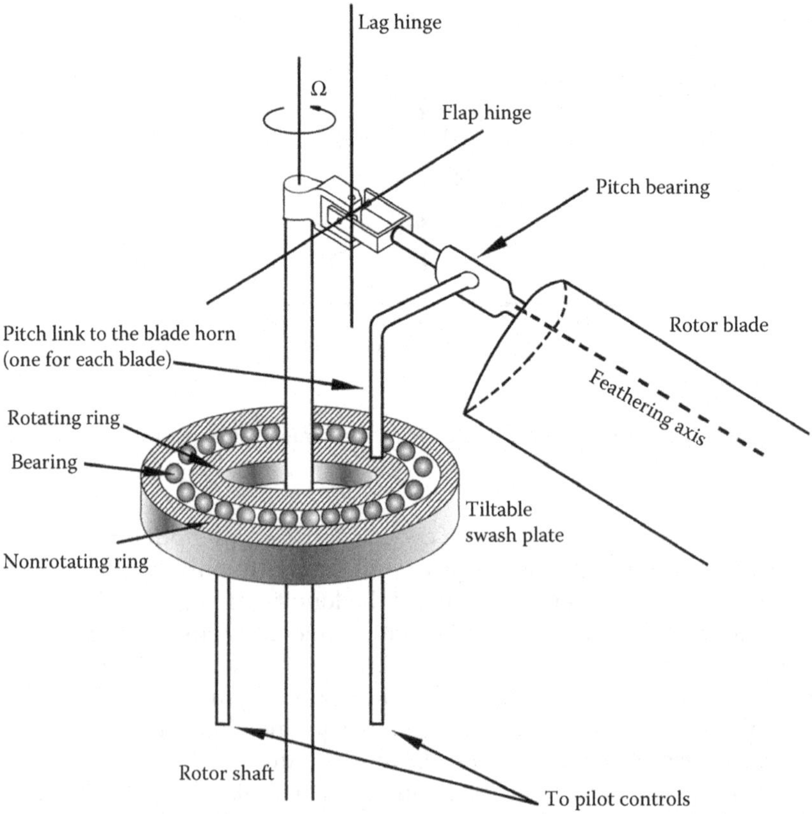

FIGURE 3.9
Swash plate mechanism.

plate up and down, which also makes the rotating part of the swash plate move up and down. This up-and-down motion of the swash plate changes the pitch angle of the blade by the same value at all azimuth locations of the rotor system. During cyclic input, the pilot tilts the nonrotating part of the swash plate about the longitudinal axis (or the lateral axis), thereby tilting the rotating part of the swash plate. In this configuration of the tilted swash plate, as the rotor blade goes around the azimuth, the pitch control link rod attached to the blade moves up and down, thereby changing the blade pitch once in a revolution.

Reference Planes

It is important to choose a reference frame for the description of the motion of the blade, control pitch angle, aerodynamic loads, etc. There are different types of reference frames (Figure 3.10), each having its own advantage in the description of the rotor blade motion.

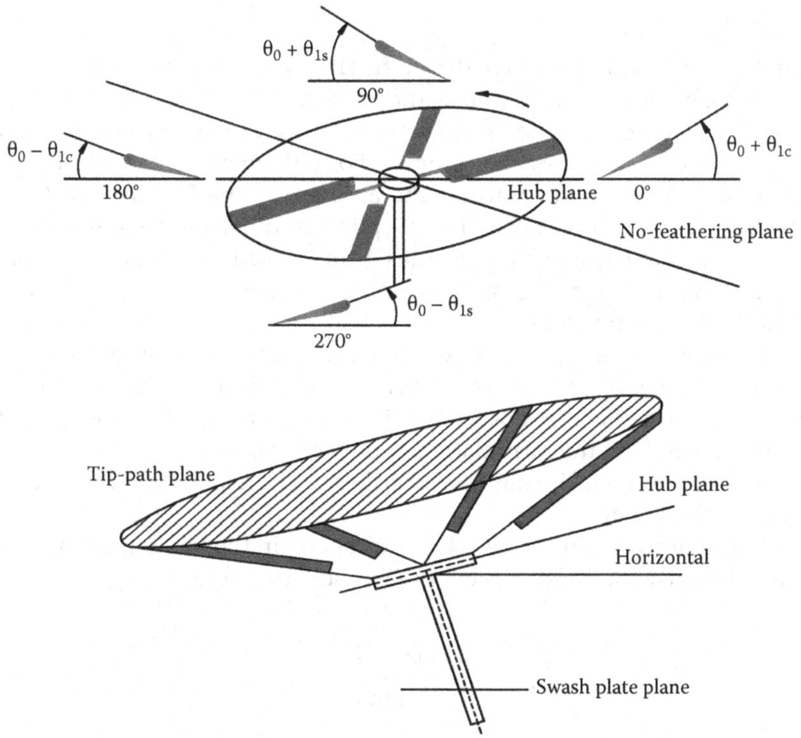

FIGURE 3.10
Various reference planes.

Tip-Path Plane

As the name suggests, it is the plane that is parallel to the plane described by the tip of the blade, particularly in the collective and first harmonic cyclic. For an observer sitting in this plane, there is no 1/rev flap motion. The orientation of the TPP with respect to any other plane defines the cyclic flap (β_{1c} and β_{1s}) angles.

No-Feathering Plane

In the no-feathering plane (NFP), there is no variation in the 1/rev pitch angle of the blade. Therefore, its orientation defines θ_{1c} and θ_{1s} with reference to any other plane.

Control Plane

The control plane (CP) represents the plane from where the pitch input is applied to the blade. This plane is the swash plate plane.

Hub Plane

The hub plane (HP) is normal to the shaft. This is the natural reference frame for defining blade motion as well as the blade loads.

In general, these reference planes are distinct, but under certain conditions, two reference planes can coincide. For a flapping rotor with no cyclic pitch control (like the tail rotor or the auto gyro), the HP and the CP are equivalent. If there is no pitch–flap coupling or other pitch sources (which will be described in a later chapter), the CP and the NFP coincide. For a feathering rotor with no flapping (such as a propeller with a cyclic pitch), the HP and the TPP are equivalent.

While analyzing helicopter rotor dynamics, one can use any reference frame for defining rotor blade velocity, blade motion, and various forces and moments. Since HP is the most convenient plane, it is used as the reference plane (Figure 3.11) in the formulation in the entire book.

Let us assume that, during forward flight, the relative air velocity V is inclined to the HP at angle α.

The horizontal component of relative airspeed in the HP is $V \cos \alpha$.

In nondimensional form, it is denoted as advance ratio

$$\mu = \frac{V \cos \alpha}{\Omega R} \tag{3.31}$$

The normal component of relative air velocity at the HP is $V \sin \alpha$.

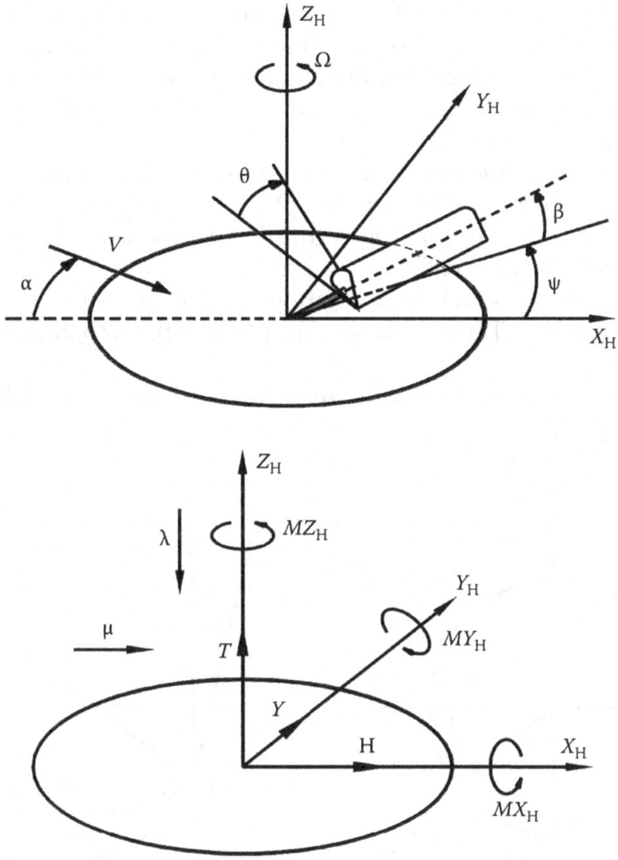

FIGURE 3.11
Velocity, blade motion, and hub loads in the reference HP.

In nondimensional form, it is denoted as inflow due to climb

$$\lambda_c = \frac{V \sin\alpha}{\Omega R} = \mu \tan\alpha$$

The total inflow must be a combination of the forward flight component and the rotor-induced flow, which is given as (from Equations 3.14–3.16)

$$\lambda = \lambda_c + \lambda_i \tag{3.32}$$

Having defined the velocity components in the HP due to helicopter forward flight, let us identify the velocity components at a typical cross section of the rotor blade at a radial distance r from the hub center and at an arbitrary azimuthal position ψ. Let us assume that the blade is centrally hinged and that it undergoes only a rigid flapping motion. For a systematic development of aerodynamic loads acting on the blade, it is essential to define the following coordinate systems (Figure 3.12).

X_H–Y_H–Z_H refers to the hub-fixed nonrotating reference coordinate system. Its origin is at the center of the hub. Axis X_H is pointing toward the tail of the helicopter, and axis Z_H is pointing vertically up. Axis Y_H completes the triad following the right-hand rule.

X_1–Y_1–Z_1 refers to the hub-fixed rotating coordinate system. Its origin is at the center of the hub. The X_1 axis is in the plane of the rotor hub, but moving with kth blade.

X_2–Y_2–Z_2 refers to the blade-fixed rotating coordinate system. Its origin is at the center of the hub.

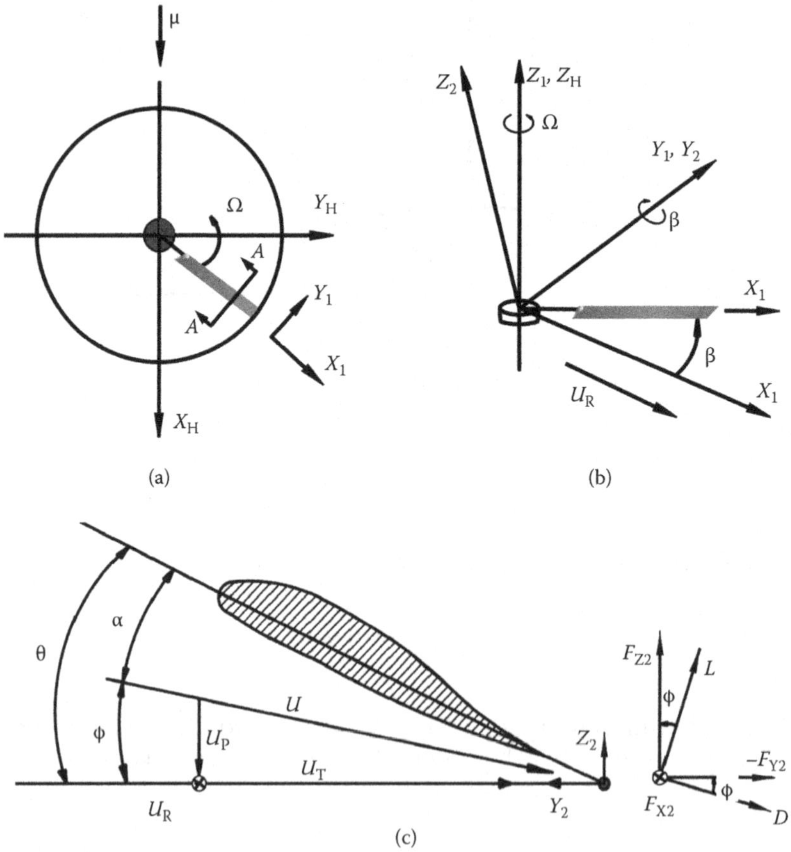

FIGURE 3.12
Coordinate systems for systematic formulation. (a) In-plane velocity components U_T and U_R. (b) U_R and blade deformation β. (c) Section A-A.

Assume that the rotor blade has a uniform cross section and that the X_2 axis is along the aerodynamic center of the blade (25% of chord). In addition, assume that the cross-sectional mass center is coincident with the aerodynamic center.

The transformation relationship between the various coordinate systems can be written as

$$\begin{Bmatrix} \vec{e}_{x_1} \\ \vec{e}_{y_1} \\ \vec{e}_{z_1} \end{Bmatrix} = \begin{bmatrix} \cos\psi_k & \sin\psi_k & 0 \\ -\sin\psi_k & \cos\psi_k & 0 \\ 0 & 0 & 1 \end{bmatrix} \begin{Bmatrix} \vec{e}_{x_H} \\ \vec{e}_{y_H} \\ \vec{e}_{z_H} \end{Bmatrix} \tag{3.33}$$

where ψ_k represents the azimuth location of the kth blade, and

$$\begin{Bmatrix} \vec{e}_{x_2} \\ \vec{e}_{y_2} \\ \vec{e}_{z_2} \end{Bmatrix} = \begin{bmatrix} \cos(-\beta_k) & 0 & -\sin(-\beta_k) \\ 0 & 1 & 0 \\ \sin(-\beta_k) & 0 & \cos(-\beta_k) \end{bmatrix} \begin{Bmatrix} \vec{e}_{x_1} \\ \vec{e}_{y_1} \\ \vec{e}_{z_1} \end{Bmatrix} = \begin{bmatrix} \cos(\beta_k) & 0 & \sin(\beta_k) \\ 0 & 1 & 0 \\ -\sin(\beta_k) & 0 & \cos(\beta_k) \end{bmatrix} \begin{Bmatrix} \vec{e}_{x_1} \\ \vec{e}_{y_1} \\ \vec{e}_{z_1} \end{Bmatrix} \tag{3.34}$$

where $(-\beta_k)$ represents the flap angle of the kth blade from HP. The negative sign denotes that it is a clockwise rotation angle.

The position vector of any arbitrary point P along the X_2 axis of the kth blade is given as

$$\vec{r}_\text{p} = r\vec{e}_{x_2} \tag{3.35}$$

Rewriting this vector in the X_1–Y_1–Z_1 system,

$$\vec{r}_\text{p} = r\vec{e}_{x_2} = r\cos\beta_k\vec{e}_{x_1} + r\sin\beta_k\vec{e}_{z_1} \tag{3.36}$$

The angular velocity of the rotor blade is taken as $\vec{\omega} = \Omega\vec{e}_{z_1}$.

The velocity of point P has two components: one due to rotation and another due to flapping motion. The velocity of point P is given by

$$\vec{v}_\text{p} = \dot{\vec{r}}_\text{p} = -r\sin\beta_k\frac{d\beta_k}{dt}\vec{e}_{x_1} + r\cos\beta_k\frac{d\beta_k}{dt}\vec{e}_{z_1} + \vec{\omega}X\vec{r}_\text{p} \tag{3.37}$$

Taking the vector cross-product, the velocity of point P can be written as

$$\vec{v}_\text{p} = -r\sin\beta_k\frac{d\beta_k}{dt}\vec{e}_{x_1} + r\cos\beta_k\frac{d\beta_k}{dt}\vec{e}_{z_1} + \Omega r\cos\beta_k\vec{e}_{y_1} \tag{3.38}$$

In the blade-fixed coordinate system, the velocity of point P can be written as

$$\vec{v}_{\mathrm{P}} = r\frac{d\beta_k}{dt}\vec{e}_{z_2} + \Omega r\cos\beta_k \vec{e}_{y_2} \tag{3.39}$$

Assuming that the flap angle β_k is a small quantity, one can make a small angle assumption for the trigonometric functions. The reduced expression for the velocity of point P can be written as

$$\vec{v}_{\mathrm{P}} \cong \Omega r\vec{e}_{y_2} + r\frac{d\beta_k}{dt}\vec{e}_{z_2} = \Omega r\vec{e}_{y_2} + r\dot{\beta}_k\vec{e}_{z_2} \tag{3.40}$$

(**NOTE**: For the sake of consistency and convenience, the time derivative of flap β_k is nondimensionalized as $\dot{\beta}_k = \Omega\frac{d\beta_k}{d\Omega t} = \Omega\frac{d\beta_k}{d\psi} = \Omega\dot{\beta}_k$. ψ is denoted as nondimensional time Ωt.) In the following, the dot symbol represents the nondimensional time derivative.

The relative air velocity at the blade section due to the motion of the helicopter and the total induced flow given in Equations 3.31 and 3.32 can be written as components along the kth blade axes system as

$$\vec{V}_{\mathrm{h}} = -\mu\Omega R\sin\psi_k\vec{e}_{y_1} + \mu\Omega R\cos\psi_k\vec{e}_{x_1} - \lambda\Omega R\vec{e}_{z_1} \tag{3.41}$$

Resolving these velocity components in the blade-fixed system, we have

$$\vec{V}_{\mathrm{h}} = -\mu\Omega R\sin\psi_k\vec{e}_{y_2} + \mu\Omega R\cos\psi_k\cos\beta_k\vec{e}_{x_2} - \mu\Omega R\cos\psi_k\sin\beta_k\vec{e}_{z_2} - \lambda\Omega R\cos\beta_k\vec{e}_{z_2}$$
$$- \lambda\Omega R\sin\beta_k\vec{e}_{x_2} \tag{3.42}$$

Invoking a small angle assumption, Equation 3.42 can be written as

$$\vec{V}_{\mathrm{h}} \cong -\mu\Omega R\sin\psi_k\vec{e}_{y_2} + \mu\Omega R\cos\psi_k\vec{e}_{x_2} - \mu\Omega R\cos\psi_k\beta_k\vec{e}_{z_2} - \lambda\Omega R\vec{e}_{z_2} - \lambda\Omega R\beta_k\vec{e}_{x_2} \tag{3.43}$$

The relative air velocity at the blade cross section at radial station r from the hub center is given as

$$\vec{V}_{\mathrm{rel}} = \vec{V}_{\mathrm{h}} - \vec{v}_{\mathrm{P}} \tag{3.44}$$

Substituting Equations 3.40 and 3.43, one obtains

$$\vec{V}_{\text{rel}} = (\mu\Omega R\cos\psi_k - \lambda\Omega R\beta_k)\vec{e}_{x_2} + (-\mu\Omega R\sin\psi_k - \Omega r)\vec{e}_{y_2} + (-\mu\Omega R\cos\psi_k\beta_k - \lambda\Omega R - \Omega r\dot{\beta}_k)\vec{e}_{z_2} \tag{3.45}$$

With reference to the airfoil cross section (Figure 3.12), using Equation 3.45, the velocity components can be expressed as follows:

Tangential velocity component at any radial location r:

$$U_{\mathrm{T}} = \Omega r + \mu\Omega R\sin\psi_k = \Omega R\left(\frac{r}{R} + \mu\sin\psi_k\right) \tag{3.46}$$

Radial velocity component along the blade:

$$U_{\mathrm{R}} = \mu\Omega R\cos\psi_k - \lambda\Omega R\,\beta_k \tag{3.47}$$

The normal velocity component at r:

$$U_{\mathrm{P}} = \lambda\Omega R + \Omega r\dot{\beta}_k + \beta_k\mu\Omega R\cos\psi_k \tag{3.48}$$

Note that λ is the total inflow, which includes the normal velocity component due to forward speed V and the induced velocity $v = \lambda_i\Omega R$, that is, $\lambda\Omega R = \mu\Omega R\tan\alpha + \lambda_i\Omega R$. Also, note that the time derivative of β_k is with respect to the nondimensional time Ωt.

Consider the two-dimensional cross section of the rotor blade. The resultant velocity of the oncoming flow can be written, by assuming that the inflow velocity U_{P} is small compared to the oncoming velocity component U_{T}, as

$$U = \sqrt{U_{\mathrm{T}}^2 + U_{\mathrm{P}}^2} \approx U_{\mathrm{T}} \tag{3.49}$$

The inflow angle is given by

$$\tan\phi = \frac{U_{\mathrm{P}}}{U_{\mathrm{T}}} \tag{3.50}$$

or, for small angles,

$$\phi \approx \frac{U_{\mathrm{P}}}{U_{\mathrm{T}}} \tag{3.51}$$

The sectional aerodynamic lift and drag acting on the airfoil are given as

$$L = \frac{1}{2}\rho U^2 C C_l \tag{3.52}$$

$$D = \frac{1}{2}\rho U^2 C C_d \tag{3.53}$$

where C is the blade chord, and C_l and C_d are the sectional lift and drag coefficients, respectively. These coefficients are functions of the Mach number and the angle of attack. Resolving these forces along the normal and in-plane directions in the reference plane (Figure 3.12),

$$F_{z2} = L\cos\phi - D\sin\phi \cong L \tag{3.54}$$

$$F_{y2} = -(L\sin\phi + D\cos\phi) \tag{3.55}$$

Note that $L\sin\phi$ represents the induced drag term. The components of these distributed aerodynamic loads along the X_1–Y_1–Z_1 axis system can be given as

$$\begin{aligned} &F_{x1} = -F_{z2}\sin\beta_k \text{ (note that radial drag effects are neglected)} \\ &F_{y1} = F_{y2} \\ &F_{z1} = F_{z2}\cos\beta_k \end{aligned} \tag{3.56}$$

Invoking a small angle assumption and noting that $L \gg D$, and neglecting radial drag effects, the aerodynamic force components can be approximated as

$$F_{z1} \cong F_{z2} \cong L \tag{3.57}$$

$$F_{y1} \cong -(L\phi + D) \tag{3.58}$$

$$F_{x1} \cong -L\beta_k \tag{3.59}$$

The aerodynamic root moment can be obtained as

$$\vec{M}_A = \vec{r} \times \vec{F} = (r\cos\beta_k\vec{e}_{x1} + r\sin\beta_k\vec{e}_{z1}) \times \vec{F} \tag{3.60}$$

The root aerodynamic moment can be expressed in component form as

$$\vec{M}_A = \left\{-F_{y1}r\sin\beta_k\vec{e}_{x1} + (F_{x1}r\sin\beta_k - r\cos\beta_k F_{z1})\vec{e}_{y1} + r\cos\beta_k F_{y1}\vec{e}_{z1}\right\} \tag{3.61}$$

Substituting in Equation 3.52, for lift coefficient as $C_l = a\alpha_e$, where a is the lift curve slope and α_e is the effective angle of attack, which is given as

$$\alpha_e = \theta - \phi \approx \theta - \frac{U_P}{U_T} \tag{3.62}$$

Using Equations 3.49, 3.52, and 3.62, the aerodynamic lift per unit span of the blade can be written as

$$F_{z1} \cong \frac{1}{2}\rho U_T^2 C\left[a\left(\theta - \frac{U_P}{U_T}\right)\right] \tag{3.63}$$

Rewriting Equation 3.63 as

$$F_{z1} \cong \frac{1}{2}\rho C[a(U_T^2\theta - U_P U_T)] \tag{3.64}$$

The velocity components can be written in nondimensional form (from Equations 3.46 and 3.48) as

$$U_T = \Omega R\bar{U}_T = \Omega R\left(\frac{r}{R} + \mu \sin\psi_k\right) \tag{3.65}$$

$$U_P = \Omega R\bar{U}_P = \Omega R\left(\lambda + \frac{r}{R}\dot{\beta}_k + \beta_k \mu \cos\psi_k\right) \tag{3.66}$$

Substituting these velocity components in Equation 3.64, the normal force (thrust) per unit length of the blade can be expressed as

$$F_{z1} = \frac{1}{2}\rho Ca(\Omega R)^2\left[\left(\frac{r}{R} + \mu\sin\psi_k\right)^2\theta - \left(\lambda + \frac{r}{R}\dot{\beta}_k + \beta_k\mu\cos\psi_k\right)\left(\frac{r}{R} + \mu\sin\psi_k\right)\right] \tag{3.67}$$

Similarly, in-plane drag force acting per unit span of the blade can be obtained using Equations 3.52, 3.53, and 3.55

$$F_{y1} = -\frac{1}{2}\rho U^2 C[C_l \sin\phi + C_d \cos\phi] \tag{3.68}$$

Using small angle assumption (Equation 3.58), the drag expression is given as

$$F_{y1} \approx \frac{1}{2}\rho U_T^2 C\left[a\left(\theta - \frac{U_P}{U_T}\right)\frac{U_P}{U_T} + C_d\right] \tag{3.69}$$

Rewriting it as

$$F_{y1} = \frac{1}{2}\rho C\left[aU_P U_T \theta - aU_P^2 + U_T^2 C_d\right] \tag{3.70}$$

Substituting the velocity components from Equations 3.65 and 3.66,

$$F_{y1} = -\frac{1}{2}\rho Ca(\Omega R)^2 \left[\begin{array}{l} \left(\lambda + \frac{r}{R}\dot{\beta}_k + \beta_k \mu \cos\psi_k\right)\left(\frac{r}{R} + \mu\sin\psi_k\right)\theta \\ -\left(\lambda + \frac{r}{R}\dot{\beta}_k + \mu\beta_k \cos\psi_k\right)^2 + \left(\frac{r}{R} + \mu\sin\psi_k\right)^2 \frac{C_d}{a} \end{array}\right] \tag{3.71}$$

The in-plane radial force per unit span can be expressed as (using Equations 3.59 and 3.67),

$$F_{x1} = -\frac{1}{2}\rho Ca(\Omega R)^2 \left[\begin{array}{l} \left(\frac{r}{R} + \mu\sin\psi_k\right)^2 \theta \\ -\left(\lambda + \frac{r}{R}\dot{\beta}_k + \beta_k \mu \cos\psi_k\right)\left(\frac{r}{R} + \mu\sin\psi_k\right) \end{array}\right]\beta_k \tag{3.72}$$

Rotor Hub Forces

Integrating the blade loads along the span results in blade root loads. Then, by summing up the loads due to all the blades in the rotor system, one can obtain the hub loads. The average value of these hub loads can be obtained by integrating over the azimuth and dividing by 2π. One can also obtain the mean and various harmonics by Fourier analysis. The hub loads in symbolic representation are given in Figure 3.11. The nondimensional form of the hub forces and moments can be written as

Thrust: $C_T = \dfrac{T}{\rho\pi R^2(\Omega R)^2}$

Longitudinal in-plane force: $C_H = \dfrac{H}{\rho\pi R^2(\Omega R)^2}$

Lateral in-plane force: $C_Y = \dfrac{Y}{\rho\pi R^2(\Omega R)^2}$

Roll moment: $M_{xH} = \dfrac{M_x}{\rho\pi R^2(\Omega R)^2 R}$

Pitch moment: $M_{yH} = \dfrac{M_y}{\rho\pi R^2(\Omega R)^2 R}$

Yaw moment (or torque): $C_Q = \dfrac{M_{zH}}{\rho\pi R^2(\Omega R)^2 R}$

It may be noted that, in obtaining the hub loads, only aerodynamic effects are considered and inertial loads due to blade motion are neglected.

Nondimensional Thrust

The nondimensional thrust is given by summing up the root loads due to all blades in the rotor system as

$$C_T = \frac{T}{\rho\pi R^2(\Omega R)^2} = \frac{1}{N}\sum_{k=1}^{N} N\int_0^R F_{z1}\,dr \frac{1}{\rho\pi R^2(\Omega R)^2} \tag{3.73}$$

where N represents the number of blades in the rotor system. It is to be noted that it is assumed that all the blades in the rotor system are identical and in Equation 3.73, N is multiplied and divided for convenience of nondimensionalization.

Substituting for F_{z1} from Equations 3.64–3.67, and assuming constant chord and airfoil section, the nondimensional thrust expression can be simplified as

$$C_T = \frac{1}{N}\sum_{k=1}^{N}\left(\sigma a\frac{1}{2}\int_0^1 (\bar{U}_T^2\theta - \bar{U}_T\bar{U}_P)\,d\bar{r}\right) \tag{3.74}$$

where σ is the solidity ratio and $\bar{r} = \dfrac{r}{R}$.

Substituting for the nondimensional velocities as

$$\bar{U}_T = \bar{r} + \mu\sin\psi_k$$

$$\bar{U}_P = \lambda + \bar{r}\dot{\beta}_k + \beta_k\mu\cos\psi_k$$

the expression for nondimensional thrust becomes

$$C_T = \frac{1}{N}\sum_{k=1}^{N}\frac{\sigma a}{2}\int_0^1\left[\begin{matrix}(\bar{r}^2 + 2\mu\bar{r}\sin\psi_k + \mu^2\sin^2\psi_k)\theta_k \\ -\{\bar{r}\lambda + \bar{r}^2\dot{\beta}_k + \bar{r}\beta_k\mu\cos\psi_k + \lambda\mu\sin\psi_k + \bar{r}\dot{\beta}_k\mu\sin\psi_k + \beta_k\mu^2\sin\psi_k\cos\psi_k\}\end{matrix}\right]d\bar{r}$$

$$C_T = \frac{1}{N}\sum_{k=1}^{N}\frac{\sigma a}{2}[\text{function of azimuth } \psi] \tag{3.75}$$

Before integrating Equation 3.75 over the azimuth to obtain the mean value of nondimensional thrust (or thrust coefficient), one must know the form of θ_k and β_k, where θ_k represents the blade pitch input and β_k represents the flap response of the kth blade.

In-Plane Drag Force Coefficient (Longitudinal)

The nondimensional in-plane drag in the longitudinal direction is obtained as a sum of the components of in-plane forces F_{y1} and F_{x1}. It is given as

$$C_H = \frac{H}{\rho\pi R^2(\Omega R)^2} = \frac{1}{N}\sum_{k=1}^{N} N\int_0^R \frac{(-F_{y1}\sin\psi_k + F_{x1}\cos\psi_k)}{\rho\pi R^2(\Omega R)^2}dr \tag{3.76}$$

Substituting for the respective force components from Equations 3.68–3.72 and using nondimensional representation, the in-plane drag force coefficient can be written as

$$C_H = \frac{1}{N}\sum_{k=1}^{N}\left(\frac{\sigma a}{2}\int_0^1\left[\left\{\bar{U}_P\bar{U}_T\theta - \bar{U}_P^2 + \bar{U}_T^2\frac{C_d}{a}\right\}\sin\psi_k - \left\{\bar{U}_T^2\theta - \bar{U}_P\bar{U}_T\right\}\beta_k\cos\psi_k\right]d\bar{r}\right) \tag{3.77}$$

Substituting the nondimensional velocity components from Equations 3.65 and 3.66

$$C_H = \frac{1}{N}\sum_{k=1}^{N}\frac{\sigma a}{2}\int_0^1\left[\begin{matrix}\left\{\begin{matrix}(\lambda + \bar{r}\dot{\beta}_k + \beta_k\mu\cos\psi_k)(\bar{r} + \mu\sin\psi_k)\theta \\ -(\lambda + \bar{r}\dot{\beta}_k + \beta_k\mu\cos\psi_k)^2 + (\bar{r} + \mu\sin\psi_k)^2\dfrac{C_d}{a}\end{matrix}\right\}\sin\psi_k \\ -\left\{(\bar{r} + \mu\sin\psi_k)^2\theta - (\lambda + \bar{r}\dot{\beta}_k + \beta_k\mu\cos\psi_k)(\bar{r} + \mu\sin\psi_k)\right\}\beta_k\cos\psi_k\end{matrix}\right]d\bar{r} \tag{3.78}$$

Rearranging the terms provides

$$C_H = \frac{1}{N}\sum_{k=1}^{N}\frac{\sigma a}{2}\int_0^1\left[\begin{matrix}(\bar{r} + \mu\sin\psi_k)^2\dfrac{C_d}{a}\sin\psi_k \\ +\left\{(\bar{r} + \mu\sin\psi_k)\theta - (\lambda + \bar{r}\dot{\beta}_k + \beta_k\mu\cos\psi_k)\right\} \\ \left\{(\lambda + \bar{r}\dot{\beta}_k + \beta_k\mu\cos\psi_k)\sin\psi_k - (\bar{r} + \mu\sin\psi_k)\beta_k\cos\psi_k\right\}\end{matrix}\right]d\bar{r} \tag{3.79}$$

This expression consists of two effects, which correspond to profile drag and induced drag. This can be expressed as

$$C_H = \frac{1}{N}\sum_{k=1}^{N}\frac{\sigma a}{2}\,[\text{profile drag term} + \text{induced drag term}]$$

The above expression can be symbolically represented as

$$C_H = C_{H0} + C_{Hi}$$

Subscript "0" stands for profile drag term and "i" stands for induced drag term.

In-Plane Force Coefficient (Lateral)

Similar to in-plane drag force, in-plane side force (or lateral force) can be obtained. The nondimensional in-plane force in the lateral direction is obtained as a sum of the components of in-plane forces F_{y1} and F_{x1}. It is given as

$$C_Y = \frac{Y}{\rho\pi R^2(\Omega R)^2} = \frac{1}{N}\sum_{k=1}^{N} N \int_0^R \frac{(F_{Y1}\cos\psi_k + F_{X1}\sin\psi_k)}{\rho\pi R^2(\Omega R)^2}\,dr \tag{3.80}$$

Substituting for the respective force components from Equations 3.68–3.72 and using nondimensional representation, the in-plane drag force coefficient can be written as

$$C_Y = \frac{1}{N}\sum_{k=1}^{N}\frac{\sigma a}{2}\int_0^1\left[-\left\{\bar{U}_P\bar{U}_T\theta - \bar{U}_P^2 + \bar{U}_T^2\frac{C_d}{a}\right\}\cos\psi_k - \left\{\bar{U}_T^2\theta - \bar{U}_P\bar{U}_T\right\}\beta_k\sin\psi_k\right]d\bar{r} \tag{3.81}$$

Substituting the nondimensional velocity components

$$C_Y = \frac{1}{N}\sum_{k=1}^{N}\frac{\sigma a}{2}\int_0^1\left[\begin{array}{l} -\left\{\begin{array}{l}(\lambda + \bar{r}\dot{\beta}_k + \beta_k\mu\cos\psi_k)(\bar{r} + \mu\sin\psi_k)\theta \\ -(\lambda + \bar{r}\dot{\beta}_k + \beta_k\mu\cos\psi_k)^2 + (\bar{r} + \mu\sin\psi_k)^2\dfrac{C_d}{a}\end{array}\right\}\cos\psi_k \\ -\left\{(\bar{r} + \mu\sin\psi_k)^2\theta - (\lambda + \bar{r}\dot{\beta}_k + \beta_k\mu\cos\psi_k)(\bar{r} + \mu\sin\psi_k)\right\}\beta_k\sin\psi_k\end{array}\right]d\bar{r} \tag{3.82}$$

Rearranging the terms provides

$$C_Y = \frac{1}{N}\sum_{k=1}^{N}\frac{\sigma a}{2}\int_0^1\left[\begin{array}{l} -(\bar{r} + \mu\sin\psi_k)^2\dfrac{C_d}{a}\cos\psi_k \\ -\left\{(\bar{r} + \mu\sin\psi_k)\theta - (\lambda + \bar{r}\dot{\beta}_k + \beta_k\mu\cos\psi_k)\right\} \\ \left\{(\lambda + \bar{r}\dot{\beta}_k + \beta_k\mu\cos\psi_k)\cos\psi_k + (\bar{r} + \mu\sin\psi_k)\beta_k\sin\psi_k\right\}\end{array}\right]d\bar{r} \tag{3.83}$$

This expression consists of two effects that correspond to profile drag and induced drag, and it is expressed as

$$C_Y = \frac{1}{N}\sum_{k=1}^{N} \frac{\sigma a}{2} \text{ [profile drag term + induced drag term]}$$

The above expression can be symbolically represented as

$$C_Y = C_{Y0} + C_{Yi}$$

C_{Y0} represents the profile drag effect and C_{Yi} represents the induced drag effect.

Moment Coefficient (C_{Mx} and C_{My})

The moment coefficients C_{Mx} and C_{My} depend on the type of attachment at the root of the blade to the hub. If the blade is centrally hinged (as shown in Figure 3.13a), no net moment is transferred to the hub. On the other hand, if the rotor is idealized as a centrally hinged, spring restrained rigid blade (Figure 3.13b), then the root moment is equal to the product of the blade flap deformation and the root spring constant K_β.

In the next chapter, various idealized models of the rotor blade will be discussed. A brief glimpse of various idealizations is given here for information. In an articulated rotor with a hinge offset *e*, the hub moment is the vector product of the blade root shear at the hinge and the hinge offset distance. If the rotor is a hingeless rotor, the hub moment is the integrated effect of the blade loads. A hingeless rotor blade can be idealized as a spring restrained offset hinged rigid blade (as shown in Figure 3.14), simulating the fundamental elastic mode of the blade.

In Figure 3.14, *e* is the hinge offset and K_β is the equivalent root spring constant representing the flexibility of the blade in flap. In this chapter, for simplicity, let us assume that the blade is idealized as a centrally hinged

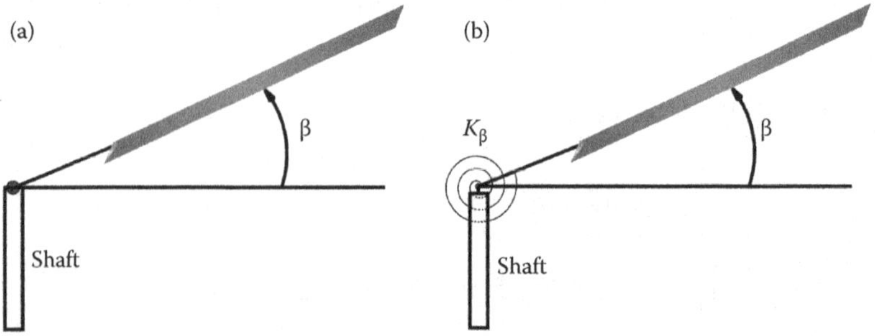

FIGURE 3.13
Rigid blade models: (a) centrally hinged model and (b) centrally hinged spring restrained model.

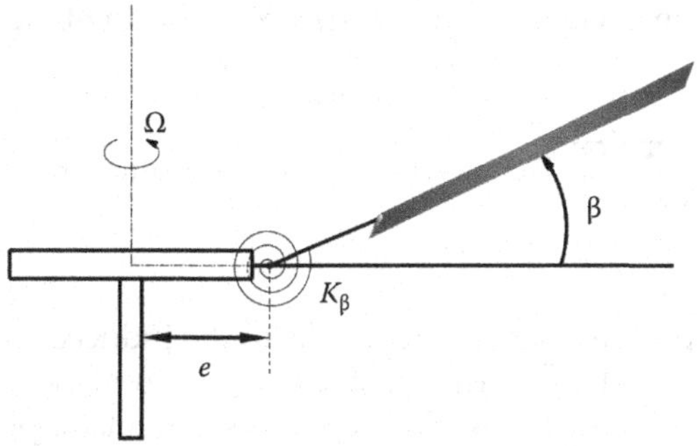

FIGURE 3.14
Idealized offset-hinged spring-restrained model of a rotor blade.

rigid blade with a spring restraint. In this case, the blade root moment is $K_\beta \beta$, where β is the flap deformation of the blade. The hub moments can be written as (refer to Figure 3.13b)

$$C_{Mx} = \frac{1}{\rho \pi R^2 (\Omega R^2) R} \sum_{k=1}^{N} K_\beta \beta_k \sin \psi_k \tag{3.84}$$

$$C_{My} = -\frac{1}{\rho \pi R^2 (\Omega R^2) R} \sum_{k=1}^{N} K_\beta \beta_k \cos \psi_k \tag{3.85}$$

Torque Coefficient (C_Q)

The torque coefficient due to all blades of the rotor system can be expressed as

$$C_Q = \frac{M_z}{\rho \pi R^2 (\Omega R)^2} = \frac{1}{N} \sum_{k=1}^{N} N \int_0^R r\{F_{Y1}\} dr \frac{1}{\rho \pi R^2 (\Omega R)^2} \tag{3.86}$$

Substituting for F_{y1} and expressing it in a nondimensional form:

$$C_Q = -\frac{1}{N} \sum_{k=1}^{N} \frac{\sigma a}{2} \int_0^1 \bar{r} \begin{bmatrix} (\lambda + \bar{r}\dot{\beta}_k + \beta_k \mu \cos \psi_k)(\bar{r} + \mu \sin \psi_k)\theta \\ -(\lambda + \bar{r}\dot{\beta}_k + \beta_k \mu \cos \psi_k)^2 + (\bar{r} + \mu \sin \psi_k)^2 \frac{C_d}{a} \end{bmatrix} d\bar{r} \tag{3.87}$$

The – sign indicates that the torque is acting in the clockwise direction. Since we are interested in the magnitude of torque, we will not carry the

– sign. The two components of drag forces contributing to the torque can be expressed as

$$C_Q = \frac{1}{N}\sum_{k=1}^{N} \frac{\sigma a}{2} \text{ [Induced drag term + profile drag term]}$$

$$C_Q = C_{Qi} + C_{Q0}$$

C_{Q0} is due to the profile drag and C_{Qi} is due to the induced drag effect.

It may be noted that the thrust coefficient (C_T), in-plane force coefficients (C_H, C_Y), and torque coefficient (C_Q) depend on the blade pitch input θ and the flap motion β. For the evaluation of these coefficients, one should know the time variation of θ and β. The time variation of β has to be obtained from the dynamics of blade motion in flap. The flap response is then substituted in the hub load expressions to obtain the total hub load (mean + time variation). The mean value is used for trimming the helicopter. The time variation provides the vibratory loads. This clearly brings out the fact that the blade dynamics is inherently coupled to the equilibrium and response analysis of the helicopter. In a more general formulation, the time response of the blade flap (β), lag ζ, and torsional φ deformations have to be included in the analysis.

In the following, let us obtain the expressions for the mean values of the various rotor hub load coefficients by assuming that the blade pitch input θ and the flap response β are known. Assuming a linear blade twist, the pitch angle θ at a radial location (r/R) of the kth blade is given as

$$\theta_k = \theta_0 + \theta_{tw}\frac{r}{R} + \theta_{1c}\cos\psi_k + \theta_{1s}\sin\psi_k \tag{3.88}$$

Similarly, the flap response β_k can be written with emphasis only on first harmonic terms as

$$\beta_k = \beta_0 + \beta_{1c}\cos\psi_k + \beta_{1s}\sin\psi_k + \text{higher harmonic terms} \tag{3.89}$$

where $\psi_k = \psi + \frac{2\pi}{N}(k-1)$ and $\psi = \Omega t$ represents the nondimensional time or the azimuthal position of the first blade, i.e., $k = 1$.

Note that from Equation 3.89, one can express the harmonic contents as (neglecting all higher harmonic terms)

$$\beta_0 = \frac{1}{N}\sum_{k=1}^{N}\beta_k$$

$$\beta_{1c} + \beta_{1c}\frac{1}{N}\sum_{k=1}^{N}\cos 2\psi_k = \frac{2}{N}\sum_{k=1}^{N}\beta_k\cos\psi_k \tag{3.90}$$

$$\beta_{1s} - \beta_{1s}\frac{1}{N}\sum_{k=1}^{N}\cos 2\psi_k = \frac{2}{N}\sum_{k=1}^{N}\beta_k\sin\psi_k$$

Substituting the expressions for pitch angle and flap angle given in Equations 3.88 and 3.89, in the various force and moment coefficients, and integrating over the radius and taking the summation over all the N blades in the rotor system, the resulting algebraic expression will contain several harmonic terms in ψ (which is equal to Ωt). These terms can be grouped into (a) constant term and (b) terms corresponding to different harmonics. The constant term represents the mean load, and the harmonics represent the vibratory loads at the hub due to all blades. The frequency of the terms treated in the flap angle expression (Equation 3.89). In general, the frequencies of the harmonics can be symbolically represented as $mN\Omega$, where $m = 1, 2, 3, \ldots$, and N is the number of blades, and Ω is the rotor angular velocity. For example, if there are two blades in the rotor system, the frequencies of the harmonics will be 2Ω, 4Ω, 6Ω, etc. (also denoted as 2/rev, 4/rev, 6/rev harmonic contents). On the other hand, if the rotor system contains four blades, the harmonics of the hub loads will be 4Ω, 8Ω, 12Ω (denoted as 4/rev, 8/rev, 12/rev harmonics), and so on. The same criterion can be applied to a rotor system with any number of blades. It is now evident that as the number of blades in the rotor system increases, the harmonics of the hub loads shift to higher frequencies.

In the following, the constant terms (or the mean values) of the force and moment coefficients as a function of advance ratio, blade pitch input, blade flap harmonics, and inflow through the rotor disc are given in closed-form analytical expressions. (Note: The derivation of the closed-form expressions can be taken as an exercise.)

Mean Thrust Coefficient C_T:

$$C_T = \frac{\sigma a}{2}\left\{\frac{\theta_0}{3}\left[1+\frac{3}{2}\mu^2\right]+\frac{\theta_{tw}}{4}(1+\mu^2)+\frac{\mu}{2}\theta_{1s}-\frac{\lambda}{2}\right\} \tag{3.91}$$

Mean In-plane Longitudinal Force Coefficient C_H:

$$C_H = \frac{\sigma a}{2}\begin{bmatrix} \theta_0\left\{\frac{\lambda\mu}{2}-\frac{\beta_{1c}}{3}\right\}+\theta_{tw}\left\{\frac{\lambda\mu}{4}-\frac{\beta_{1c}}{4}\right\} \\ +\theta_{1s}\left\{\frac{\lambda}{4}-\frac{\mu\beta_{1c}}{4}\right\}-\theta_{1c}\frac{\beta_0}{6} \\ +\frac{3}{4}\lambda\beta_{1c}+\beta_{1s}\frac{\beta_0}{6}+\frac{\mu}{4}\left\{\beta_0^2+\beta_{1c}^2\right\} \end{bmatrix}+\frac{\sigma a}{2}\left\{\frac{\mu}{2}\frac{C_{d0}}{a}\right\} \tag{3.92}$$

Mean In-plane Lateral Force Coefficient C_Y:

$$C_Y = -\frac{\sigma a}{2}\begin{bmatrix} \theta_0\left\{\frac{3}{4}\mu\beta_0+\frac{\beta_{1s}}{3}\left(1+\frac{3}{2}\mu^2\right)\right\}+\theta_{tw}\left\{\frac{\mu\beta_0}{2}+\frac{\beta_{1s}}{4}(1+\mu^2)\right\} \\ +\theta_{1c}\left\{\frac{\lambda}{4}+\frac{1}{4}\mu\beta_{1c}\right\}+\theta_{1s}\left\{\frac{\beta_0}{6}(1+3\mu^2)+\frac{1}{2}\mu\beta_{1s}\right\} \\ -\frac{3}{2}\lambda\mu\beta_0+\beta_0\beta_{1c}\left(\frac{1}{6}-\mu^2\right)-\frac{3}{4}\lambda\beta_{1s}-\frac{\mu}{4}\beta_{1c}\beta_{1s} \end{bmatrix} \tag{3.93}$$

Mean Torque Coefficient: $C_Q = C_{Mz}$:

$$C_Q = -\frac{\sigma a}{2}\left[\frac{C_{d0}}{4a}[1+\mu^2]+\theta_0\frac{\lambda}{3}+\theta_{tw}\frac{\lambda}{4}\right.$$
$$+\theta_{1c}\left\{\frac{\beta_{1s}}{8}+\frac{\mu\beta_0}{6}+\frac{\mu^2}{16}\beta_{1s}\right\}+\theta_{1s}\left\{-\frac{\beta_{1c}}{8}+\frac{\lambda\mu}{4}+\frac{\mu^2}{16}\beta_{1c}\right\} \tag{3.94}$$
$$\left.-\frac{\lambda^2}{2}-\frac{\beta_{1c}^2}{8}-\frac{\beta_{1s}^2}{8}-\mu^2\left\{\frac{\beta_0^2}{4}+\frac{3}{16}\beta_{1c}^2+\frac{1}{16}\beta_{1s}^2\right\}-\frac{\mu\beta_{1s}\beta_0}{3}-\frac{\mu\lambda\beta_{1c}}{2}\right]$$

Mean Roll Moment Coefficient:

$$C_{Mx} = \frac{N}{2}\frac{1}{\rho\pi R^2(\Omega R^2)R}K_\beta\beta_{1s} \tag{3.95}$$

Mean Pitch Moment Coefficient:

$$C_{My} = -\frac{N}{2}\frac{1}{\rho\pi R^2(\Omega R^2)R}K_\beta\beta_{1c} \tag{3.96}$$

(NOTE: Reverse flow and radial drag effects have been neglected in obtaining these expressions.)

In addition to the mean hub loads in forward flight, one can obtain an expression for the power required for flight.

Power in Forward Flight

With all the power transmitted to the main rotor through the shaft, we have $P = \Omega Q$. It can be shown that the nondimensional power coefficient is equal to the nondimensional torque coefficient. Though the expression for the nondimensional torque coefficient has been derived and given in Equation 3.94, let us try to obtain a different form of the expression for C_Q starting from a basic expression. The mean torque coefficient can also be obtained by taking the mean load over the azimuth due to a single blade and multiplied by the number of blades. It can be expressed as (from Equation 3.86). (Note: subscript "k" is removed to denote that the first blade is treated as the reference single blade.)

$$C_Q = \frac{M_z}{\rho(\pi R)^2(\Omega R)^2 R} = \frac{1}{2\pi}\int_0^{2\pi} N\int_0^R r\{F_{y1}\}dr\frac{1}{\rho\pi R^2(\Omega R)^2 R}d\psi \tag{3.97}$$

Changing the integral to nondimensional form:

$$C_Q = \frac{1}{2\pi}\int_0^{2\pi} \frac{NR^2C}{\rho\pi R^2(\Omega R)^2 RC}\int_0^1 \bar{r}\{F_{y1}\}d\bar{r}\,d\psi \tag{3.98}$$

Substituting for the solidity ratio, the above integral equation can be written as

$$C_Q = \frac{1}{2\pi}\int_0^{2\pi} \frac{\sigma}{\rho C(\Omega R)^2}\int_0^1 \bar{r}\{F_{y1}\}d\bar{r}\,d\psi \tag{3.99}$$

Substituting the force term from Equation 3.68, the integral can be written as

$$C_Q = -\frac{1}{2\pi}\int_0^{2\pi} \frac{\sigma a}{aC}\int_0^1 \frac{1}{\rho(\Omega R)^2}\left\{\frac{1}{2}\rho U^2 C[C_l \sin\phi + C_d \cos\phi]\right\}\bar{r}\,d\bar{r}\,d\psi \tag{3.100}$$

Assuming a small flap angle, from Equations 3.52–3.54 and 3.56, one can obtain an expression for the lift coefficient term as

$$F_{z1} = \frac{1}{2}\rho U^2 C[C_l \cos\phi - C_d \sin\phi]$$

$$C_l \sin\phi = \frac{F_{z1}}{(1/2)\rho U^2 C}\frac{\sin\phi}{\cos\phi} + C_d \frac{\sin^2\phi}{\cos\phi} \tag{3.101}$$

Substituting Equation 3.101 in C_Q expression of Equation 3.100,

$$C_Q = -\frac{1}{2\pi}\int_0^{2\pi} \frac{\sigma a}{aC}\int_0^1 \frac{1}{\rho(\Omega R)^2}\left\{\frac{1}{2}\rho U^2 C\left[\frac{F_{z1}}{(1/2)\rho U^2 C}\tan\phi + C_d\frac{1}{\cos\phi}\right]\right\}\bar{r}\,d\bar{r}\,d\psi$$

(3.102)

Substituting for $\tan\phi = \dfrac{\bar{U}_P}{\bar{U}_T}$ and $\cos\phi = \dfrac{\bar{U}_T}{\bar{U}}$

$$C_Q = -\frac{1}{2\pi}\int_0^{2\pi} \frac{\sigma a}{aC}\int_0^1 \frac{1}{\rho(\Omega R)^2}\left\{F_{z1}\frac{\bar{U}_P}{\bar{U}_T} + (1/2)\frac{\rho U^2}{\bar{U}_T} C\bar{U}C_d\right\}\bar{r}\,d\bar{r}\,d\psi \tag{3.103}$$

Note that the – sign indicates that the torque is acting in the clockwise direction. Since we are interested in the magnitude, we can drop the – sign. Equation 3.103 can be written as a sum of induced (C_{Qi}) and drag term (C_{Q0}):

$$C_Q = C_{Qi} + C_{Q0} \tag{3.104}$$

Similarly, let us formulate an expression for mean C_H in terms of F_{z1} (from Equations 3.55, 3.56, and 3.76):

$$C_H = \frac{H}{\rho\pi R^2(\Omega R)^2} = \frac{1}{2\pi}\int_0^{2\pi}\frac{N}{\rho\pi R^2(\Omega R)^2}\int_0^R \{-F_{Y1}\sin\psi + F_{x1}\cos\psi\}\,dr\,d\psi \tag{3.105}$$

Using Equations 3.57, 3.59, and 3.68, and nondimensionalizing Equation 3.105

$$C_H = \frac{1}{2\pi}\int_0^{2\pi}\frac{NRC}{\rho\pi R^2(\Omega R^2)C}\int_0^1\left\{\frac{1}{2}\rho U^2 C[C_l\sin\phi + C_d\cos\phi]\sin\psi - F_{z1}\beta\cos\psi\right\}d\bar{r}\,d\psi \tag{3.106}$$

Substituting Equation 3.101, Equation 3.106 can be written as

$$C_H = \frac{1}{2\pi}\int_0^{2\pi}\frac{\sigma a}{aC}\frac{1}{\rho(\Omega R^2)}\int_0^1\left\{\frac{1}{2}\rho U^2 C\left[\frac{F_{z1}}{\frac{1}{2}\rho U^2 C}\tan\phi + C_d\frac{1}{\cos\phi}\right]\sin\psi - F_{z1}\beta\cos\psi\right\}d\bar{r}\,d\psi \tag{3.107}$$

Substituting for $\tan\phi$ and $\cos\phi$ and rearranging the terms

$$\begin{aligned} C_H = \frac{1}{2\pi}\int_0^{2\pi}\frac{\sigma a}{aC}\int_0^1 &\left\{\frac{1}{\rho(\Omega R^2)}\left[\frac{\bar{U}_P}{\bar{U}_T}\sin\psi - \beta\cos\psi\right]F_{z1}\right. \\ &\left. + \frac{1}{\rho(\Omega R^2)}\frac{1}{2}\rho U^2 C\frac{\bar{U}}{\bar{U}_T}C_d\sin\psi\right\}d\bar{r}d\psi \end{aligned} \tag{3.108}$$

Equation 3.108 can be written as a combination of the induced term (C_{Hi}) and the drag term (C_{H0}):

$$C_H = C_{Hi} + C_{H0} \tag{3.109}$$

Let us write

$$\begin{aligned} C_Q &= C_{Qi} + C_{Q0} + \mu C_{Hi} + \mu C_{H0} - \mu C_H = C_P \\ C_Q &= C_{Qi} + \mu C_{Hi} - \mu C_H + (C_{Q0} + \mu C_{H0}) = C_P \end{aligned} \tag{3.110}$$

Let us evaluate the various terms of C_Q:

$$C_{Qi} + \mu C_{Hi} = \frac{1}{2\pi}\int_0^{2\pi} \frac{\sigma a}{aC} \int_0^1 \frac{1}{\rho(\Omega R^2)} \left\{ \bar{r}\frac{\bar{U}_P}{U_T} + \frac{\bar{U}_P}{U_T}\mu \sin\psi - \mu\beta\cos\psi \right\} F_{z1} d\bar{r} d\psi$$

(3.111)

Noting from Equation 3.65 that $\bar{U}_T = \bar{r} + \mu \sin\psi$ and $\bar{r} = \frac{r}{R}$

$$C_{Qi} + \mu C_{Hi} = \frac{1}{2\pi}\int_0^{2\pi} \frac{\sigma a}{\sigma C} \int_0^1 \frac{1}{\rho(\Omega R^2)} \{\bar{U}_P - \mu\beta\cos\psi\} F_{z1}\, d\bar{r}\, d\psi \qquad (3.112)$$

Substituting from Equation 3.66, $\bar{U}_P = \lambda + \bar{r}\dot{\beta} + \mu\beta\cos\psi$

$$C_{Qi} + \mu C_{Hi} = \frac{1}{2\pi}\int_0^{2\pi} \frac{\sigma\alpha}{\alpha C} \int_0^1 \frac{1}{\rho(\Omega R)^2} \{\lambda + \bar{r}\dot{\beta}\} F_{z1}\, d\bar{r}\, d\psi \qquad (3.113)$$

Changing the order of integration and integrating over the azimuth for average quantity, the second term becomes zero because

$$\frac{1}{2\pi}\int_0^{2\pi} \dot{\beta} F_{z1}\, d\psi = \frac{1}{2\pi}\int_0^{2\pi} F_{z1}\, d\beta = 0 \qquad (3.114)$$

This term represents the work done on the blade section by the periodic force F_{z1} during one revolution. Under steady-state condition, the periodic motion of the blade requires that total work done per revolution is zero. The proof is given in the note below.

BRIEF NOTE: The flap dynamic equation of a rotating blade can be written as (this will be discussed in the next chapter)

$$\ddot{\beta} + v^2\beta = \gamma \int_0^1 \bar{r}\frac{F_z}{ac} d\bar{r}$$

Evaluating the average work done by the aerodynamic load during one revolution, i.e.,

$$W = \frac{1}{2\pi}\int_0^{2\pi}\int_0^1 \frac{\bar{r}\dot{\beta}F_z}{ac}\,d\bar{r}\,d\psi = \frac{1}{2\pi}\int_0^{2\pi}\dot{\beta}\left\{\int_0^1 \bar{r}\frac{F_z}{ac}\,d\bar{r}\right\}d\psi$$

$$W = \frac{1}{2\pi}\int_0^{2\pi}\frac{\dot{\beta}}{\gamma}\left[\ddot{\beta}+\upsilon^2\beta\right]d\psi$$

$$W = \frac{1}{2\pi}\int_0^{2\pi}\frac{1}{2\gamma}\frac{d}{d\psi}\left[\dot{\beta}^2+\upsilon^2\beta^2\right]d\psi$$

Since the term $\dot{\beta}^2+\upsilon^2\beta^2$ is periodic, the integral is zero.

Therefore, Equation 3.113 can be written as

$$C_{Qi}+\mu C_{Hi} = \frac{1}{2\pi}\int_0^{2\pi}\frac{\sigma a}{aC}\int_0^1 \lambda\frac{F_{z1}}{\rho(\Omega R)^2}\,d\bar{r}\,d\psi \tag{3.115}$$

This term can be written as

$$C_{Qi}+\mu C_{Hi} = \int_A \lambda\, dC_T \text{ area integral over the rotor disk} \tag{3.116}$$

where dC_T is the mean thrust per unit area of the rotor disk.

The total power coefficient can be written as

$$C_P = C_Q = \int \lambda\, dC_T - \mu C_H + (C_{Q0}+\mu C_{H0}) \tag{3.117}$$

This expression can be further simplified by representing the total induced flow λ in terms of helicopter drag and lift. From Equation 3.14, the total inflow is given as

$$\lambda = \mu \tan\alpha + \lambda_i$$

For small angle, the total inflow can be written as

$$\lambda = \lambda_i + \mu\tan\alpha \cong \lambda_i + \mu\alpha \tag{3.118}$$

Figure 3.15 shows an idealized representation of forces acting on the helicopter in the longitudinal direction. The velocity of the helicopter is V, and

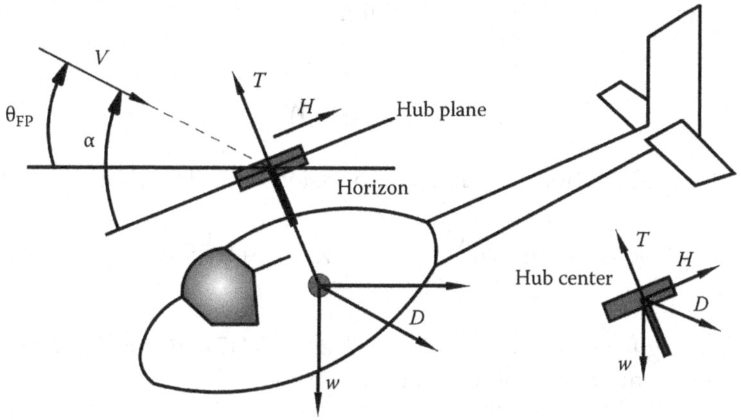

FIGURE 3.15
Idealized representation of helicopter loads in the longitudinal direction.

it is inclined with respect to the horizontal plane by the angle θ_{FP} representing the flight path angle. The reference plane is taken as a hub plane, which is inclined with respect to the flight path by angle α. The forces acting in the vertical plane are shown in Figure 3.15. The thrust T is acting normal to the hub plane, and the longitudinal force H is acting at the hub center. W is the weight of the helicopter, and D is the helicopter drag force acting along the wind velocity direction. From Figure 3.15, the climb velocity of the helicopter can be defined as

$$V_{\text{climb}} = V \sin \theta_{FP} = V_C \tag{3.119}$$

The nondimensional climb velocity can be written as (assuming small angle)

$$\lambda_C = \frac{V_C}{\Omega R} = \frac{V \sin \theta_{FP}}{\Omega R} \approx \mu \theta_{FP} \tag{3.120}$$

Force equilibrium condition requires that

$$T \cos(\alpha - \theta_{FP}) + H \sin(\alpha - \theta_{FP}) = W + D \sin \theta_{FP} \tag{3.121}$$

$$T \sin(\alpha - \theta_{FP}) - H \cos(\alpha - \theta_{FP}) = D \cos \theta_{FP} \tag{3.122}$$

Assuming small angles and neglecting the drag force in comparison to weight W, one can write the above two equations as

$$\begin{aligned} T &\approx W \\ H + D &\approx T(\alpha - \theta_{FP}) \end{aligned} \tag{3.123}$$

Rearranging and nondimensionalizing

$$\alpha = \theta_{FP} + \frac{C_H}{C_T} + \frac{D}{W} \tag{3.124}$$

Substituting Equation 3.124 in the expression for inflow (Equation 3.118):

$$\lambda = \lambda_i + \mu\theta_{FP} + \mu\frac{C_H}{C_T} + \mu\frac{D}{W} = \lambda_i + \lambda_C + \mu\frac{C_H}{C_T} + \mu\frac{D}{W} \tag{3.125}$$

Substituting for λ in C_P expression (Equation 3.117), the total power can be written as a combination of four components as

$$C_P = \int \lambda_i \, dC_T + \lambda_c C_T + \mu C_H + \mu\frac{D}{W}C_T - \mu C_H + (C_{Q0} + \mu C_{H0}) \tag{3.126}$$

Equation 3.126 can be written as

$$C_P = C_{Pi} + C_{Pc} + C_{Pp} + C_{P0} \tag{3.127}$$

where C_{Pi} is the induced power, C_{Pc} is the climb power, C_{Pp} is the parasite power to overcome the drag of the helicopter, and C_{P0} is the rotor profile power to turn the rotor in air.

The profile power expression can be simplified to obtain a closed-form expression. The profile power expression is given as

$$C_{P0} = C_{Q0} + \mu C_{H0} \tag{3.128}$$

Substituting in Equation 3.128, the individual expressions from Equations 3.103, 3.104, 3.108, and 3.109,

$$\begin{aligned} C_{P0} &= \frac{\sigma a}{ac}\int_0^1 \frac{1}{\rho(\Omega R)^2}\frac{1}{2}\rho\frac{U^2}{\bar{U}_T}c\bar{U}C_d\,\bar{r}\,d\bar{r} \\ &+ \frac{\sigma a}{ac}\mu\int_0^1 \frac{1}{\rho(\Omega R)^2}\frac{1}{2}\rho\frac{U^2}{\bar{U}_T}c\bar{U}C_d\sin\psi\,d\bar{r} \end{aligned} \tag{3.129}$$

Combining the two terms and simplifying, one obtains

$$C_{P0} = \frac{\sigma C_d}{2}\int_0^1 \left\{\bar{r}\frac{\bar{U}^3}{\bar{U}_T} + \mu\sin\psi\frac{\bar{U}^3}{\bar{U}_T}\right\}d\bar{r} \tag{3.130}$$

Assuming that $\bar{U} \approx \bar{U}_T$, the profile drag power can be written as

$$C_{P0} \approx \frac{\sigma C_d}{2} \int_0^1 (\bar{r} + \mu \sin\psi)^3 \, d\bar{r} \tag{3.131}$$

The mean value of C_{P0} can be obtained by averaging over the azimuth,

$$C_{P0} = \frac{\sigma C_d}{2} \frac{1}{2\pi} \int_0^{2\pi} \int_0^1 (\bar{r} + \mu \sin\psi)^3 \, d\bar{r} \, d\psi \tag{3.132}$$

Evaluating the integral, the expression for profile drag can be obtained as

$$C_{P0} = \frac{\sigma C_d}{8} [1 + 3\mu^2] \tag{3.133}$$

With the inclusion of reverse flow and radial drag effects, a further increase in C_{P0} is observed. The coefficient of μ^2 can be more than 3.

The rotor-induced power is given as $C_{Pi} = \int \lambda_i \, dC_T$ where $dC_T = \sigma a \frac{1}{\rho(\Omega R^2)} \left(\frac{F_z}{ac} \right) d\bar{r}$.

For uniform inflow, the induced power can be written as $C_{Pi} = \lambda_i C_T$. For forward speeds above $\mu \geq 0.1$, a good approximation for induced flow can be expressed as $\lambda_i \approx \kappa \frac{C_T}{2\mu}$ taking into account tip loss, nonuniform inflow, and other effects. The induced power loss can be expressed as

$$C_{Pi} \cong \kappa \frac{C_T^2}{2\mu} \tag{3.134}$$

The parasite power term can be approximated as $C_{Pp} = \mu \frac{D}{W} C_T \approx \frac{VD}{\rho \pi R^2 (\Omega R)^3}$ by equating the total weight of the helicopter to rotor thrust T. If the helicopter drag is written in terms of equivalent flat plate area f, then it can be expressed as $D = \frac{1}{2} \rho V^2 f \, C_{DF}$, where $C_{DF} = 1.0$. The parasite power term can now be written as

$$C_{Pp} = \frac{1}{2} \frac{V \rho V^2 f}{\rho \pi R^2 (\Omega R)^3} \cong \frac{1}{2} \mu^3 \frac{f}{\pi R^2} \tag{3.135}$$

In general, for most of the helicopters, the value of $\frac{f}{\pi R^2}$ is of order of 0.01.

In summary, the rotor power coefficient in forward flight may be written, as by combining all the individual contributions,

$$C_P \cong \frac{\kappa C_T^2}{2\mu} + \frac{\sigma C_d}{8}[1+3\mu^2] + \frac{1}{2}\mu^3\frac{f}{A} + \lambda_c C_T \tag{3.136}$$

The rotor power has four components, namely, induced power, profile drag power of the rotor, parasite drag power of helicopter, and climb power.

Figure 3.16 shows a typical variation of power with forward speed for the level flight of a helicopter. It can be seen that as the forward speed increases from hover, the power initially decreases up to a forward speed of about μ = 0.15, and then the power increases drastically with forward speed. The initial decrease in power requirement is mainly due to the reduction in induced flow through the main rotor as the helicopter increases its speed from hover. As the speed increases, the parasite drag power of the helicopter increases in cubic power of forward speed. The profile drag power shows a quadratic variation with forward speed. Therefore, at high forward speeds, the power required is mainly to overcome fuselage drag, whereas during hover, the power required is mainly the induced power to lift the weight of the helicopter.

The variation of power with forward speed is an important parameter required in the selection of a suitable engine for the helicopter. The excess power over the power required for level flight is the available power that can be used for climbing the helicopter. The power curve also indicates the maximum speed that the helicopter can fly, which is dependent on the power rating of the engine.

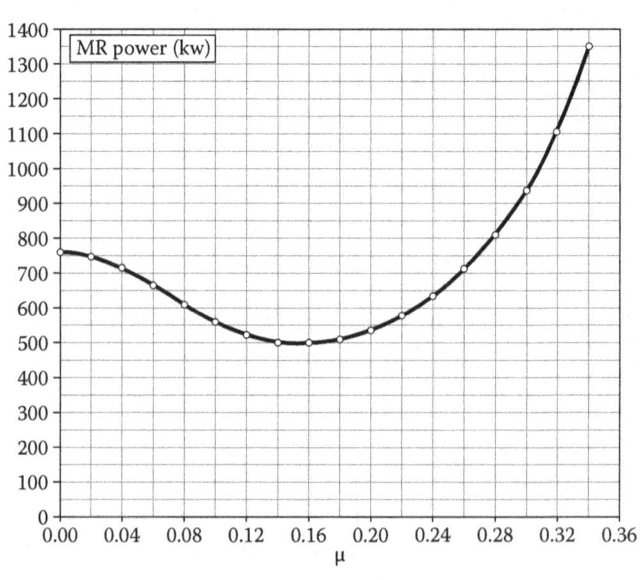

FIGURE 3.16
Variation of power with forward speed during level flight.

4

Rotor Blade Flapping Motion: Simple Model

It has been clearly brought out in Chapter 3 that, during forward flight, the hub loads are dependent on the blade pitch input and the flap response of the blade. The mean hub load expressions have been derived by assuming that the flap response can be represented by a first harmonic variation. Therefore, to complete the solution, one must evaluate the flap response of the blade to a given blade pitch input. The input quantities are collective, lateral, and longitudinal cyclic pitch inputs (θ_0, θ_{1c}, θ_{1s}), and the output is the flap response of the blade (i.e., harmonics of the flap response β_0, β_{1c}, β_{1s}). These response quantities depend on the configuration of the blade model. Initially, a simplified model will be used to highlight the essential features of the problem; later, additional features will be included in the idealized blade model. In the simplified case, the rotor blade is idealized as a rigid beam hinged at the center of the hub, as shown in Figure 4.1. The rotor blade is assumed to undergo only flap motion.

Let us derive the equation of the motion of the rotor blade, starting from first principles. The equation of the motion of the blade can be obtained by making the total moment about the flap hinge at the root 0.

$$(\bar{M}_{\mathrm{I}})_{\mathrm{Flap}} + (\bar{M}_{\mathrm{ext}})_{\mathrm{Flap}} = 0 \tag{4.1}$$

where $\bar{M}_{\mathrm{I}}$ is the inertia moment and $\bar{M}_{\mathrm{ext}}$ is the external aerodynamic moment. The subscript indicates the component of the moment vector about the flap axis.

The rotor blade is idealized as a rigid beam element undergoing only flapping motion. The position vector of any point P on the kth blade, of an N bladed rotor system, in the deformed state can be written as (Figure 4.1)

$$\vec{r}_{\mathrm{p}} = r\vec{e}_{x2} = r\cos\beta_k\vec{e}_{x1} + r\sin\beta_k\vec{e}_{z1} \tag{4.2}$$

The angular velocity of the rotor blade is taken as $\vec{\omega} = \Omega\vec{e}_{z1}$. It is assumed that the angular velocity of the rotor is a constant, and hence, the angular acceleration $\dfrac{d\bar{\omega}}{dt} = 0$.

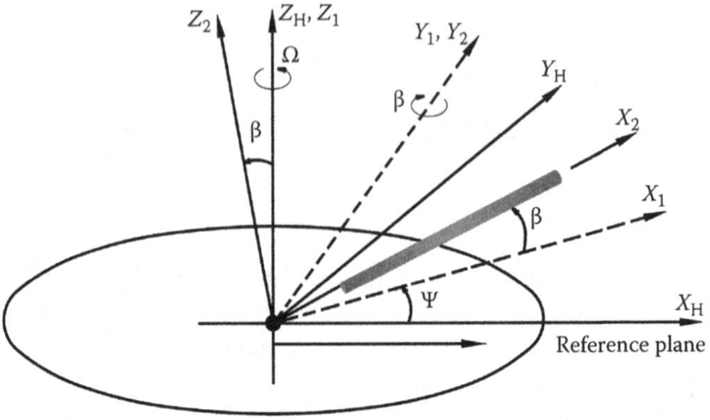

FIGURE 4.1
Centrally hinged rigid blade model.

The absolute velocity of point P has two components: one due to rotation and another due to the flapping motion in the rotating frame. The absolute velocity of point P is given by

$$\vec{v}_p = \left(\frac{d\vec{r}_p}{dt}\right)_{rel} + \vec{w}\, X\, \vec{r}_p = -r\sin\beta_k \frac{d\beta_k}{dt}\vec{e}_{x1} + r\cos\beta_k \frac{d\beta_k}{dt}\vec{e}_{z1} + \vec{\omega}\, X\, \vec{r}_p \tag{4.3}$$

Evaluating the vector cross-product, the velocity of point P can be written as

$$\vec{v}_p = -r\sin\beta_k \frac{d\beta_k}{dt}\vec{e}_{x1} + r\cos\beta_k \frac{d\beta_k}{dt}\vec{e}_{z1} + \Omega r\cos\beta_k \vec{e}_{y1} \tag{4.4}$$

The absolute acceleration of point P is given by

$$\vec{a}_p = \left(\frac{d^2\vec{r}_p}{dt^2}\right)_{rel} + \frac{d\bar{\omega}}{dt} X\, \vec{r}_p + 2\bar{\omega}\, X \left(\frac{d\vec{r}_p}{dt}\right)_{rel} + \bar{\omega}\, X(\bar{\omega}\, X\, \bar{r}_p) \tag{4.5}$$

Substituting various terms, after time differentiation, the absolute acceleration of point P can be written as

$$\vec{a}_p = \left(-r\cos\beta_k \left\{\frac{d\beta_k}{dt}\right\}^2 - r\sin\beta_k \frac{d^2\beta_k}{dt^2}\right)\vec{e}_{x1} + \left(-r\sin\beta_k \left\{\frac{d\beta_k}{dt}\right\}^2 + r\cos\beta_k \frac{d^2\beta_k}{dt^2}\right)\vec{e}_{z1}$$
$$+ \left(-2\Omega r\sin\beta_k \frac{d\beta_k}{dt}\right)\vec{e}_{y1} + (-\Omega^2 r\cos\beta_k)\vec{e}_{x1} \tag{4.6}$$

Rearranging the terms,

$$\vec{a}_{\mathrm{p}}=\left(-r\cos\beta_k\left\{\frac{d\beta_k}{dt}\right\}^2-r\sin\beta_k\frac{d^2\beta_k}{dt^2}-\Omega^2r\cos\beta_k\right)\vec{e}_{\mathrm{x}1}+\left(-2\Omega r\sin\beta_k\frac{d\beta_k}{dt}\right)\vec{e}_{\mathrm{y}1}$$
$$+\left(-r\sin\beta_k\left\{\frac{d\beta_k}{dt}\right\}^2+r\cos\beta_k\frac{d^2\beta_k}{dt^2}\right)\vec{e}_{\mathrm{z}1} \tag{4.7}$$

The inertia force acting on a blade element of length dr is given as $(-\rho dr\,\vec{a}_{\mathrm{p}})$, where ρ is the mass per unit length of the blade. The inertia moment about the root is given as

$$\bar{M}_{\mathrm{I}}=\int_0^R \vec{r}_{\mathrm{p}}X(-\rho dr\,\vec{a}_{\mathrm{p}}) \tag{4.8}$$

Substituting for the position vector from Equation 4.2 and taking the vector cross-product, the component of inertia moment about the flap axis $\bar{e}_{\mathrm{y}1}$ can be obtained. It is given as

$$(M_{\mathrm{I}})_{\mathrm{Flap}}=\int_0^R \rho\left(r^2\frac{d^2\beta_k}{dt^2}+\Omega^2r^2\sin\beta_k\cos\beta_k\right)dr \tag{4.9}$$

Integrating the term over the length of the blade and the inertia moment in flap motion can be expressed as

$$(M_{\mathrm{I}})_{\mathrm{Flap}}=I_{\mathrm{b}}\left(\frac{d^2\beta_k}{dt^2}+\Omega^2\sin\beta_k\cos\beta_k\right) \tag{4.10}$$

where $I_{\mathrm{b}}=\int_0^R \rho r^2\,dr$ is the mass moment of inertia of the blade about the flap hinge at the center of the hub.

Invoking a small angle assumption for flap angle β_k, the inertia moment in the flap can be simplified as

$$(M_{\mathrm{I}})_{\mathrm{Flap}}\cong I_{\mathrm{b}}\left(\frac{d^2\beta_k}{dt^2}+\Omega^2\beta_k\right) \tag{4.11}$$

The external flap moment due to the distributed aerodynamic lift acting on the blade can be written, using Equations 3.54 and 3.60, as

$$(M_{\text{ext}})_{\text{Flap}} = \int_0^R (r\vec{e}_{x2} \times F_{z2}\vec{e}_{z2})dr = \int_0^R (-rF_{z2})dr \tag{4.12}$$

where F_{z2} is the lift per unit length acting on the blade, which is also equal to F_{z1} (Equation 3.57).

Combining the inertia and the aerodynamic effects (Equation 4.1), the flap equation can be written as

$$I_b\left(\frac{d^2\beta_k}{dt^2} + \Omega^2\beta_k\right) = \int_0^R rF_{z1}\,dr \tag{4.13}$$

Substitute in Equation 4.13 for F_{z1} from Equation 3.67 as

$$F_{z1} = \frac{1}{2}\rho aC(\Omega R)^2\left[\left(\frac{r}{R} + \mu\sin\psi_k\right)^2\theta - \left(\lambda + \frac{r}{R}\dot{\beta}_k + \beta_k\mu\cos\psi_k\right)\left(\frac{r}{R} + \mu\sin\psi_k\right)\right] \tag{4.14}$$

In Equation 4.13, nondimensionalize the time derivative term on the LHS as $\frac{d^2\beta_k}{dt^2} = \Omega^2\ddot{\beta}_k$ and the integral on the RHS with respect to the rotor radius R. After integrating RHS over the length of the blade, the flap equation can be written in symbolic form as

$$\ddot{\beta}_k + \beta_k = \frac{\rho aCR^4}{I_b}M_{\text{Flap}} = \gamma M_{\text{Flap}} \tag{4.15}$$

where $\gamma = \frac{\rho aCR^4}{I_b}$ is denoted as the Lock number, which represents the ratio of the aerodynamic effect to the inertia effect of the blade. Typically, γ is of the order 8 to 10 for articulated rotors and 5 to 7 for hingeless blades. The term M_{Flap} is given as

$$M_{\text{Flap}} = \int_0^1 \frac{1}{2}\bar{r}\left\{\begin{matrix}(\bar{r}^2 + 2\mu\bar{r}\sin\psi_k + \mu^2\sin^2\psi_k)\theta \\ -(\lambda\bar{r} + \lambda\mu\sin\psi_k + \bar{r}^2\dot{\beta}_k + \dot{\beta}_k\bar{r}\mu\sin\psi_k + \beta_k\bar{r}\mu\sin\psi_k + \beta_k\mu^2\sin\psi_k\cos\psi_k)\end{matrix}\right\}d\bar{r} \tag{4.16}$$

where $\bar{r} = \frac{r}{R}$ represents the nondimensional radial variable.

Assume that the blade pitch angle consists of the pilot input and a linear geometric twist of the blade:

$$\theta = \theta_{\mathrm{I}} + \theta_{\mathrm{tw}} \frac{r}{R} \tag{4.17}$$

Substituting Equation 4.17 in Equation 4.16 and assuming uniform inflow (λ), after integration over the length of the blade, the aerodynamic moment term can be obtained as

$$\begin{aligned} M_{\mathrm{Flap}} = \theta_{\mathrm{I}} & \left\{ \frac{1}{8} + \frac{\mu}{3} \sin \psi_k + \frac{\mu^2}{4} \sin^2 \psi_k \right\} \\ & + \theta_{\mathrm{tw}} \left\{ \frac{1}{10} + \frac{\mu}{4} \sin \psi_k + \frac{\mu^2}{6} \sin^2 \psi_k \right\} \\ & - \lambda \left\{ \frac{1}{6} + \frac{\mu}{4} \sin \psi_k \right\} \\ & - \dot{\beta}_k \left\{ \frac{1}{8} + \frac{\mu}{6} \sin \psi_k \right\} \\ & - \beta_k \mu \cos \psi_k \left\{ \frac{1}{6} + \frac{\mu}{4} \sin \psi_k \right\} \end{aligned} \tag{4.18}$$

Equation 4.18 can be symbolically written as

$$M_{\mathrm{Flap}} = M_{\theta} \theta_{\mathrm{I}} + M_{\theta_{\mathrm{tw}}} \theta_{\mathrm{tw}} + M_{\lambda} \lambda + M_{\dot{\beta}} \dot{\beta}_k + M_{\beta} \beta_k \tag{4.19}$$

The aerodynamic flap moment at the blade root is dependent on blade pitch input (θ_{I}), the blade geometric twist (θ_{tw}), the rotor inflow (λ), the blade flap response (β_k), and the time derivative of the flap response ($\dot{\beta}_k$). The time derivative term $\dot{\beta}_k$ represents aerodynamic damping in the flap mode.

The blade pitch control input (θ_{I}) given by the pilot can be mathematically expressed as

$$\theta_{\mathrm{I}} = \theta_0 + \theta_{1c} \cos \psi_k + \theta_{1s} \sin \psi_k \tag{4.20}$$

It may be noted that the steady-state periodic flap response is represented as a Fourier series approximation containing several harmonics. For the sake of simplicity, but without losing the physics of the problem, let us assume that the flap response is represented by the following truncated Fourier series.

$$\beta_k = \beta_0 + \beta_{1c} \cos \psi_k + \beta_{1s} \sin \psi_k + \text{higher harmonic terms} \tag{4.21}$$

For the given pitch input, the flap response β_k can be obtained by either (1) the harmonic balance method or (2) the operator method (very similar to the evaluation of Fourier coefficients). The flapping equation is operated by the following integrals:

$$\frac{1}{2\pi}\int_0^{2\pi}(\text{Flap Equation})\,d\psi_k = 0 \tag{4.22}$$

$$\frac{1}{\pi}\int_0^{2\pi}(\text{Flap Equation})\cos\psi_k\,d\psi_k = 0 \tag{4.23}$$

$$\frac{1}{\pi}\int_0^{2\pi}(\text{Flap Equation})\sin\psi_k\,d\psi_k = 0 \tag{4.24}$$

Performing the above operations, we obtain a set of three algebraic equations in terms of the harmonic coefficients of β_k and θ_I. Solving these equations, one can obtain the coefficients β_0, β_{1c}, β_{1s} in terms of θ_0, θ_{1c}, θ_{1s}. Performing these three integral operations on the left-hand side of the flap equation (Equation 4.15), and using Equation 4.21, one obtains

$$\frac{1}{2\pi}\int_0^{2\pi}(\ddot{\beta}_k+\beta_k)\,d\psi_k = \frac{1}{2\pi}\int_0^{2\pi}\beta_0\,d\psi_k = \beta_0 \tag{4.25}$$

$$\frac{1}{\pi}\int_0^{2\pi}(\ddot{\beta}_k+\beta_k)\cos\psi_k\,d\psi_k = \frac{1}{\pi}\int_0^{2\pi}\beta_0\cos\psi_k\,d\psi_k = 0 \tag{4.26}$$

$$\frac{1}{\pi}\int_0^{2\pi}(\ddot{\beta}_k+\beta_k)\sin\psi_k\,d\psi_k = \frac{1}{\pi}\int_0^{2\pi}\beta_0\sin\psi_k\,d\psi_k = 0 \tag{4.27}$$

Apply the operator to the right-hand side of the flap equation (Equation 4.15). After neglecting all the higher harmonic contents of the flap response, and using Equations 4.18 and 4.20, the following set of three algebraic equations are obtained:

$$\beta_0 = \gamma\left\{\frac{\theta_0}{8}(1+\mu^2)+\frac{\theta_{tw}}{10}\left(1+\frac{5}{6}\mu^2\right)+\frac{\mu}{6}\theta_{1s}-\frac{\lambda}{6}\right\} \tag{4.28}$$

$$0 = \frac{\theta_{1c}}{8}\left(1+\frac{1}{2}\mu^2\right)-\frac{\beta_{1s}}{8}-\frac{\mu\beta_0}{6}-\frac{\mu^2}{16}\beta_{1s} \tag{4.29}$$

$$0 = \frac{\theta_{1s}}{8}\left(1 + \frac{3}{2}\mu^2\right) + \theta_0\frac{\mu}{3} + \theta_{tw}\frac{\mu}{4} - \frac{\lambda\mu}{4} + \frac{\beta_{1c}}{8} - \beta_{1c}\frac{\mu^2}{16} \tag{4.30}$$

The inflow can also be defined in the no-feathering plane (NFP). Note that the inflow λ is defined in the hub plane, as shown in Figure 4.2.

On the other hand, if the inflow is defined with respect to the NFP (as shown in Figure 4.2), Equations 4.28 to 4.30 can be written in a modified form. Inflow with respect to NFP is given by (using a small angle assumption)

$$\lambda_{NFP} \cong \lambda - \mu\theta_{1s} \tag{4.31}$$

Using this modified definition, Equations 4.28 to 4.30 can be written as

$$\beta_0 = \gamma\left\{\frac{\theta_{0.8}}{8}(1+\mu^2) - \frac{\theta_{tw}}{60}\mu^2 - \frac{\lambda_{NFP}}{6}\right\} \tag{4.32}$$

$$\beta_{1s} - \theta_{1c} = \frac{-(4/3)\mu\beta_0}{1 + \frac{1}{2}\mu^2} \tag{4.33}$$

$$\beta_{1c} + \theta_{1s} = \frac{-(8/3)\mu[\theta_{0.75} - (3/4)\lambda_{NFP}]}{1 - \frac{1}{2}\mu^2} \tag{4.34}$$

where $\theta_{0.8} = \theta_0 + 0.8\theta_{tw}$ and $\theta_{0.75} = \theta_0 + 0.75\theta_{tw}$ represent the blade pitch angle at radial locations 0.8*R* and 0.75*R*, respectively.

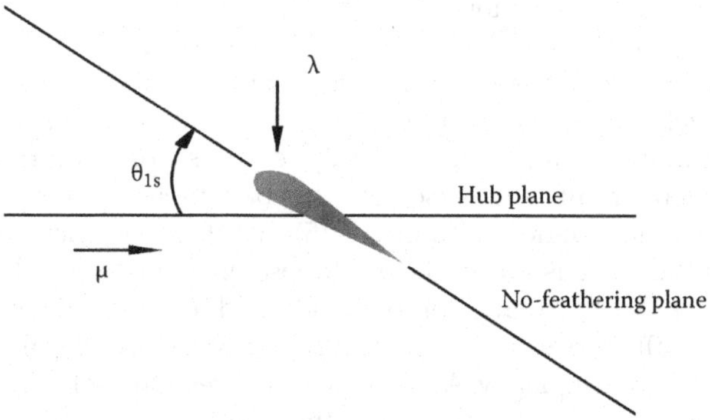

FIGURE 4.2
Flow directions in the hub plane and the NFP.

The three equations, Equations 4.28 to 4.30 (or Equations 4.32–4.34), can be solved for the harmonics of flap motion β_0, β_{1c}, β_{1s}, due to the blade pitch input and the inflow.

Based on these equations, certain interesting observations can be made.

1. The rotor coning angle β_0 is proportional to the pitch angle and inflow (Equation 4.32). Since blade loading $\frac{C_T}{\sigma}$ is related to the blade pitch angle and the inflow, one can state that the coning angle is proportional to the blade loading.
2. The rotor coning angle is also proportional to the Lock number (Equation 4.32). For the normal operating range of $\mu \leq 0.35$, the coning angle remains almost a constant.
3. The first harmonics β_{ic} and β_{is} are proportional to the advance ratio μ and also to the blade loading $\frac{C_T}{\sigma}$.

Typically, β_0, β_{1c}, and β_{1s} are of the order of few degrees. Reducing Equations 4.28 to 4.30 to hovering condition ($\mu = 0$) results in

$$\beta_0 = \gamma\left\{\frac{\theta_0}{8} + \frac{\theta_{tw}}{10} - \frac{\lambda}{6}\right\} \tag{4.35}$$

$$\beta_{1s} - \theta_{1c} = 0 \tag{4.36}$$

$$\beta_{1c} + \theta_{1s} = 0 \tag{4.37}$$

It is interesting to note that a cosine harmonic pitch input (θ_{1c}) produces a sine harmonic flap (β_{1s}) of the same magnitude and a sine harmonic input (θ_{1s}) produces a cosine harmonic flap (β_{1c}). This is an important result of flap dynamics to the blade pitch input in the helicopter rotor system. It implies that forward tilting of the rotor tip path plane (β_{1c}) requires a blade pitch input θ_{1s}, which is denoted as a longitudinal cyclic input. Similarly, for the lateral tilting of the rotor tip path plane, β_{1s} requires a pitch input θ_{1c}, denoted as a lateral cyclic input. The reason for this behavior is as follows. It is evident from the flap equation (Equation 4.15) that the nondimensional natural frequency of the blade is 1/rev. The excitation force due to aerodynamic load has a fundamental harmonic component (i.e., due to the blade pitch input θ_I; Equation 4.20) whose frequency is also 1/rev. Hence, the flap motion is excited at resonant frequency. At resonance, the response of a single-degree-of-freedom system always lags behind the input by a phase of 90° (that is, one quarter of a cycle means an azimuth angle of 90°). The proof is given in the following.

Using Equation 4.18, the flap equation (Equation 4.15) under hovering condition can be written as (assuming zero twist)

$$\ddot{\beta}_k + \beta_k = \gamma\left\{\frac{1}{8}\theta_I - \frac{\lambda}{6} - \frac{\dot{\beta}_k}{8}\right\} \tag{4.38}$$

Rearranging the terms

$$\ddot{\beta}_k + \frac{\gamma}{8}\dot{\beta}_k + \beta_k = \frac{\gamma}{8}\theta_I - \frac{\gamma}{6}\lambda \tag{4.39}$$

For harmonic variation of the input θ_I, the flap response β_k can be obtained easily. Assuming that λ is a constant, it will affect only the flap coning angle (Equation 4.35). One can notice from Equation 4.39 that the rotor blade flap equation in hover is essentially a second-order damped system. Assuming $\theta_I = \bar{\theta}\cos[n(\psi_k + \psi_0)]$, the steady-state flap response can be written as $\beta_k = \bar{\beta}\cos[n(\psi_k + \psi_o) - \Delta\psi]$. Note that n represents the ratio of excitation frequency to the natural frequency of the system, which, in this case, is 1/rev. Solving for the magnitude and the phase angle of β_k, one obtains

$$\frac{\bar{\beta}}{\bar{\theta}} = \left|\frac{\gamma/8}{\left\{\left(n\frac{\gamma}{8}\right)^2 + (1-n^2)^2\right\}^{1/2}}\right| \tag{4.40}$$

$$\Delta\psi = \tan^{-1}\left\{\frac{n\frac{\gamma}{8}}{1-n^2}\right\} = 90° + \tan^{-1}\left\{\frac{n^2-1}{n\frac{\gamma}{8}}\right\} \tag{4.41}$$

For $n = 1$, the phase lag is $\Delta\psi = 90°$.

For example, if $\theta_I = \theta_{1c}\cos\psi_k$, then $\beta_k = \bar{\beta}\cos[\psi_k - 90°] = \bar{\beta}\sin\psi_k = \beta_{1s}\sin\psi_k$. A cosine harmonic pitch input produces a sine harmonic in flap motion. Similarly, a sine harmonic pitch input produces a cosine harmonic flap motion.

It is important to recognize that these results are valid only for a centrally hinged rotor blade. For a general configuration of blades such as hingeless or bearingless blades, there will be a cross-coupling between the sine and the cosine harmonics of the pitch input to the longitudinal (β_{1c}) and the lateral (β_{1s}) flap response of the blade. The phase lag will be less than 90° because the natural frequency in the flap will be more than 1/rev. Therefore, in the design of the rotor control system, this phase lag must be properly accounted for in the hardware of the swash plate system. Even though the phase lag between the pilot pitch input and the blade flap response can vary with forward speed μ, the control system is usually designed for a single value of a phase angle since the variation in the phase between the pitch control input and the flap response in forward speed is not very significant.

Flap Motion with a Centrally Placed Root Spring

In the previous section, the rotor blade is modeled as a rigid blade hinged at the center of the hub. It is noted that the natural frequency of the rotating blade in the flap is observed to be exactly equal to 1/rev. To represent the realistic rotor blade having the natural frequency more than 1/rev, one can modify the rotor blade model by including a linear torsional spring at the root, as shown in Figure 4.3. The spring constant is taken as K_β. This model can simulate a hingeless rotor blade configuration. In the undeformed state, the rotor blade is having a precone angle β_p, as shown in Figure 4.3. The influence of the precone angle on the blade root load will be brought out in the following discussion.

Introduction of a root spring introduces a resisting moment in the flap direction, which is directly proportional to the flap deformation. Inclusion of the flap restoring moment due to the root spring alters the flap equation (Equation 4.13). The modified flap equation can be written as

$$I_b\left(\frac{d^2\beta_k}{dt^2}+\Omega^2\beta_k\right)+K_\beta(\beta_k-\beta_p)=\int_0^R rF_{z1}\,dr \tag{4.42}$$

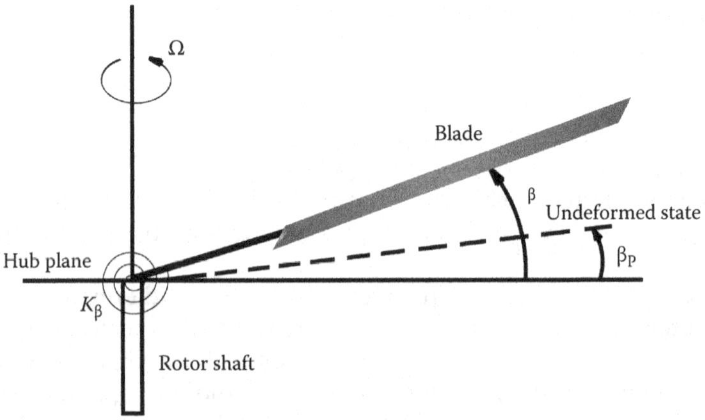

FIGURE 4.3
Rigid rotor blade with centrally hinged spring restrained idealization.

Note that the flap angle β_k is defined with respect to the hub plane. Following the procedure described in the previous section, substituting for F_{z1}, performing the integration, and nondimensionalizing, the flap equation becomes

$$\ddot{\beta}_k + \left(1 + \frac{K_\beta}{I_b \Omega^2}\right)\beta_k - \frac{K_\beta}{I_b \Omega^2}\beta_p = \gamma M_F \tag{4.43}$$

From Equation 4.43, it can be noted that the nondimensional rotating natural frequency of the blade in the flap mode is given as

$$\bar{\omega}_{RF}^2 = 1 + \frac{K_\beta}{I_b \Omega^2} = 1 + \bar{\omega}_{NR}^2 \tag{4.44}$$

where the nonrotating natural frequency in the flap is defined as $\bar{\omega}_{NR}^2 = \dfrac{K_\beta}{I_b \Omega^2}$.

M_F is the aerodynamic moment coefficient as given earlier in Equation 4.18. Applying the operator method, assuming that the flap response contains only up to first harmonics, the LHS of the flap equation (Equation 4.43) can be written as

$$\frac{1}{2\pi}\int_0^{2\pi}\left\{\ddot{\beta}_k + \bar{\omega}_{RF}^2\beta_k - (\bar{\omega}_{RF}^2 - 1)\beta_p\right\}d\psi_k = \bar{\omega}_{RF}^2\beta_0 - (\bar{\omega}_{RF}^2 - 1)\beta_p \tag{4.45}$$

$$\frac{1}{\pi}\int_0^{2\pi}\left\{\ddot{\beta}_k + \bar{\omega}_{RF}^2\beta_k - (\bar{\omega}_{RF}^2 - 1)\beta_p\right\}\cos\psi_k\, d\psi_k = (\bar{\omega}_{RF}^2 - 1)\beta_{1c} \tag{4.46}$$

$$\frac{1}{\pi}\int_0^{2\pi}\left\{\ddot{\beta}_k + \bar{\omega}_{RF}^2\beta_k - (\bar{\omega}_{RF}^2 - 1)\beta_p\right\}\sin\psi_k\, d\psi_k = (\bar{\omega}_{RF}^2 - 1)\beta_{1s} \tag{4.47}$$

Performing the integral operation on the right-hand side of Equation 4.43, and equating to the corresponding LHS coefficients, one obtains the following equations for flap harmonics:

$$\bar{\omega}_{RF}^2\beta_0 = (\bar{\omega}_{RF}^2 - 1)\beta_p + \gamma\left\{\frac{\theta_{0.8}}{8}(1+\mu^2) - \frac{\theta_{tw}}{60}\mu^2 - \frac{\lambda_{NFP}}{6}\right\} \tag{4.48}$$

$$(\bar{\omega}_{RF}^2 - 1)\beta_{1c} = \gamma\left\{\frac{1}{8}(\theta_{1c} - \beta_{1s})\left(1 + \frac{1}{2}\mu^2\right) - \frac{\mu}{6}\beta_0\right\} \tag{4.49}$$

$$(\bar{\omega}_{RF}^2 - 1)\beta_{1s} = \gamma\left\{\frac{1}{8}(\theta_{1s} + \beta_{1c})\left(1 - \frac{1}{2}\mu^2\right) + \frac{\mu}{3}\theta_{0.75} - \frac{\mu}{4}\lambda_{NFP}\right\} \tag{4.50}$$

From Equation 4.31, $\lambda_{NFP} = \lambda - \mu\theta_{1s}$.
From Equation 4.48, the coning angle can be expressed as

$$\beta_0 = \frac{(\bar{\omega}_{RF}^2 - 1)}{\omega_{RF}^2}\beta_p + \frac{\gamma}{\bar{\omega}_{RF}^2}\left\{\frac{\theta_{0.8}}{8}(1+\mu^2) - \frac{\theta_{tw}}{60}\mu^2 - \frac{\lambda_{NFP}}{6}\right\} \quad (4.51)$$

Rewriting the expression for coning angle as

$$\beta_0 = \frac{(\bar{\omega}_{RF}^2 - 1)}{\omega_{RF}^2}\beta_p + \frac{1}{\bar{\omega}_{RF}^2}\beta_{ideal} \quad (4.52)$$

where β_{ideal} is defined as the coning angle when $\bar{\omega}_{RF} = 1$, and is given by

$$\beta_{ideal} = \gamma\left\{\frac{\theta_{0.8}}{8}(1+\mu^2) - \frac{\theta_{tw}}{60}\mu^2 \frac{\lambda_{NFP}}{6}\right\} \quad (4.53)$$

The effect of the precone angle is to reduce the steady-state deformation in the flap angle. The mean hinge moment due to the flap deformation of the spring at the root is given as follows (using Equation 4.52). (Note: From Figure 4.3, that flap angle is measured from the hub plane not from the initial undeformed state with a precone angle β_p.)

$$M_F = K_\beta(\beta_0 - \beta_p) = K_\beta\left\{\frac{(\bar{\omega}_{RF}^2 - 1)}{\omega_{RF}^2}\beta_p + \frac{1}{\bar{\omega}_{RF}^2}\beta_{ideal} - \beta_p\right\} \quad (4.54)$$

Simplifying Equation 4.54, the mean hinge moment in flap motion can be written as

$$M_F = K_\beta(\beta_0 - \beta_p) = \left\{\frac{1}{\bar{\omega}_{RF}^2}\beta_{ideal} - \frac{1}{\bar{\omega}_{RF}^2}\beta_p\right\} \quad (4.55)$$

Substituting for K_β in terms of nondimensional rotating flap natural frequency $\bar{\omega}_{RF}^2$, from Equation 4.44, the flap hinge moment becomes

$$M_F = I_b\Omega^2\left\{\frac{(\bar{\omega}_{RF}^2 - 1)}{\bar{\omega}_{RF}^2}\right\}(\beta_{ideal} - \beta_p) \quad (4.56)$$

The mean hinge moment is nonzero when $\bar{\omega}_{RF}^2 > 1$, unless the precone angle is chosen to be equal to β_{ideal}, i.e., ($\beta_p = \beta_{ideal}$). When $\beta_p = \beta_{ideal}$, the equilibrium coning angle will be $\beta_0 = \beta_{ideal}$ (from Equation 4.52), and the root spring K_β does not get strained. Hence, the mean hub moment is 0. This indicates that a proper choice of the precone will reduce the blade root moment, thereby reducing the hub moment. The ideal value of the precone depends on the rotor loading and the operating condition. Hence, the precone can have a value equal to β_{ideal} only for a particular operating condition. Normally, the value of the precone in practical rotor blades will be about 2°.

Considering hovering condition ($\mu = 0$), and simplifying the response equations (Equations 4.49 and 4.50) for β_{1c} and β_{1s},

$$\frac{(\bar{\omega}_{RF}^2 - 1)}{\gamma/8}\beta_{1c} + \beta_{1s} = \theta_{1c} \tag{4.57}$$

$$\frac{(\bar{\omega}_{RF}^2 - 1)}{\gamma/8}\beta_{1s} - \beta_{1c} = \theta_{1s} \tag{4.58}$$

Solving for β_{1c} and β_{1s}, one obtains

$$\beta_{1c} = \frac{-\theta_{1s} + \dfrac{(\bar{\omega}_{RF}^2 - 1)}{\gamma/8}\theta_{1c}}{1 + \left(\dfrac{(\bar{\omega}_{RF}^2 - 1)}{\gamma/8}\right)^2} \tag{4.59}$$

$$\beta_{1s} = \frac{\theta_{1c} + \dfrac{(\bar{\omega}_{RF}^2 - 1)}{\gamma/8}\theta_{1s}}{1 + \left(\dfrac{(\bar{\omega}_{RF}^2 - 1)}{\gamma/8}\right)^2} \tag{4.60}$$

Comparing Equations 4.59 and 4.60 with Equations 4.36 and 4.37, it can be seen that the effect of having $\bar{\omega}_{RF}^2 > 1$ is to introduce longitudinal flapping β_{1c} due to lateral cyclic pitch θ_{1c} and lateral flapping β_{1c} due to longitudinal cyclic pitch θ_{1s}.

Assuming $\theta = \bar{\theta}\cos(\psi + \psi_0)$ and $\beta = \bar{\beta}\cos(\psi + \psi_0 - \Delta\psi)$, the magnitude and the phase of the flap response become

$$\frac{\bar{\beta}}{\bar{\theta}} = \left[1 + \left\{\frac{(\bar{\omega}_{RF}^2 - 1)}{\gamma/8}\right\}^2\right]^{-1/2} \tag{4.61}$$

where $\bar{\beta} = \left\{\beta_{1c}^2 + \beta_{1s}^2\right\}^{1/2}$ and $\bar{\theta} = \left\{\theta_{1c}^2 + \theta_{1s}^2\right\}^{1/2}$

$$\tan \Delta\psi = \frac{\gamma/8}{(\bar{\omega}_{RF}^2 - 1)} \tag{4.62}$$

Rewriting the phase difference or the phase angle between the pitch input and the flap response as

$$\tan \Delta\psi = \tan(90° - \theta) = \frac{1}{\tan\theta} = \frac{\gamma/8}{(\bar{\omega}_{RF}^2 - 1)} \tag{4.63}$$

$$\Delta\psi = 90° - \theta = 90° - \tan^{-1}\left\{\frac{(\bar{\omega}_{RF}^2 - 1)}{\gamma/8}\right\} \tag{4.64}$$

It is evident from these expressions that increasing the flap frequency $\bar{\omega}_{RF}$ more than 1 reduces the phase lag in the flap response from 90°. When $\bar{\omega}_{RF} = 1.15$ and $\gamma = 8$, the amplitude $\bar{\beta}/\bar{\theta}$ is reduced by about 5%, but the phase lag reduces to 72° from 90°, as compared to a centrally hinged rotor blade. This phase change constitutes a coupling of the lateral and the longitudinal response of the tip path plane with respect to blade pitch inputs. As far as the control of the helicopter by the pilot is concerned, this coupling is partially reduced by compensating the phase lag between the control plane and the Tip Path Plane (TPP). The control system geometry is modified so that the rotor still responds with purely longitudinal TPP tilt due to θ_{1s}. Note that forward speed has an influence on phase shift. Moreover, it is not the same in both axes of the cyclic. Therefore, the control rigging to compensate for the lateral–longitudinal coupling is a function of forward speed, and it is not the same in both lateral and longitudinal axes. Since this variation is not very large, a single value of phase is used for the geometric design of the control system so that it is reasonably satisfactory over the entire speed range of the helicopter.

In the rotating frame, the blade root moment due to flap deflection is given as (replacing the spring constant in terms of blade inertia and rotating natural frequency in the flap; Equation 4.44)

$$M_F = K_\beta(\beta_k - \beta_p) = I_b\Omega^2(\bar{\omega}_{RF}^2 - 1)(\beta_k - \beta_p) \tag{4.65}$$

The mean value of the pitch and roll moments acting at the rotor hub, due to all the blades, can be obtained by summing over all the blades and averaging

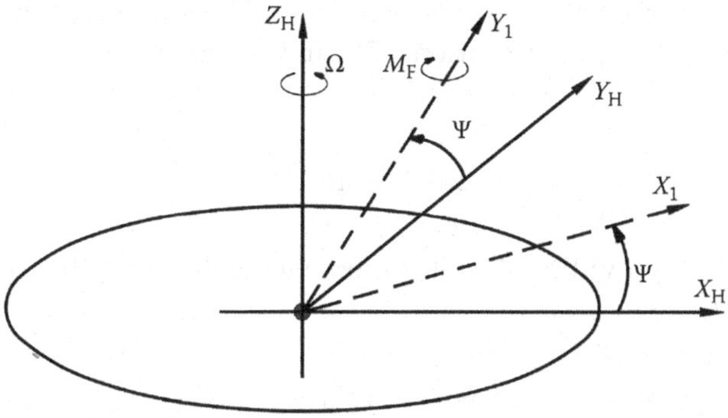

FIGURE 4.4
Flap moment M_F.

over the azimuth (see Figure 4.4 for the transformation of the blade root flap moment to the hub moment in pitch and roll).

$$\text{Pitch moment:} \quad M_{yH} = -\frac{N}{2\pi}\int_0^{2\pi} M_F \cos\psi_k \, d\psi_k \tag{4.66}$$

$$\text{Roll moment:} \quad M_{xH} = \frac{N}{2\pi}\int_0^{2\pi} M_F \sin\psi_k \, d\psi_k \tag{4.67}$$

Substituting for M_F from Equation 4.65 and for β_k from Equation 4.21 and integrating, the pitch and roll moments can be written as

$$M_{yH} = -\frac{N}{2} I_b\Omega^2(\bar{\omega}_{RF}^2 - 1)\beta_{1c} \tag{4.68}$$

$$M_{xH} = \frac{N}{2} I_b\Omega^2(\bar{\omega}_{RF}^2 - 1)\beta_{1s} \tag{4.69}$$

The pitching and rolling moment coefficients at the hub can be expressed in nondimensional form as

$$C_{MyH} = \frac{M_{yH}}{\rho\pi R^2(\Omega R)^2 R} \tag{4.70}$$

Substituting for M_{yH} from Equation 4.68, the pitch moment coefficient becomes

$$C_{MyH} = -\frac{NI_b\Omega^2}{2\rho\pi R^2\Omega^2 R^2 R}(\bar{\omega}_{RF}^2 - 1)\beta_{1c} \tag{4.71}$$

Multiplying and dividing by the blade area *CR* (assuming constant chord) and the lift curve slope *a*, Equation 4.71 can be written as

$$C_{\mathrm{MyH}} = -\frac{NCR}{2\pi R^2}\frac{I_b}{\rho CaR^4}a(\bar{\omega}_{\mathrm{RF}}^2 - 1)\beta_{1c} \tag{4.72}$$

Equation 4.72, providing the pitch moment coefficient at the rotor hub, can also be written as

$$\frac{2C_{\mathrm{MyH}}}{\sigma a} = -\frac{(\bar{\omega}_{\mathrm{RF}}^2 - 1)}{\gamma}\beta_{1c} \tag{4.73}$$

Similarly, the roll moment coefficient can be expressed as

$$\frac{2C_{\mathrm{MxH}}}{\sigma a} = \frac{(\bar{\omega}_{\mathrm{RF}}^2 - 1)}{\gamma}\beta_{1s} \tag{4.74}$$

These expressions show that the moment-generating capability of the rotor system is dependent on the rotating natural frequency in the flap, and the moment is greatly increased when $\bar{\omega}_{\mathrm{RF}} > 1$. The articulated rotor provides fewer hub moments because the rotating flap natural frequency is less than that of the hingeless or bearingless rotor configurations.

The pitch and roll moments about the fuselage center of gravity (c.g.) consist of two components. One is due to the direct hub moment (given in Equations 4.73 and 4.74) and the other is due to the hub in-plane forces *H* and *Y*. In the following, a derivation is given, providing a simple expression for the pitch and roll moments about the fuselage c.g. for hovering condition. The final expression clearly brings out the effect of the flap natural frequency in generating control moments about the fuselage c.g.

Writing the thrust coefficient and the in-plane force coefficients under hovering condition as (from Equations 3.91–3.94)

$$C_T = \frac{\sigma a}{2}\left[\frac{\theta_0}{3} + \frac{\theta_{tw}}{4} - \frac{\lambda}{2}\right] \tag{4.75}$$

$$C_H = \frac{\sigma a}{2}\left[\frac{-\theta_0}{3}\beta_{1c} - \frac{\theta_{tw}}{4}\beta_{1c} + \lambda\frac{\theta_{1s}}{4} - \theta_{1c}\frac{\beta_0}{6} + \frac{3}{4}\lambda\beta_{1c} + \beta_{1s}\frac{\beta_0}{6}\right] \tag{4.76}$$

$$C_Y = -\frac{\sigma a}{2}\left[\frac{\theta_0}{3}\beta_{1s} + \frac{\theta_{tw}}{4}\beta_{1s} + \lambda\frac{\theta_{1c}}{4} + \theta_{1s}\frac{\beta_0}{6} - \frac{3}{4}\lambda\beta_{1s} + \beta_{1c}\frac{\beta_0}{6}\right] \tag{4.77}$$

Assuming $\bar{\omega}_{RF} \approx 1.0$, then from Equations 4.57 and 4.58, $\beta_{1s} \approx \theta_{1c}$ and $-\beta_{1c} \approx \theta_{1s}$. Using these relations, the hub in-plane free coefficients (Equations 4.76 and 4.77) can be simplified as

$$C_H = \frac{\sigma a}{2}\left[\frac{\theta_0}{3} + \frac{\theta_{tw}}{4} - \frac{\lambda}{2}\right]\{-\beta_{1c}\} \quad (4.78)$$

$$C_Y = \frac{\sigma a}{2}\left[\frac{\theta_0}{3} + \frac{\theta_{tw}}{4} - \frac{\lambda}{2}\right]\{-\beta_{1s}\} \quad (4.79)$$

In other words,

$$C_H = -C_T\beta_{1c} \quad (4.80)$$

$$C_Y = -C_T\beta_{1s} \quad (4.81)$$

It is now evident that the in-plane hub forces are due to the tilt of the thrust vector with respect to the hub plane, which is achieved by tilting the rotor tip path plane due to the cyclic flap. From Figure 4.5, it can be seen that the moment about the c.g. of the fuselage will be due to the direct hub moment and the moment due to the hub in-plane forces. When the c.g. of the helicopter

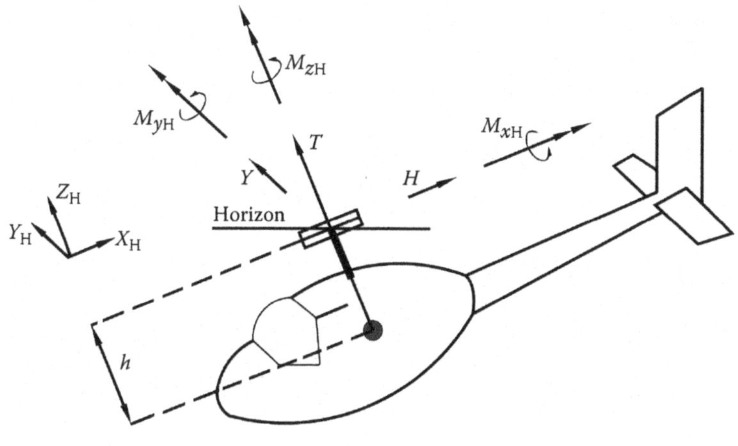

FIGURE 4.5
Hub loads.

is at a height (h/R) directly below the hub, the moment about the c.g. can be written as

$$\begin{Bmatrix} \dfrac{-2C_{My}}{\sigma a} \\ \dfrac{2C_{Mx}}{\sigma a} \end{Bmatrix}_{CG} = \left(\frac{\bar{\omega}_{RF}^2 - 1}{\gamma} + \frac{2C_T}{\sigma a} \frac{h}{R} \right) \begin{Bmatrix} \beta_{1c} \\ \beta_{1s} \end{Bmatrix} \tag{4.82}$$

Several interesting observations can be made from Equation 4.82. (1) Direct hub moment (first term on the right side) is dependent on flap frequency, and this effect is 0 when the flap frequency is 1/rev. (2) Direct hub moment is independent of rotor thrust or rotor load factor. (3) For hingeless rotors, the direct hub moment (first term) may be two to four times greater than the effect due to the thrust tilt term (second term). This simple derivation clearly brings out the effect of flap natural frequency on the generation of control moment by the rotor system. The moment expressions become more complex in forward flight due to the inclusion of advance ratio μ. The above expression provides a clear insight on the control moment–generating capability of the rotor system in helicopters.

5

Helicopter Trim (or Equilibrium) Analysis

In the previous chapters, it is shown that rotor forces and moments are functions of flap response and blade pitch input. In the most general case, these forces and moments will also be functions of blade lag, torsion, and perturbational hub motion. During steady flight, the trim condition or equilibrium of the helicopter requires that the mean values of the forces and moments acting at the center of the mass of the helicopter must be 0. Since there are six equations of equilibrium, which are three force and three moment equations, one can solve for six unknown quantities satisfying the equilibrium equations. For hover and level flight conditions, the six unknown quantities are the collective pitch (θ_0) input of the main rotor, the cyclic pitch inputs (θ_{1c}, θ_{1s}) of the main rotor, the tail rotor collective pitch (θ_{TR}), the pitch attitude (α), and the roll attitude (Φ) of the helicopter. It may be recognized that the rotor loads are influenced by the blade response and the rotor inflow; hence, it becomes necessary to solve for the rotor inflow and the blade response equations. They form the intermediate stage in the solution procedure.

In the following, a step-by-step procedure for the trim analysis of a helicopter in steady-level flight is presented. In this example, the helicopter model is highly simplified. Empennage control surfaces are not included. The helicopter is assumed to be flying at steady forward speed V. There is no side-slip velocity. Tail rotor is assumed to generate only thrust. Only the first harmonic flap response of the main rotor is considered. This analysis is sometimes referred to as "flap trim" since only the flap motion of the main rotor is considered in rotor dynamics.

Let us first define the relevant coordinate systems that are shown in Figure 5.1.

X_{ea}–Y_{ea}–Z_{ea} is the earth-fixed inertial coordinate system, with Z_{ea} pointing toward the earth's center.

X_b–Y_b–Z_b is the body fixed coordinate system with origin at the center of gravity (c.g.) of the fuselage. X_b is pointing toward the nose, Y_b is pointing toward the starboard side, and Z_b is pointing downward.

X_H–Y_H–Z_H is the hub-fixed nonrotating coordinate system with its origin at the center of the main rotor hub. X_H is pointing toward the tail, Y_H is pointing toward the starboard side, and Z_H is pointing vertically upward.

Let us now define all the relevant quantities shown in Figure 5.1. The velocity vector V of the helicopter is in the body-fixed X_b–Z_b plane. The angle θ_{FP} represents the flight path angle defined as the angle between the horizon and the velocity vector. Note that X_H and Z_H are parallel to X_b and Z_b

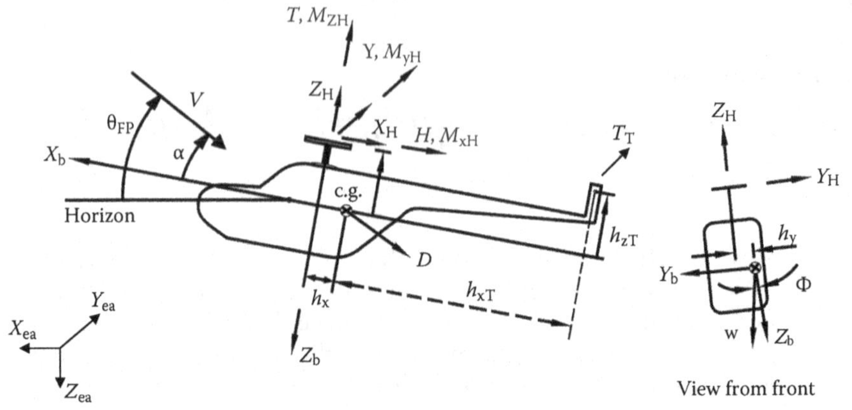

FIGURE 5.1
Forces and moments acting on a helicopter.

respectively, but pointing in opposite directions. The angles Θ and Φ are the equilibrium pitch and the roll attitude, respectively, of the helicopter, and angle α represents the angle between the body X_b axis and the velocity vector V of the helicopter. The weight of the helicopter is acting down at the c.g. of the helicopter. The drag force acting on the fuselage due to forward flight is assumed to act at the c.g. along the direction of the relative air velocity vector, as shown in Figure 5.1. The mean loads due to the main rotor system are acting at the main rotor hub center as shown. The tail rotor is assumed to provide only thrust force as shown. The coordinates of the main rotor hub center and the tail rotor thrust location are given respectively as (h_x, h_y, h_z) and $(h_{xT}, 0, h_{zT})$. It may be noted that one can add forces due to other lifting surfaces such as horizontal tail and vertical fin, which are neglected in the present formulation.

The transformation matrix relating the earth-fixed system and the body-fixed system at the fuselage c.g. is obtained by using Euler angle transformation. The earth axis system is first rotated counterclockwise about the Y_{ea} axis through an angle Θ representing the pitch angle and then followed by a counterclockwise rotation about the rotated X_{ea} axis through an angle Φ representing the roll angle so that the new orientation is along the body-fixed fuselage c.g. coordinate system. The relationship between the earth-fixed system and the fuselage axis system can be given as

$$\begin{Bmatrix} X_b \\ Y_b \\ Z_b \end{Bmatrix} = \begin{bmatrix} 1 & 0 & 0 \\ 0 & \cos\Phi & \sin\Phi \\ 0 & -\sin\Phi & \cos\Phi \end{bmatrix} \begin{bmatrix} \cos\Theta & 0 & -\sin\Theta \\ 0 & 1 & 0 \\ \sin\Theta & 0 & \cos\Theta \end{bmatrix} \begin{Bmatrix} X_{ea} \\ Y_{ea} \\ Z_{ea} \end{Bmatrix} \tag{5.1}$$

Equation 5.1 can be written as

$$\begin{Bmatrix} X_b \\ Y_b \\ Z_b \end{Bmatrix} = \begin{bmatrix} \cos\Theta & 0 & -\sin\Theta \\ \sin\Phi\ \sin\Theta & \cos\Phi & \sin\Phi\ \cos\Theta \\ \cos\Phi\ \sin\Theta & -\sin\Phi & \cos\Phi\ \cos\Theta \end{bmatrix} \begin{Bmatrix} X_{ea} \\ Y_{ea} \\ Z_{ea} \end{Bmatrix} \tag{5.2}$$

From Figure 5.1, note that, when the angles are small, the pitch angle $\Theta = (\theta_{FP} - \alpha)$.

The force and moment equilibrium equations are written in the c.g. fixed fuselage coordinate system (X_b–Y_b–Z_b). Hence, all the forces and moments acting on the helicopter are first transferred to the c.g. of the helicopter in the body-fixed fuselage axis system, and the equilibrium equations are written in the body coordinate system. In the following, all the relevant equations for the trim analysis of the helicopter in forward flight are given.

Assuming uniform inflow model, the rotor inflow equation is given as

1. Inflow equation

$$\lambda = \mu \tan\alpha + \frac{C_T}{2\sqrt{\mu^2 + (\mu \tan\alpha + \lambda_i)^2}} \tag{5.3}$$

or can also be written as

$$\lambda = \mu \tan\alpha + \frac{C_T}{2\sqrt{\mu^2 + \lambda}} \tag{5.4}$$

Resolving the forces and moments along the fuselage axis system, the six equilibrium equations can be written as follows.

2. Thrust equation: vertical force perpendicular to the hub plane

$$-T + W\cos(\theta_{FP} - \alpha)\cos\Phi + D\sin\alpha = 0 \tag{5.5}$$

3. Horizontal force equation: longitudinal direction

$$-H - D\cos\alpha - W\sin(\theta_{FP} - \alpha) = 0 \tag{5.6}$$

4. Side force equation: starboard side

$$Y + T_T + W\sin\Phi\cos(\theta_{FP} - \alpha) = 0 \tag{5.7}$$

5. Roll moment equation

$$-M_{xH} - Yh_z - T_T h_{zT} - Th_y = 0 \tag{5.8}$$

6. Pitch moment equation

$$M_{yH} + Th_x - Hh_z = 0 \tag{5.9}$$

7. Yawing moment

$$-M_{zH} + Yh_x + Hh_y + T_T h_{xT} = 0 \tag{5.10}$$

In the above equilibrium equations, the expression for tail rotor thrust can be taken as similar to the expression for main rotor thrust, except for the fact that tail rotor has only collective pitch angle and that all the parameters in that expression must correspond to tail rotor parameters. Since the main rotor hub loads depend on the flap harmonics of the blade β_0, β_{1c}, β_{1s}, these harmonics have to be obtained before solving the above equilibrium equations.

The step-by-step procedure for trim analysis is given in the following.

The given quantities are V, θ_{FP}, W, D, h_x, h_y, h_z, h_{xT}, h_{zT}, β_p, $\bar{\omega}_{RF}$, θ_{tw}, γ, R, Ω.

Step 1: Assume α, and let $T \approx W$ evaluate the advance ratio μ.

Using inflow equation (Equation 5.4), solve for the total inflow λ.

Step 2: Assuming θ_0, θ_{1c}, and θ_{1s} and using the following equations representing the equilibrium deformation of the blade in the flap mode, solve for β_0, β_{1c}, and β_{1s} (using Equations 4.31, and 4.48 to 4.50).

$$\beta_0 = \frac{\bar{\omega}_{RF}^2 - 1}{\bar{\omega}_{RF}^2}\beta_p + \frac{\gamma}{\bar{\omega}_{RF}^2}\left\{\frac{\theta_o}{8}(1+\mu^2) + \frac{\theta_{tw}}{10}\left(1+\frac{5}{6}\mu^2\right) + \frac{\mu}{6}\theta_{1s} - \frac{\lambda}{6}\right\} \tag{5.11}$$

$$\beta_{1c} = \frac{\gamma}{\bar{\omega}_{RF}^2 - 1}\left\{\frac{1}{8}(\theta_{1c} - \beta_{1s})\left(1+\frac{1}{2}\mu^2\right) - \frac{\mu}{6}\beta_0\right\} \tag{5.12}$$

$$\beta_{1s} = \frac{\gamma}{\bar{\omega}_{RF}^2 - 1}\left\{\frac{1}{8}\theta_{1s}\left(1+\frac{3}{2}\mu^2\right) + \theta_0\frac{\mu}{3} + \theta_{tw}\frac{\mu}{4} - \frac{\lambda\mu}{4} + \frac{\beta_{1c}}{8} - \frac{\beta_{1c}\mu^2}{16}\right\} \tag{5.13}$$

Step 3: Knowing λ, β_0, β_{1c}, and β_{1s}, solve the six equilibrium equations (Equations 5.4 to 5.10) for θ_0, θ_{1c}, θ_{1s}, α, Φ, and θ_{TR}.

Go to step 1.

Iterate until convergence is achieved.

Step 4: Evaluate the power required using the product of the rotor torque and the rotor angular speed of the main rotor system. An approximate evaluation of the tail rotor power can be obtained from the tail rotor thrust.

Repeat the calculation for different values of forward speed.

Step 5: Plot flap response.

Plot the variation of θ_0, θ_{1c}, θ_{1s}, α, Φ, and θ_{TR} and β_0, β_{1c}, and β_{1s} as a function of advance ratio μ.

A highly simplified sample trim problem is given in the following.

Sample Trim Problem

Perform a trim procedure to evaluate the variation of collective and cyclic input to the main rotor blades, the fuselage pitch attitude, and the inflow ratio with the advance ratio (μ = 0, 0.05, 0.1, 0.15, 0.2, and 0.3) using the hub loads and the flap response (up to the first harmonic only) for the given helicopter data:

- Weight coefficient of the helicopter: 0.0032
- Lift-curve slope: 2π
- Blade profile drag coefficient: 0.0079
- Twist: 0
- Fuselage aerodynamic moments: 0.0
- Density of air: 1.225 kg/m^3
- Solidity ratio: 0.1
- Number of blades: 4
- Lock number: 12
- Rotating flap-natural frequency: 1.1/rev
- Equivalent flat plate area: $0.037/\pi$ (nondimensionalized with respect to [w.r.t.] rotor disk area)
- Location of the hub centre from c.g. (h_x, h_y, h_z): (0, 0, –0.426) (nondimensionalized w.r.t. rotor radius) (c.g. is directly below the rotor shaft)
- Location of the tail rotor hub centre from c.g. (h_{xT}, h_{yT}, h_{zT}): (–1.2, 0, 0) (nondimensionalized w.r.t. main rotor radius)
- Flight path angle: 0

Assume that the rotor side force and the main rotor torque are balanced by the tail rotor thrust. Hence, neglect the side force equation and the yaw equation. Use the remaining four equilibrium equations (Equations 5.5, 5.6, 5.8, and 5.9) to obtain the four quantities θ_0, θ_{1c}, θ_{1s}, and α. (Please note that, in a general case, all the six equilibrium equations are solved for the six unknowns θ_0, θ_{1c}, θ_{1s}, α, Φ, and θ_{TR}.) In the following, a clear description of the procedure in obtaining the converged solution of this simple problem is presented.

Procedures

Step 1: Solve for hover ($\mu = 0$):

$$C_T = C_W = 0.0032$$

$$\lambda = \sqrt{\frac{C_T}{2}} = 0.04$$

$$C_T = \frac{\sigma a}{2}\left[\frac{\theta_o}{3} - \frac{\lambda}{2}\right] \Rightarrow \theta_o = 0.091 \text{ rad}$$

$$\theta_{1s} = 0;\ \theta_{1c} = 0$$

$$\beta_o = 0.046 \text{ rad}$$

$$\beta_{1s} = 0;\ \beta_{1c} = 0$$

$$\alpha = 0$$

Step 2: Use the above values as an initial guess to solve for successive μ (say, $\mu = 0.05$), as follows:

- Initial guess:

$$\theta_o = 0.091,\ \theta_{1s} = 0,\ \theta_{1c} = 0,\ \alpha = 0$$

- Assume $C_T = C_W$, and use the following inflow equation to solve for λ:

$$\lambda = \mu \tan\alpha + \frac{C_T}{2\sqrt{\mu^2 + \lambda^2}}$$

- Solve for β_0, β_{1s}, β_{1c} using the following equations (for 0 twist and no preflap):

$$\beta_0 = \frac{\gamma}{\varpi_{RF}^2}\left\{\frac{\theta_o}{8}(1+\mu^2) + \frac{\mu}{6}\theta_{1s} - \frac{\lambda}{6}\right\}$$

$$\beta_{1s} = \frac{\gamma}{\varpi_{RF}^2 - 1}\left\{\frac{1}{8}\theta_{1s}\left(1 + \frac{3}{2}\mu^2\right) + \theta_0\frac{\mu}{3} + -\frac{\lambda\mu}{4} + \frac{\beta_{1c}}{8} - \frac{\beta_{1c}\mu^2}{16}\right\}$$

$$\beta_{1c} = \frac{\gamma}{\varpi_{RF}^2 - 1}\left\{\frac{1}{8}(\theta_{1c} - \beta_{1s})\left(1 + \frac{1}{2}\mu^2\right) - \frac{\mu}{6}\beta_0\right\}$$

- Compute C_T, C_H, C_Y, C_{MxH}, and C_{MyH} using the following expressions (from Equations 3.91 through 3.93, 4.73, and 4.74):

$$C_T = \frac{\sigma a}{2}\left[\frac{\theta_0}{3}\left[1+\frac{3}{2}\mu^2\right]+\frac{\mu}{2}\theta_{1s}-\frac{\lambda}{2}\right]$$

$$C_H = \frac{\sigma a}{2}\left[\theta_0\left\{\frac{\lambda\mu}{2}-\frac{\beta_{1c}}{3}\right\}+\theta_{1s}\left\{\frac{\lambda}{4}-\frac{\mu\beta_{1c}}{4}\right\}-\theta_{1c}\frac{\beta_0}{6}+\frac{3}{4}\lambda\beta_{1c}\right.$$
$$\left.+\beta_{1s}\frac{\beta_0}{6}+\frac{\mu}{4}\left\{\beta_0^2+\beta_{1c}^2\right\}+\frac{\mu C_{Do}}{2a}\right]$$

$$C_Y = -\frac{\sigma a}{2}\left[\theta_0\left\{\frac{3}{4}\mu\beta_0+\frac{\beta_{1s}}{3}\left(1+\frac{3}{2}\mu^2\right)\right\}+\theta_{1c}\left\{\frac{\lambda}{4}+\frac{1}{4}\mu\beta_{1c}\right\}\right.$$
$$+\theta_{1s}\left\{\frac{\beta_0}{6}(1+3\mu^2)+\frac{1}{2}\mu\beta_{1s}\right\}-\frac{3}{2}\lambda\mu\beta_0$$
$$\left.+\beta_0\beta_{1c}\left(\frac{1}{6}-\mu^2\right)-\frac{3}{4}\lambda\beta_{1s}-\frac{\mu}{4}\beta_{1c}\beta_{1s}\right]$$

$$C_{MyH} = -\frac{\sigma a}{2}\left[\frac{\left(\bar{\omega}_{RF}^2-1\right)}{\gamma}\right]\beta_{1c}$$

$$C_{MxH} = \frac{\sigma a}{2}\left[\frac{\left(\bar{\omega}_{RF}^2-1\right)}{\gamma}\right]\beta_{1s}$$

- Solve the four equilibrium equations for θ_o, θ_{1s}, θ_{1c}, and α using the Newton–Raphson method, as follows:
  - The four equilibrium equations: (roll angle $\phi = 0$ and using the given data)

    Thrust equation: $-T + W\cos\alpha + D\sin\alpha = 0$

    Drag equation: $-H + W\sin\alpha - D\cos\alpha = 0$

    Roll moment equation: $-M_{xH} - Yh_z = 0$

    Pitch moment equation: $M_{yH} - Hh_z = 0$
  - In nondimensional form using given data:

$$C_T - C_w\cos\alpha - 0.0059\,\mu^2\,\frac{\sin\alpha}{(\cos\alpha)^2} = 0$$

$$C_H - C_w\sin\alpha - 0.0059\,\mu^2\,\frac{1}{\cos\alpha} = 0$$

$$0.426\,C_Y - C_{MxH} = 0$$

$$0.426\,C_H - C_{MyH} = 0$$

- Using the expressions of C_T, C_H, and C_Y, let

$$
\begin{aligned}
C_T &= a_1\theta_o + b_1\theta_{1s} + c_1\theta_{1c} + d_1 \\
C_H &= a_2\theta_o + b_2\theta_{1s} + c_2\theta_{1c} + d_2 \\
C_Y &= a_3\theta_o + b_3\theta_{1s} + c_3\theta_{1c} + d_3
\end{aligned}
$$

where

$$
\begin{aligned}
a_1 &= \frac{\sigma a}{6}\left[1 + \frac{3}{2}\mu^2\right] \\
b_1 &= \frac{\sigma a \mu}{4} \\
c_1 &= 0 \\
d_1 &= -\frac{\sigma a \lambda}{4}
\end{aligned}
$$

$$
\begin{aligned}
a_2 &= \frac{\sigma a}{2}\left[\frac{\lambda\mu}{2} - \frac{\beta_{1c}}{3}\right] \\
b_2 &= \frac{\sigma a}{2}\left[\frac{\lambda}{4} - \frac{\mu\beta_{1c}}{4}\right] \\
c_2 &= -\frac{\sigma a \beta_o}{12} \\
d_2 &= \frac{\sigma a}{2}\left[\frac{3}{4}\lambda\beta_{1c} + \frac{\beta_{1s}\beta_o}{6} + \frac{\mu}{4}\left(\beta_o^2 + \beta_{1c}^2\right) + \frac{\mu C_{Do}}{2a}\right]
\end{aligned}
$$

$$
\begin{aligned}
a_3 &= \frac{\sigma a}{2}\left[\frac{3}{4}\mu\beta_o + \frac{\beta_{1s}}{3}\left(1 + \frac{3}{2}\mu^2\right)\right] \\
b_3 &= \frac{\sigma a}{2}\left[\frac{\beta_o}{6}(1 + 3\mu^2) + \frac{\mu\beta_{1s}}{2}\right] \\
c_3 &= \frac{\sigma a}{2}\left[\frac{\lambda}{4} + \frac{\mu\beta_{1c}}{4}\right] \\
d_3 &= \frac{\sigma a}{2}\left[-\frac{3}{2}\lambda\mu\beta_o + \beta_o\beta_{1c}\left(\frac{1}{6} - \mu^2\right) - \frac{3}{4}\lambda\beta_{1s} - \frac{\mu}{4}\beta_{1c}\beta_{1s}\right]
\end{aligned}
$$

- Using the equations above, define $f_i = f_i(\theta_o, \theta_{1s}, \theta_{1c}\alpha)$ as follows:

$$f_1 = (a_1\theta_o + b_1\theta_{1s} + c_1\theta_{1c} + d_1) - C_w \cos\alpha - 0.0059\mu^2 \frac{\sin\alpha}{(\cos\alpha)^2} = 0$$

$$f_2 = (a_2\theta_o + b_2\theta_{1s} + c_2\theta_{1c} + d_2) - C_w \sin\alpha + 0.0059\mu^2 \frac{1}{\cos\alpha} = 0$$

$$f_3 = 0.426(a_3\theta_o + b_3\theta_{1s} + c_3\theta_{1c} + d_3) - C_{MxH} = 0$$

$$f_4 = 0.426(a_2\theta_o + b_2\theta_{1s} + c_2\theta_{1c} + d_2) + C_{MyH} = 0$$

- Solve the simultaneous equations using the Newton–Raphson method:

$$f_{1o} + \Delta\theta_o\, \partial f_1/\partial\theta_o + \Delta\theta_{1s}\, \partial f_1/\partial\theta_{1s} + \Delta\theta_{1c}\, \partial f_1/\partial\theta_{1c} + \Delta\alpha\, \partial f_1/\partial\alpha = 0$$
$$f_{2o} + \Delta\theta_o\, \partial f_2/\partial\theta_o + \Delta\theta_{1s}\, \partial f_2/\partial\theta_{1s} + \Delta\theta_{1c}\, \partial f_2/\partial\theta_{1c} + \Delta\alpha\, \partial f_2/\partial\alpha = 0$$
$$f_{3o} + \Delta\theta_o\, \partial f_3/\partial\theta_o + \Delta\theta_{1s}\, \partial f_3/\partial\theta_{1s} + \Delta\theta_{1c}\, \partial f_3/\partial\theta_{1c} + \Delta\alpha\, \partial f_3/\partial\alpha = 0$$
$$f_{4o} + \Delta\theta_o\, \partial f_4/\partial\theta_o + \Delta\theta_{1s}\, \partial f_4/\partial\theta_{1s} + \Delta\theta_{1c}\, \partial f_4/\partial\theta_{1c} + \Delta\alpha\, \partial f_4/\partial\alpha = 0$$

Assuming small α:

$$f_{1o} + \Delta\theta_o[a_1] + \Delta\theta_{1s}[b_1] + \Delta\theta_{1c}[c_1] + \Delta\alpha[0.0032\alpha - 0.0059\mu^2] = 0$$
$$f_{2o} + \Delta\theta_o[a_2] + \Delta\theta_{1s}[b_2] + \Delta\theta_{1c}[c_2] + \Delta\alpha[-0.0032 - 0.0059\mu^2\alpha] = 0$$
$$f_{3o} + \Delta\theta_o[0.426a_3] + \Delta\theta_{1s}[0.426b_3] + \Delta\theta_{1c}[0.426c_3] + \Delta\alpha[0] = 0$$
$$f_{4o} + \Delta\theta_o[0.426a_2] + \Delta\theta_{1s}[0.426b_2] + \Delta\theta_{1c}[0.426c_2] + \Delta\alpha[0] = 0$$

In matrix form

$$\underbrace{\begin{bmatrix} [a_1] & [b_1] & [c_1] & [0.0032\alpha - 0.0059\mu^2] \\ [a_2] & [b_2] & [c_2] & [-0.0032 - 0.0059\mu^2\alpha] \\ [0.426a_3] & [0.426b_3] & [0.426c_3] & 0 \\ [0.426a_2] & [0.426b_2] & [0.426c_2] & 0 \end{bmatrix}}_{\text{Jacobian}} \begin{Bmatrix} \Delta\theta_o \\ \Delta\theta_{1s} \\ \Delta\theta_{1c} \\ \Delta\alpha \end{Bmatrix} = -\begin{Bmatrix} f_{1o} \\ f_{2o} \\ f_{3o} \\ f_{4o} \end{Bmatrix}$$

$$\begin{Bmatrix} \Delta\theta_o \\ \Delta\theta_{1s} \\ \Delta\theta_{1c} \\ \Delta\alpha \end{Bmatrix} = -\begin{bmatrix} [a_1] & [b_1] & [c_1] & [0.0032\alpha - 0.0059\mu^2] \\ [a_2] & [b_2] & [c_2] & [-0.0032 - 0.0059\mu^2\alpha] \\ [0.426a_3] & [0.426b_3] & [0.426c_3] & 0 \\ [0.426a_2] & [0.426b_2] & [0.426c_2] & 0 \end{bmatrix}^{-1} \begin{Bmatrix} f_{1o} \\ f_{2o} \\ f_{3o} \\ f_{4o} \end{Bmatrix}$$

- Obtain the new approximations as follows:

$$\theta_o = \theta_o + \Delta\theta_o / k$$
$$\theta_{1s} = \theta_{1s} + \Delta\theta_{1s} / k$$
$$\theta_{1c} = \theta_{1c} + \Delta\theta_{1c} / k$$
$$\alpha = \alpha + \Delta\alpha / k$$

NOTE: If convergence is not obtained with $k = 1$, then k can be taken as 2,4,... etc., to get the desired convergence.

- Repeat this process by replacing the new approximations with the previous guess until we get the values to the desired accuracy.

Results

The results are given in Figures 5.2 to 5.5. The observations made are as follows:

1. The fuselage pitch attitude (α) increases progressively with forward speed (Figure 5.2).
2. The collective pitch (θ_o) decreases from its hover value to a minimum at a moderate forward speed. This is due to the benefit of translational velocity component on the lifting capability of the rotor. The collective pitch then increases with forward speed (Figure 5.2).
3. The inflow variation is similar to that of the collective pitch (Figure 5.3).
4. The longitudinal cyclic pitch (θ_{1s}) increases (sign reversed) with the forward speed to tilt the rotor forward to overcome the drag of the helicopter fuselage (or, in other words, to counteract the effect of β_{1c}) (Figure 5.2).
5. The variation in longitudinal cyclic pitch (θ_{1s}) along with the forward speed effect of the increased oncoming velocity in the advancing side and the reduced oncoming velocity in the retreating side results in a net increase in longitudinal flapping (β_{1c}) with forward speed (Figure 5.4).
6. The collective flap angle does not show much variation with forward speed (Figure 5.4).
7. The blade coning causes a lateral flapping response. When a blade passes over the front of the disk, the effects of the forward flight

cause an increase in the angle of attack because it acts as an upwash. At the rear of the disk, the angle of attack is decreased. This causes a lateral tilt of the disk as shown by the variation of lateral flapping (β_{1s}) with forward speed (Figure 5.4). This angle is very small.

8. The lateral cyclic pitch (θ_{1c}) increases gradually with forward speed to counteract the lateral disc tilt caused by coning. Actually, an additional amount of lateral cyclic will be required to balance the tail rotor thrust. However, this effect is ignored in the present exercise (Figure 5.2).
9. The flapping response of the blade is of the form

$$\beta_k = \beta_o + \beta_{1c} \cos \psi_k + \beta_{1s} \sin \psi_k$$

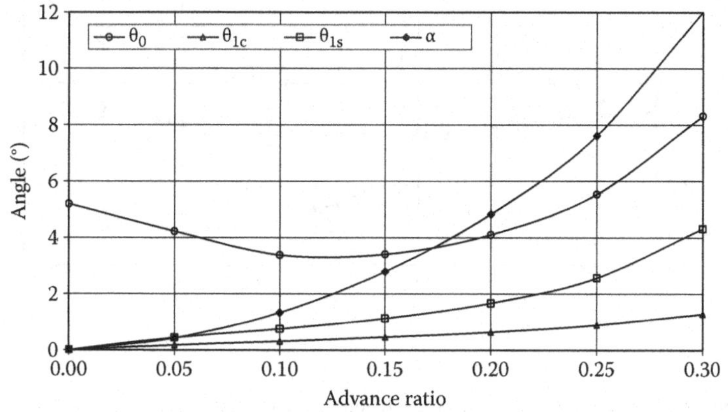

FIGURE 5.2
Variation of control input and pitch attitude with forward speed.

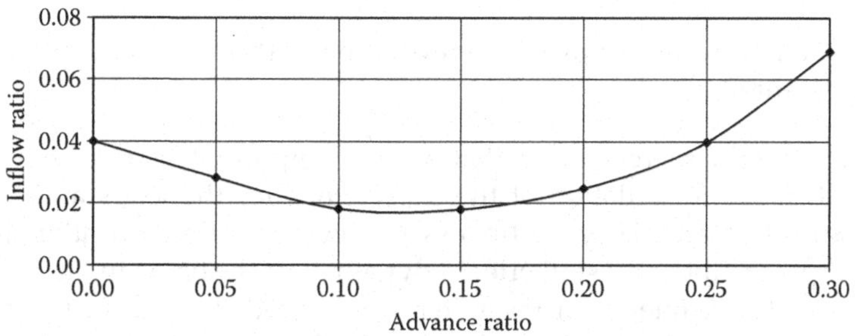

FIGURE 5.3
Variation of total inflow with forward speed.

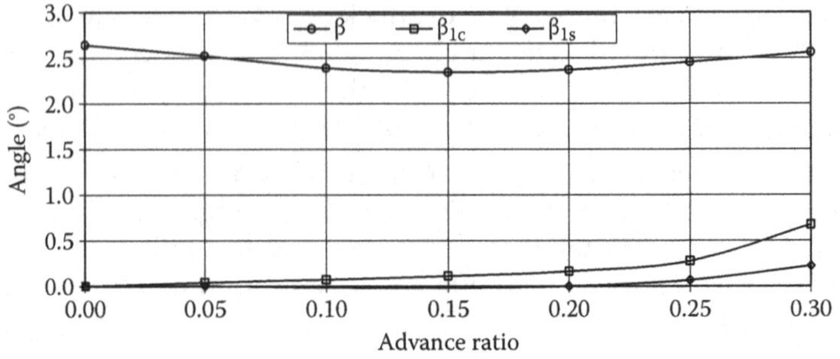

FIGURE 5.4
Variation of flap harmonic with forward speed.

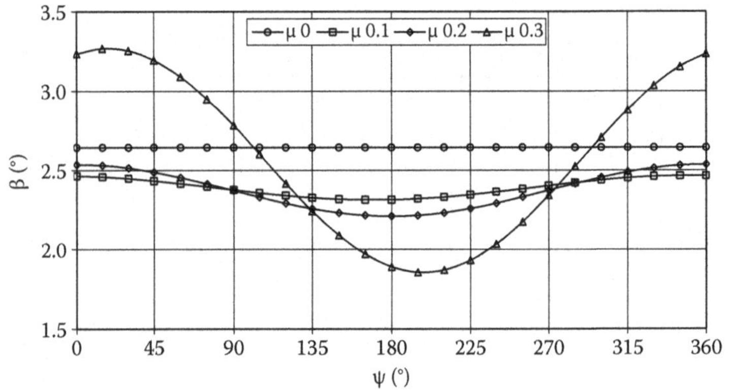

FIGURE 5.5
Flap response over one rotor revolution for different forward speeds.

where, for the kth blade,

$$\psi_k = \frac{2\pi}{N}[k-1]+\psi$$

The flap response over one revolution is shown in Figure 5.5 for different forward speeds.

It is important to recognize that several important aspects have been neglected in the formulation of the expressions for the loads and the flap response of the blade. They are tip loss, root cutout, inflow variation, reverse flow, compressibility and stall effects, dynamics of the blade in lag, torsional modes, etc. The influence of these items will need to be considered while analyzing the complete problem. In general, these items will quantitatively modify the trim values but will not affect the qualitative nature of the results, except in the case of lateral cyclic pitch (θ_{1c}) in low forward speeds.

Reverse Flow

During forward flight, an area on the retreating side of the rotor disk experiences a velocity of the airflow directed from the trailing edge to the leading edge. This region is known as the "reverse flow region," and it can be computed as follows.

The component of forward speed normal to the rotor blade cross section at any azimuth angle is given by

$$V \cos \alpha \sin \psi = \mu \Omega R \sin \psi$$

The relative air velocity due to blade motion is Ωr.

The condition for reverse flow is $(\Omega r + \mu \Omega R \sin \psi) \leq 0$.

In nondimensional form, it is given as $\bar{r} + \mu \sin \psi \leq 0$.

The boundary of the reverse flow region is given by the equation $r + \mu R \sin \psi = 0$.

From Figure 5.6,

$$OA = r = OB \cos \theta = OB \cos (270 - \psi) = OB\,[-\sin \psi] = \mu R \sin \psi$$

Since $\sin \psi$ is negative for $\psi > 180°$, the reverse flow region occurs in the retreating side of the rotor. Hence, the diameter of the reverse flow region is μR. The ratio of the reverse flow region to the total disk area is $\frac{\mu^2}{4}$. Since the root cutout extends up to 15% to 25% of the rotor radius, it will more or less cover the reverse flow region. Therefore, the reverse flow region can be neglected in the analysis. However, at high advance ratio, the reverse flow may become significant, and it may be included in calculating the aerodynamic loads on the blade. Near the reverse flow boundary (i.e., outside the reverse flow region), there will be a significant separated and radial flow disturbance.

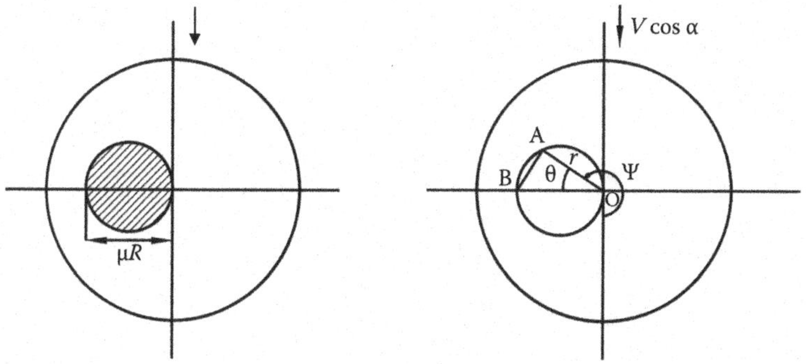

FIGURE 5.6
Reverse flow region in forward flight.

6

Isolated Rotor Blade Dynamics

Helicopter rotor blades are long slender beams undergoing axial, lag, flap, and torsional deformations. Figure 6.1 shows the deformation of an elastic blade model. A detailed analysis of blade dynamics requires the formulation of coupled equations of motion. These equations are nonlinear due to the inclusion of moderate deformation effects, involving nonlinear strain–displacement relationships. Formulation of the equations of motion of the rotor blade has been a topic of research since the 1970s. Earlier theories were restricted to treating the dynamics of isotropic blades; later, in the late 1980s and early 1990s, beam theories suitable for composite rotor blades were formulated. Research efforts are also directed to the development of a multidisciplinary optimization of composite rotor blades.

For a fundamental understanding of rotor blade dynamics, one can formulate an idealized model of the blade. In the following, a simple model of the rotor blade is formulated by idealizing the blade as a rigid blade having a spring restraint and a root offset. The blade is assumed to be uniform. The root springs represent the stiffness of the blade in the flap, lag, and torsional modes (Figure 6.2). This model, although relatively simple, captures the essential features of the blade dynamics and its aeroelastic behavior. This model is equally valid for both articulated and hingeless rotor blades. The limitation of this model is that it represents only the fundamental vibratory mode of the blade. Therefore, this type of blade model may not be suitable for vibration analysis of the helicopter, where the participation of higher modes of the blades is significant.

To have a fundamental understanding, the dynamics of the blade will be analyzed independently for the flap, lead–lag, and torsion modes (in an uncoupled manner), taking one degree of freedom at a time. Such an analysis will bring out not only the essential features of the blade dynamics, but also the constraints that have to be considered in the design of a rotor blade.

Isolated Blade Flap Dynamics in Uncoupled Mode

Consider a rotor blade undergoing only flap deflection. The blade is idealized as a uniform rigid beam with a root offset e and a root spring K_β, as shown in Figure 6.3. The rotor blade is assumed to have a precone angle β_P.

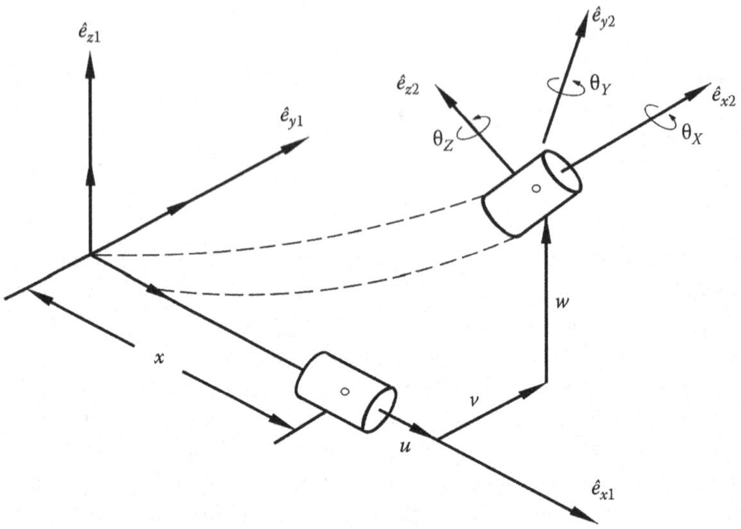

FIGURE 6.1
General deformation of a rotor blade.

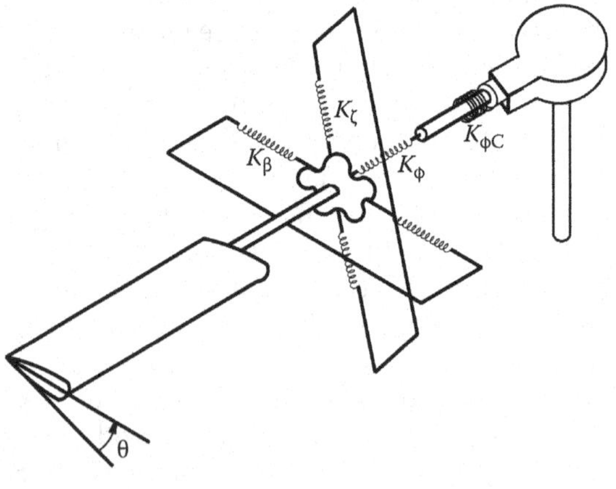

FIGURE 6.2
Idealized model of the rotor blade.

To describe the motion of the blade and also for a consistent formulation of the dynamics of the rotor blade, several coordinate systems are required. These coordinate systems are shown in Figures 6.3 and 6.4.

$\hat{e}_{xH}$, $\hat{e}_{yH}$, $\hat{e}_{zH}$ represent the unit vectors of the hub-fixed nonrotating coordinate system, with origin at the center of the hub.

$\hat{e}_{x1}$, $\hat{e}_{y1}$, $\hat{e}_{z1}$ are the unit vectors of the hub-fixed rotating coordinate system, with origin at the center of the hub. It may be noted that unit vectors $\hat{e}_{zH}$ and $\hat{e}_{z1}$ are parallel and coincident vectors.

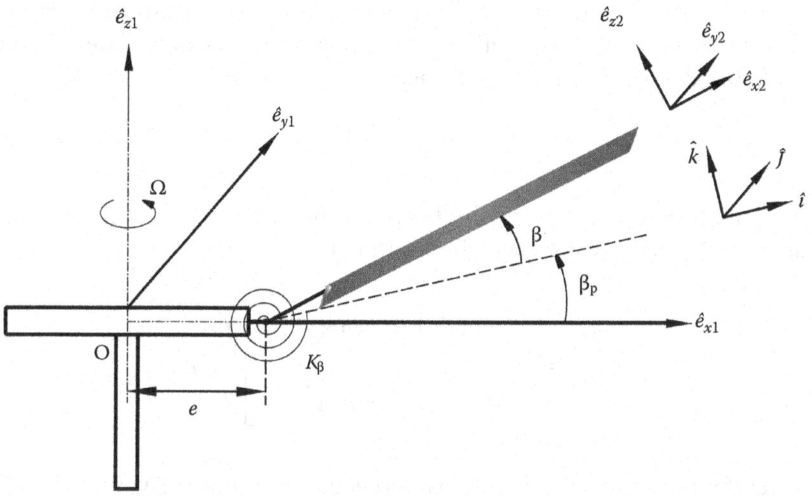

FIGURE 6.3
Idealized rotor blade in flap motion.

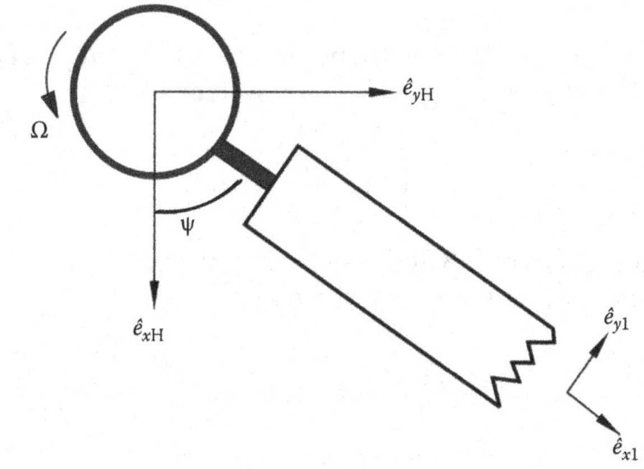

FIGURE 6.4
Hub-fixed nonrotating and rotating coordinate system.

$\hat{\imath}, \hat{\jmath}, \hat{k}$ are the unit vectors along the undeformed state of the rotating blade, with origin at the hinge offset location.

$\hat{e}_{x2}, \hat{e}_{y2}, \hat{e}_{z2}$ are the blade-fixed rotating system in the deformed state of the blade, with origin at the hinge offset location.

It may be noted that the flap deformation β of the blade is defined with respect to the undeformed state of the blade having a precone angle β_P. The rotor is operating at a constant angular velocity Ω about $\hat{e}_{z1}$. It may be noted that, in this formulation, the hub is fixed and it does not have any motion.

Assuming that the mass of the blade is distributed uniformly along the blade reference axis $\hat{e}_x$, the position vector of an arbitrary mass point P on the blade in the deformed state is given as

$$\bar{r}_{\mathrm{p}} = e\hat{e}_{x1} + r\hat{e}_{x2} \tag{6.1}$$

The transformation relationship between the unit vectors along hub-fixed rotating coordinate systems can be written as

$$\begin{Bmatrix} \hat{e}_{x1} \\ \hat{e}_{y1} \\ \hat{e}_{z1} \end{Bmatrix} = \begin{bmatrix} \cos(\beta+\beta_{\mathrm{p}}) & 0 & -\sin(\beta+\beta_{\mathrm{p}}) \\ 0 & 1 & 0 \\ \sin(\beta+\beta_{\mathrm{p}}) & 0 & \cos(\beta+\beta_{\mathrm{p}}) \end{bmatrix} \begin{Bmatrix} \hat{e}_{x2} \\ \hat{e}_{y2} \\ \hat{e}_{z2} \end{Bmatrix} \tag{6.2}$$

Using the transformation, the position vector of point P given in Equation 6.1 can be written as

$$\vec{r}_{\mathrm{p}} = e\hat{e}_{x1} + r\cos(\beta+\beta_{\mathrm{p}})\hat{e}_{x1} + r\sin(\beta+\beta_{\mathrm{p}})\hat{e}_{z1} \tag{6.3}$$

The absolute velocity of point P can be obtained as (noting that the position vector is defined in the rotating coordinate system)

$$\vec{v}_{\mathrm{p}} = \left\{\dot{\vec{r}}_{\mathrm{p}}\right\}_{\mathrm{rel}} + \Omega\hat{e}_{z1} X \vec{r}_{\mathrm{p}} \tag{6.4}$$

Differentiating Equation 6.3 and substituting the respective quantities, the absolute velocity of point P can be written as

$$\vec{v}_{\mathrm{p}} = -r\sin(\beta+\beta_{\mathrm{p}})\frac{d\beta}{dt}\hat{e}_{x1} + r\cos(\beta+\beta_{\mathrm{p}})\frac{d\beta}{dt}\hat{e}_{z1} + \Omega\hat{e}_{z1} X\vec{r}_{\mathrm{p}} \tag{6.5}$$

Equation 6.5 can be expanded as

$$\vec{v}_{\mathrm{p}} = -r\sin(\beta+\beta_{\mathrm{p}})\frac{d\beta}{dt}\hat{e}_{x1} + r\cos(\beta+\beta_{\mathrm{p}})\frac{d\beta}{dt}\hat{e}_{z1} + \Omega[e + r\cos(\beta+\beta_{\mathrm{p}})]\hat{e}_{y1} \tag{6.6}$$

The absolute acceleration of the mass point P can be obtained from the expression (note that Ω is a constant)

$$\vec{a}_p = \left(\frac{d^2\vec{r}_p}{dt^2}\right)_{\mathrm{rel}} + \frac{d\bar{w}}{dt} X\vec{r}_p + 2\bar{w} X\left(\frac{d\vec{r}_p}{dt}\right)_{\mathrm{rel}} + \bar{w} X\left(\bar{w} X\vec{r}_p\right) \tag{6.7}$$

(Note: $\bar{\omega} = \Omega\hat{e}_{z1}$)

Differentiating Equation 6.3 and substituting various quantities, the absolute acceleration of mass point P can be expressed as

$$\vec{a}_{\mathrm{P}} = \left\langle -r\sin(\beta+\beta_{\mathrm{p}})\frac{d^2\beta}{dt^2} - r\cos(\beta+\beta_{\mathrm{p}})\left(\frac{d\beta}{dt}\right)^2 - \Omega^2\{e + r\cos(\beta+\beta_{\mathrm{p}})\}\right\rangle \hat{e}_{x1} + \left\langle -2\Omega r\frac{d\beta}{dt}\sin(\beta+\beta_{\mathrm{p}})\right\rangle \hat{e}_{y1} + \left\langle r\cos(\beta+\beta_{\mathrm{p}})\frac{d^2\beta}{dt^2} - r\left(\frac{d\beta}{dt}\right)^2 \sin(\beta+\beta_{\mathrm{p}})\right\rangle \hat{e}_{z1} \tag{6.8}$$

The inertia force of elemental mass at point P is given as $-m\vec{a}_{\mathrm{P}}\,dr$, where m is the mass per unit length of the blade.

The inertia moment about the flap hinge can be obtained by taking moment about the hinge.

$$M_{\mathrm{I}} = \int_0^{R-e} \left(r\cos(\beta+\beta_{\mathrm{p}})\hat{e}_{x1} + r\sin(\beta+\beta_{\mathrm{p}})\hat{e}_{z1}\right) X\{-m\vec{a}_{\mathrm{P}}\}\,dr \tag{6.9}$$

Substituting the acceleration term from Equation 6.8 and taking cross-product, the inertia moment about the hinge point can be obtained, which is given as

$$M_I = \int_0^{R-e} m\Bigg\{\left\langle -2\Omega r^2 \frac{d\beta}{dt}\sin^2(\beta+\beta_{\mathrm{p}})\hat{e}_{x1}\right\rangle + < r^2\frac{d^2\beta}{dt^2}\cos^2(\beta+\beta_{\mathrm{p}}) - r^2\left(\frac{d\beta}{dt}\right)^2 \sin(\beta+\beta_{\mathrm{p}})\cos(\beta+\beta_{\mathrm{p}}) + r^2\frac{d^2\beta}{dt^2}\sin^2(\beta+\beta_{\mathrm{p}}) + r^2\left(\frac{d\beta}{dt}\right)^2 \sin(\beta+\beta_{\mathrm{p}})\cos(\beta+\beta_{\mathrm{p}}) + \Omega^2[e + r\cos(\beta+\beta_{\mathrm{p}})]r\sin(\beta+\beta_{\mathrm{p}}) > \hat{e}_{y1} + 2\Omega r^2\frac{d\beta}{dt}\sin(\beta+\beta_{\mathrm{p}})\cos(\beta+\beta_{\mathrm{p}})\hat{e}_{z1}\Bigg\}dr \tag{6.10}$$

Considering only the flap moment (i.e., the component of the moment about the $\hat{e}_{y1}$ axis), the inertia moment in the flap mode can be written as

$$M_{\beta I} = \int_0^{R-e} m\left\{r^2 \frac{d^2\beta}{dt^2} + \Omega^2\left\langle e\,r\sin(\beta+\beta_p) + r^2\sin(\beta+\beta_p)\cos(\beta+\beta_p)\right\rangle\right\} dr \tag{6.11}$$

Integrating over the length of the blade, the inertia moment in flap mode can be written as

$$M_{\beta I} = I_b \frac{d^2\beta}{dt^2} + \Omega^2 MX_{c.g.}\, e\sin(\beta+\beta_p) + \Omega^2 I_b \sin(\beta+\beta_p)\cos(\beta+\beta_p) \tag{6.12}$$

The term I_b represents the mass moment of inertia of the blade about the flap hinge, and $MX_{c.g.}$ represents the first moment of the mass about the flap hinge. For a blade with a uniform mass distribution, $I_b = m\frac{(R-e)^3}{3}$ and $MX_{c.g.} = m\frac{(R-e)^2}{2}$.

When the angles β and β_p are small, one can make the approximation as

$$\sin(\beta + \beta_p) \approx (\beta + \beta_p)$$

$$\cos(\beta + \beta_p) \approx 1$$

Using these small angle approximation and uniform mass distribution of the blade, the flap inertia moment about the flap hinge (Equation 6.12) can be written as

$$M_{\beta I} = m\frac{(R-e)^3}{3}\frac{d^2\beta}{dt^2} + \Omega^2 me\frac{(R-e)^2}{2}(\beta+\beta_p) + m\Omega^2\frac{(R-e)^3}{3}(\beta+\beta_p) \tag{6.13}$$

Rearranging the terms, and nondimensionalizing the time derivative term, the inertia moment in the flap can be written as (note that, for convenience, the symbol 'dot' is used in the following for the nondimensional time derivative)

$$M_{\beta I} = m\,\frac{l^3}{3}\left\{\Omega^2\;\ddot{\beta} + \left(\frac{3e}{2l} + 1\right)\Omega^2\;\beta\right\} + \frac{ml^3}{3}\Omega^2\left(\frac{3e}{2l} + 1\right)\beta_P \tag{6.14}$$

where $l = (R - e)$ is the length of the rotor blade from the hinge offset.

Next, let us evaluate the flap moment due to the aerodynamic load acting on the uniform untwisted blade. The aerodynamic load acting on the blade can be obtained using either unsteady aerodynamics due to the Theodorsen theory or quasi-steady approximation of the Theodorsen theory or quasi-static approximation. In quasi-static approximation, the aerodynamic lift is evaluated based on the instantaneous angle of attack at every cross section of the blade. In the following, the quasi-static formulation is used for convenience. Assuming that the blade has zero pretwist, the aerodynamic lift (L) and drag (D) forces per unit length of the blade can be written as (Figure 6.5)

$$L = \frac{1}{2}\rho U^2 C a\left(\theta_{con} + \frac{U_P}{U_T}\right) \tag{6.15}$$

$$D = \frac{1}{2}\rho U^2 C C_{d0} \tag{6.16}$$

where θ_{con} is the control pitch input at the cross section of the blade. The resultant velocity and the induced angle are given as

$$U = \sqrt{U_P^2 + U_T^2} \approx U_T \tag{6.17}$$

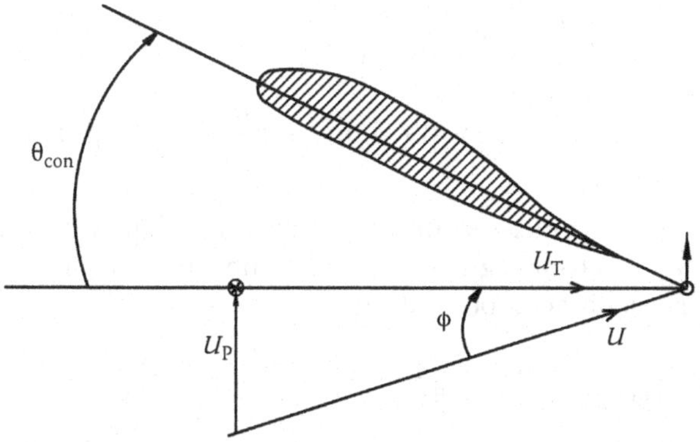

FIGURE 6.5
Flow directions and effective angle of attack at a typical cross section.

$$\tan\Phi \approx \Phi = \frac{U_P}{U_T} \tag{6.18}$$

where U_P and U_T are the relative air velocity components in normal and tangential directions to the airfoil cross section of the blade, as shown in Figure 6.5.

The components of velocity have to be defined in the blade-fixed $(\hat{e}_{x2}, \hat{e}_{y2}, \hat{e}_{z2})$, coordinate system, in the deformed state of the blade, as shown in Figure 6.3.

The net relative velocity of airflow at the blade cross section is due to two components:

1. The velocity due to the forward speed of the vehicle and the induced flow
2. The velocity due to the blade motion

The net relative air velocity vector can be written as

$$\vec{V}_{\text{net}} = \vec{V}_F - \vec{v}_P \tag{6.19}$$

$\vec{V}_F$ is the free stream velocity in the hub plane due to the forward speed of the vehicle and rotor inflow (as shown in Figure 3.2), which is expressed as

$$\vec{V}_F = V\cos\alpha\,\hat{e}_{xH} - (V\sin\alpha + \nu)\hat{e}_{zH} = \Omega R\,(\mu\,\hat{e}_{xH} - \lambda\,\hat{e}_{zH}) \tag{6.20}$$

where V is the forward velocity of the vehicle, ν is the induced velocity, and α is the angle of tilt of the rotor hub plane. In nondimensional form, μ represents the advance ratio and λ represents the total inflow through the rotor disk.

Velocity due to blade motion $\vec{v}_P$ is given in Equation 6.6.

$$\begin{aligned}\vec{v}_P = &-r\,\frac{d\beta}{dt}\sin(\beta + \beta_p)\hat{e}_{x1}\\ &+ \Omega\,(e + r\cos(\beta + \beta_p))\hat{e}_{y1} + r\frac{d\beta}{dt}\cos(\beta + \beta_p)\hat{e}_{z1}\end{aligned} \tag{6.21}$$

Transforming the velocity components given in Equation 6.20, along the hub-fixed $(\hat{e}_{x1}, \hat{e}_{y1}, \hat{e}_{z1})$ rotating system, and combining with Equation 6.21, the net relative air velocity can be written as

$$\begin{aligned}\vec{V}_{net} = &\,\Omega R\left\langle \mu\cos\psi\hat{e}_{x1} - \mu\sin\psi\hat{e}_{y1} - \lambda\hat{e}_{z1}\right\rangle\\ &+ r\frac{d\beta}{dt}\sin(\beta + \beta_p)\hat{e}_{x1} - \Omega\left\{e + r\cos(\beta + \beta_p)\right\}\hat{e}_{y1} - r\frac{d\beta}{dt}\cos(\beta + \beta_p)\hat{e}_{z1}\end{aligned} \tag{6.22}$$

Simplifying Equation 6.22 and transforming the velocity components to the blade-fixed ($\hat{e}_{x2}$, $\hat{e}_{y2}$, $\hat{e}_{z2}$) rotating system, using Equation 6.2, the net relative air velocity experienced at the cross section of the rotor blade can be written as

$$\begin{aligned}\vec{V}_{Net} &= \left\langle \Omega R\mu\cos\psi\cos(\beta+\beta_p) - \Omega R\lambda\sin(\beta+\beta_p)\right\rangle \hat{e}_{x2}\\ &+\left\langle -\mu\Omega R\sin\psi - \Omega\left\{e + r\cos(\beta+\beta_p)\right\}\right\rangle \hat{e}_{y2}\\ &+\left\langle -\mu\Omega R\cos\psi\sin(\beta+\beta_p) - r\frac{d\beta}{dt} - \lambda\Omega R\cos(\beta+\beta_p)\right\rangle \hat{e}_{z2}\end{aligned} \tag{6.23}$$

Assuming small angles, and identifying the respective velocity components, the tangential and normal relative air velocities at the airfoil cross section of the rotor blade can be written as

$$U_T = -\{-\mu\Omega\, R\sin\psi - \Omega\,\{e + r\}\} \tag{6.24}$$

$$U_P = -\mu(\beta+\beta_p)\Omega R\cos\psi - r\frac{d\beta}{dt} - \lambda\Omega R \tag{6.25}$$

The aerodynamic force along $\hat{e}_{z2}$ direction acting at the cross section of the rotor blade is given as

$$F_z = L\cos\phi + D\sin\phi$$

Assuming ϕ to be small

$$F_z \approx L = \frac{1}{2}\rho U_T^2 Ca\left(\theta_{con} + \frac{U_P}{U_T}\right)$$

Substituting the velocity components, the force per unit length can be written as

$$F_z = \frac{1}{2}\rho Ca\left\{\begin{array}{c}\left\langle \mu\Omega R\sin\psi + \Omega(e+r)\right\rangle^2\theta_{con}\\ -\left\langle \mu\Omega R\sin\psi + \Omega(e+r)\right\rangle\left\langle \mu\Omega R(\beta+\beta_p)\cos\psi + r\dfrac{d\beta}{dt} + \lambda\Omega R\right\rangle\end{array}\right\} \tag{6.26}$$

Nondimensionalizing various quantities as $\bar{e} = \frac{e}{R}$, $\bar{r} = \frac{r}{R}$, and converting the time derivative flap term with respect to nondimensional time $\psi = \Omega t$ (denoted by "dot"), the aerodynamic lift force per unit length can be written as

$$F_z = \frac{1}{2}\rho(\Omega R)^2 Ca \left\{ \begin{array}{c} \left\langle \mu \sin\psi + \bar{e} + \bar{r} \right\rangle^2 \theta_{con} \\ -\left\langle \mu \sin\psi + \bar{e} + \bar{r} \right\rangle \left\langle \mu(\beta + \beta_p)\cos\psi + \bar{r}\dot{\beta} + \lambda \right\rangle \end{array} \right\} \tag{6.27}$$

Taking moment about the flap hinge and integrating over the length of the blade, the aerodynamic flap moment can be obtained as

$$M_{\beta A} = \int_0^{R-e} F_z r\, dr \tag{6.28}$$

Nondimensionalizing the integral,

$$M_{\beta A} = R^2 \int_0^{1-\bar{e}} F_z \bar{r}\, d\bar{r} \tag{6.29}$$

Substituting for F_z from Equation 6.27 and integrating over the length of the blade, the aerodynamic flap moment can be obtained. While integrating over the length of the rotor blade, it is assumed that the rotor inflow λ is constant over the blade span. The aerodynamic moment is given as

$$M_{\beta A} = \frac{1}{2}\rho(\Omega R)^2 aCR^2 \left\{ \begin{array}{l} \left\langle \frac{\bar{l}^4}{4} + 2\bar{e}\frac{\bar{l}^3}{3} + 2\mu\sin\psi\frac{\bar{l}^3}{3} + \bar{e}^2\frac{\bar{l}^2}{2} + 2\bar{e}\mu\sin\psi\frac{\bar{l}^2}{2} + \mu^2\sin^2\psi\frac{\bar{l}^2}{2} \right\rangle \theta_{con} \\ -\lambda\left\langle \frac{\bar{l}^3}{3} + \bar{e}\frac{\bar{l}^2}{2} + \mu\sin\psi\frac{\bar{l}^2}{2} \right\rangle \\ -\dot{\beta}\left\langle \frac{\bar{l}^4}{4} + \bar{e}\frac{\bar{l}^3}{3} + \mu\sin\psi\frac{\bar{l}^3}{3} \right\rangle \\ -\beta\mu\cos\psi\left\{ \frac{\bar{l}^3}{3} + \bar{e}\frac{\bar{l}^2}{2} + \mu\sin\psi\frac{\bar{l}^2}{2} \right\} \\ -\beta_p\mu\cos\psi\left\{ \frac{\bar{l}^3}{3} + \bar{e}\frac{\bar{l}^2}{2} + \mu\sin\psi\frac{\bar{l}^2}{2} \right\} \end{array} \right\}$$

where

$$\bar{l} = (R-e)/R = 1-\bar{e} \tag{6.30}$$

Combining the inertia moment (Equation 6.14), the aerodynamic moment (Equation 6.30), and the elastic moment of the root spring due to flap deformation, the flap dynamic equation can be formulated. Nondimensionalizing time derivatives, combining various terms and rearranging them in order, the flap equation can be written as

$$\begin{aligned}
&\ddot{\beta}+\frac{\gamma}{2}\left\{\frac{\bar{l}^4}{4}+\bar{e}\frac{\bar{l}^3}{3}+\mu\sin\psi\frac{\bar{l}^3}{3}\right\}\dot{\beta}+\left\{1+\frac{3}{2}\frac{\bar{e}}{\bar{l}}+\frac{K_\beta}{I_b\Omega^2}\right\}\beta\\
&+\frac{\gamma}{2}\left\{\mu\cos\psi\left\langle\frac{\bar{l}^3}{3}+\bar{e}\frac{\bar{l}^2}{2}+\mu\sin\psi\frac{\bar{l}^2}{2}\right\rangle\right\}\beta=-\beta_p\left\{1+\frac{3}{2}\frac{\bar{e}}{\bar{l}}\right\}\\
&+\frac{\gamma}{2}\left\{\left\langle\frac{\bar{l}^4}{4}+2\bar{e}\frac{\bar{l}^3}{3}+\bar{e}^2\frac{\bar{l}^2}{2}+\mu\sin\psi\left\{2\frac{\bar{l}^3}{3}+2\bar{e}\frac{\bar{l}^2}{2}\right\}+\mu^2\sin^2\psi\frac{\bar{l}^2}{2}\right\rangle\theta_{con}\right\}\\
&-\lambda\left\langle\frac{\bar{l}^3}{3}+\bar{e}\frac{\bar{l}^2}{2}+\mu\sin\psi\frac{\bar{l}^2}{2}\right\rangle-\beta_p\left\langle\mu\cos\psi\left\{\frac{\bar{l}^3}{3}+\bar{e}\frac{\bar{l}^2}{2}+\mu\sin\psi\frac{\bar{l}^2}{2}\right\}\right\rangle\Bigg\}
\end{aligned}\tag{6.31}$$

where $\gamma=\frac{\rho aCR^4}{I_b}$ is the Lock number and $I_b=\frac{ml^3}{3}$ is the blade mass moment of inertia about the flap hinge.

Let us first discuss about the form of the flap equation in forward flight, given in Equation 6.31. The equation of motion is a linear differential equation with time-varying periodic coefficients associated with the advance ratio μ. In addition, it is evident from the equation that, for a first harmonic blade pitch input (θ_{con}), the flap response of the blade will have higher harmonics.

There is no closed solution for Equation 6.31. It may be noted that this equation was formulated neglecting reverse flow, stall, and compressibility effects. In the early days of helicopter development, considerable attention was paid to this equation since its solution provides an important concept on the stability of periodic systems. The stability of the solution is analyzed by using the Floquet–Liapunov theory. Another approximate approach was to use multiblade coordinate transformation, which converts the lower-order periodic systems into constant coefficient terms, but higher-order periodic terms will still remain in the equation. Therefore, after performing the multiblade coordinate transformation, the remaining higher harmonic coefficients are neglected, and the stability of the system is analyzed using the approximate constant coefficient equation.

One can obtain, for a given initial condition, the time response of the homogeneous part of the flap equation (Equation 6.31), by numerical integration scheme. For the sake of academic interest, the homogenous part of Equation 6.31 is solved for two values of advance ratio. They are $\mu = 0.5$ and 1.4. The initial condition is taken as $\beta(0) = 0.06$ and $\dot{\beta}(0) = 0.0$. Figure 6.6 shows the flap response for these two different values of advance ratio. It can be seen from the response that, for $\mu = 0.5$, the flap response shows a damped behavior, but for $\mu = 1.4$, flap response shows an undamped oscillatory behavior. For

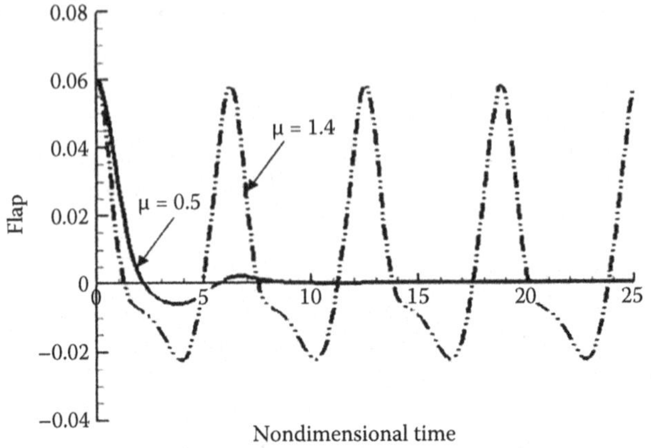

FIGURE 6.6
Flap response of a rotor blade for a given initial condition at different advance ratios.

small values of advance ratio, the response is stable, and for high values of advance ratio, the flap response becomes unstable. Fortunately, for the case of real helicopters, the value of advance ratio is less than $\mu = 0.4$. Therefore, even though the flap equation of a rotor blade is having periodic coefficients, the flap motion of the blade is well damped in the operating speed of the helicopters.

Considering hovering condition, Equation 6.31 can be simplified by setting an advance ratio $\mu = 0$. The simplified flap equation applicable for hovering condition can be written as

$$\ddot{\beta}+\frac{\gamma}{2}\left(\frac{\bar{l}^4}{4}+e\frac{\bar{l}^3}{3}\right)\dot{\beta}+\left(1+\frac{3}{2}\frac{\bar{e}}{\bar{l}}+\frac{K_\beta}{I_b\Omega^2}\right)\beta$$
$$=\frac{\gamma}{2}\left\{\left\langle\frac{\bar{l}^4}{4}+2\bar{e}\frac{\bar{l}^3}{3}+\bar{e}^2\frac{\bar{l}^2}{2}\right\rangle\theta_{\text{con}}-\lambda\left\langle\frac{\bar{l}^3}{3}+\bar{e}^2\frac{\bar{l}^2}{2}\right\rangle\right\}-\beta_p\left\{\left(1+\frac{3}{2}\frac{\bar{e}}{\bar{l}}\right)\right\} \quad (6.32)$$

Equation 6.32 resembles the dynamics of a single-degree-of-freedom spring–mass–damper system. The coefficient of β corresponds to the stiffness effect, the coefficient of $\dot{\beta}$ represents the damping, and the $\ddot{\beta}$ term corresponds to the inertia effect.

A spring–mass–damper system can be represented by the dynamical equation.

$$\ddot{X}+\frac{c}{m}\dot{X}+\frac{k}{m}X=\frac{F}{m}(t) \quad (6.33)$$

Rewriting Equation 6.33, by assuming that the system is an underdamped system,

$$\ddot{X} + 2\varsigma\omega_n \dot{X} + \omega_n^2 X = \bar{F}(t) \tag{6.34}$$

where the natural frequency (ω_n^2) and the damping ratio (ζ) are defined as $\frac{k}{m} = \omega_n^2$ and $\frac{c}{m} = \frac{c}{c_c}\frac{c_c}{m} = \varsigma 2\omega_n$ with $c_c = 2\sqrt{km}$ and $\varsigma = \frac{c}{c_c}$.

Comparing term by term between Equations 6.32 and 6.34, the nondimensional rotating natural frequency in the flap mode ($\bar{\omega}_{\text{RF}}$) is given by

$$\omega_n = \bar{\omega}_{\text{RF}} = \frac{\omega_{\text{RF}}}{\Omega} = \sqrt{1 + \frac{3}{2}\frac{\bar{e}}{1-\bar{e}} + \frac{K_\beta}{I_b\,\Omega^2}} \tag{6.35}$$

where the nondimensional nonrotating flap frequency is given by

$$\bar{\omega}_{\text{NRF}} = \sqrt{\frac{K_\beta}{I_b\,\Omega^2}} \tag{6.36}$$

From Equation 6.35, it can be seen that the effect of root spring K_β is to increase the rotating natural frequency in the flap mode. Similarly, the hinge offset also increases the flap natural frequency. Note that the rotating natural frequency in the flap mode is always greater than 1/rev. Generally, for articulated rotor with a hinge offset of $\bar{e} = 0.02$ to 0.05, the nondimensional rotating flap natural frequency is about 1.015 to 1.04. For hingeless rotor blades, the nondimensional rotating flap natural frequency is around 1.09 to 1.1/rev, and such a high natural frequency corresponds to a large hinge offset of an equivalent articulated blade ($\bar{e} \simeq 0.11 \sim 0.12$). It is shown earlier that a high value of flap natural frequency results in large control moments (or hub moments). Large control moment provides a large margin for center of gravity (c.g.) movement in the helicopter for trim purposes, in addition to providing good maneuvering capability. However, large control moment also results in higher vibratory loads from the rotor to be transmitted to the fuselage.

The damping in the flap mode is given by (comparing Equations 6.32 and 6.34),

$$2\varsigma\omega_n = \frac{\gamma}{8}\left(\bar{l}^4 + \frac{4}{3}\bar{e}\,\bar{l}^3\right) \tag{6.37}$$

Since $\omega_n = \bar{\omega}_{RF}$, the damping ratio in the flap mode can be written as

$$\varsigma = \frac{\gamma}{16\bar{\omega}_{RF}}\left(\bar{l}^4 + \frac{4}{3}\bar{e}\,\bar{l}^3\right) \tag{6.38}$$

For small values of flap hinge offset $\bar{e}$, the length of the blade $\bar{l} \approx 1$. Now, for typical values of Lock number γ (in the range of 6–8), and for the nondimensional rotating flap natural frequency $\bar{\omega}_{RF}$ close to unity, the damping ratio ς = 30% ~ 50%. This is a very high value of damping for any vibrating system, and this damping is due to aerodynamics. Because of high damping, the flap motion of a rotor blade is a well-damped mode, and hence, the flap response reaches its steady-state value, to a given step input in pitch angle, in a very short time, that is, within the time taken for one rotor revolution. In practice, the actual value of damping in the flap mode may be around 25%, which, itself, is very high. The damped natural frequency of the blade in the flap mode $\left(\omega_d = \omega_n\sqrt{1-\zeta^2}\right)$ can be less than 1/rev.

Pitch–Flap (δ_3) Coupling

Pitch–flap coupling is employed in a rotor blade to introduce a pitch change during flap motion. This coupling, known as δ_3 coupling, can be introduced geometrically in articulated blades using a kinematic arrangement in the hinge, as shown in Figure 6.7. In hingeless and bearingless rotors, during operation, because of elastic deformation, this coupling is always present.

Mathematically, the effect of pitch–flap coupling is represented as a change in blade pitch angle due to flap motion by the expression

$$\Delta\theta = -K_{P\beta}\beta \tag{6.39}$$

where $\Delta\theta$ is the change in pitch angle due to flap motion, β is the flap angle, and $K_{P\beta}$ is the pitch–flap coupling parameter.

From Figure 6.7, it can be noted that the blade rotation about the hinge is denoted as q. The component of rotation q, defining the flap motion of the blade, is $\beta = q \cos \delta_3$, and the component of q along the blade axis representing pitch change is $\Delta\theta = -q \sin \delta_3$. Replacing q in terms of flap angle β, the change in blade pitch angle due to flap deflection becomes $\Delta\theta = -\beta \tan \delta_3$. Hence, it is evident that $K_{P\beta} = \tan \delta_3$. For positive values of pitch–flap coupling, a flap-up motion produces a decrease in the pitch angle (nose down) of the blade, and a negative value of pitch–flap coupling increases the pitch angle (nose-up change) for flap up motion.

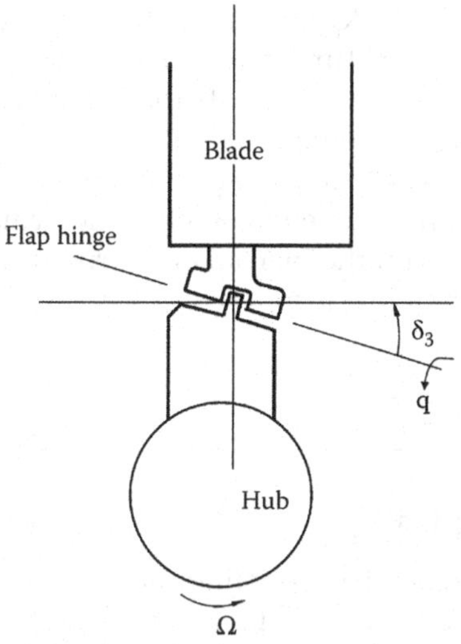

FIGURE 6.7
Geometric arrangement showing pitch–flap coupling.

The effect of this pitch–flap coupling is to increase the stiffness of the blade in flap motion, which, in turn, increases the flap frequency. The pitch–flap coupling modifies the flap dynamics. For simplicity, the influence of pitch–flap coupling on the flap equation of a blade in hover is given below.

Earlier, the equation of motion was derived for flapping motion without the δ_3 coupling. The effect of this coupling can be introduced easily (in a quasistatic manner) by replacing θ_{con} by the modified pitch angle $(\theta_{con} - K_{P\beta}\beta)$ in the flap equation given in Equation 6.32. Assuming $K_{P\beta}$ to be positive, the flap equation for a blade with pitch–flap coupling under hovering condition can be written as

$$\ddot{\beta} + \frac{\gamma}{2}\left(\frac{\bar{l}^4}{4} + e\frac{\bar{l}^3}{3}\right)\dot{\beta} + \left(1 + \frac{3}{2}\frac{\bar{e}}{\bar{l}} + \frac{K_\beta}{I_b\Omega^2} \right.$$
$$\left. + \underline{\frac{\gamma}{2}\left\{\frac{\bar{l}^4}{4} + 2e\frac{\bar{l}^3}{3} + \bar{e}^2\frac{\bar{l}^2}{2}\right\}K_{p\beta}}\right)\beta$$
$$= \frac{\gamma}{2}\left\{\frac{\bar{l}^4}{4} + 2\bar{e}\frac{\bar{l}^3}{3} + \bar{e}^2\frac{\bar{l}^2}{2}\theta_{con} - \lambda\frac{\bar{l}^3}{3} + \bar{e}\frac{\bar{l}^2}{2}\right\} - \beta_p\left\{1 + \frac{3}{2}\frac{\bar{e}}{\bar{l}}\right\} \quad (6.40)$$

From Equation 6.40, it is seen that the effect of $K_{p\beta}$ is to modify the stiffness term (i.e., coefficient of β). Since the additional term depends on Lock number γ, this underlined term is also referred to as aerodynamic stiffness due to the

δ_3 effect. A positive value of $K_{p\beta}$ increases the flap natural frequency, and a negative $K_{p\beta}$ decreases the natural frequency in the flap mode. For positive values of $K_{p\beta}$, because of the increased stiffness, the flap response is reduced. Generally, this coupling is geometrically introduced in tail rotors to reduce its flap response. The typical value of $K_{p\beta}$ is about 1.0. For hingeless and bearingless rotors, this coupling depends on blade deformation in the lag mode, and therefore, it depends on the inertial and structural characteristics of the blade and on the aerodynamic loads.

Pitch–Lag (δ_1) Coupling

Similar to pitch–flap geometric coupling, one can also form pitch–lag coupling. For an articulated rotor blade, the pitch–lag coupling can be introduced by kinematic arrangement, and the pitch change due to lag motion can be expressed as

$$\Delta\theta = -K_{P\varsigma}\varsigma \tag{6.41}$$

For positive values of $K_{P\varsigma}$, lag back motion (i.e., for negative lag deformation, $(-\varsigma)$) introduces a pitch-down attitude, and for negative values of $K_{P\varsigma}$, lag back motion produces a nose-up attitude of the blade. The pitch–lag coupling is known to have a very strong influence on the damping in lag mode. Combining both δ_3 and δ_1 effects, the net change in the pitch angle of the blade due to flap and lag motions can be written as

$$\Delta\theta = K_{P\varsigma}\varsigma - K_{P\beta}\beta \tag{6.42}$$

(It should be pointed out that the sign convention for $K_{P\varsigma}$ may differ from one reference to another. Hence, one must be consistent while formulating the equations of motion.)

In practical design of rotor blades, (δ_1) coupling is not introduced in the construction of the blade. However, these (δ_1 and δ_3) couplings are effectively introduced in an elastic blade due to their deformation in the flap and lag modes.

Structural Flap–Lag Coupling

Even though helicopter blades are long, slender flexible beams, for the purpose of analysis, they can be idealized as a rigid blade with root springs. These idealized models provide information about the essential features of helicopter blade dynamics and coupled rotor/body dynamics, and hence, these models are widely used both in the design and the fundamental understanding of the dynamics of the rotor blades. One of the major limitations of this model is that only the fundamental modes of the blade are represented. Therefore, for loads and vibration analysis, these models are not suitable. Since most of the blade flexibility is concentrated near the root of the blade, deformation of the blade takes place about these flexible locations, which form the virtual hinge. An idealized model of a blade undergoing flap and lag motions is shown in Figure 6.8.

The rotational springs K_β and $K\varsigma$ represent the equivalent spring constants representing the flexibility of the blade in the flap and lag deformations, respectively, and they are orthogonal to each other. Angle θ represents the blade pitch angle. Note that, depending on the location of the pitch bearing, during pitch angle change, the spring assembly can rotate through the same angle (if the pitch bearing is in-board of the spring assembly, say, at location A in Figure 6.8) or it need not rotate at all (if the pitch bearing is out-board of the spring assembly at location B in Figure 6.8). However, one can theoretically form a general approach such that when the blade rotates through a pitch angle θ, the spring assembly rotates through an angle $R\theta$. When $R = 1$, the flap–lag coupling is in unity, and when $R = 0$, there is no flap–lag coupling. This parameter R simulates the condition on the location of pitch bearing with respect to flap–lag hinges; if the pitch bearing is in-board of the spring assembly, the value of $R = 1$; if the pitch bearing is out-board of the spring assembly, the value of $R = 0$.

The formulation of the relationship between the flap–lag deformation of the spring and their respective moments can be obtained by considering a

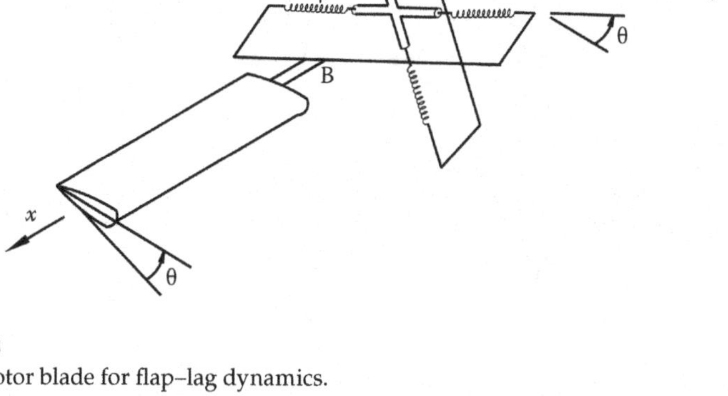

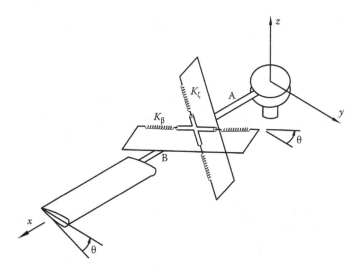

FIGURE 6.8
Idealized rotor blade for flap–lag dynamics.

flap rotation β and a lag rotation ζ of the spring assembly. Let us now formulate the relationship between flap and lag moments, and flap and lag deformations. Consider the case where the pitch bearing is in-board of the spring assembly. For a given pitch angle θ, the spring assembly will also rotate through the same angle θ, as shown in Figure 6.8. The coordinate system (x, y, z) represents the system before pitch input. The coordinate system (x_1, y_1, z_1) represents the system after rotating through angle θ about the x (or x_1) axis.

Let us assume that flap deformation is represented by an angular rotation $-\beta$ about the y axis (note that the negative sign denotes clockwise rotation and the positive value of β refers to the flap-up deformation) and the lag deformation is represented by an angular rotation ζ about the z axis (the positive value of ζ represents the lead-forward deformation of the blade), as shown in Figure 6.9.

Assuming that the rotation angles are very small, the deformation of the spring assembly can be written as follows:

The spring K_β will be rotated through an angle $-\beta \cos\theta + \zeta \sin\theta$ about the y_1 axis.

The spring K_ζ will be rotated through an angle $\zeta \cos\theta + \beta \sin\theta$ about the z_1 axis.

The applied moment about the y_1 axis is $-M_{y1} = K_\beta\,(-\beta \cos\theta + \zeta \sin\theta)$.

The applied moment about the z_1 axis is $M_{z1} = K_\zeta\,(\zeta \cos\theta + \beta \sin\theta)$.

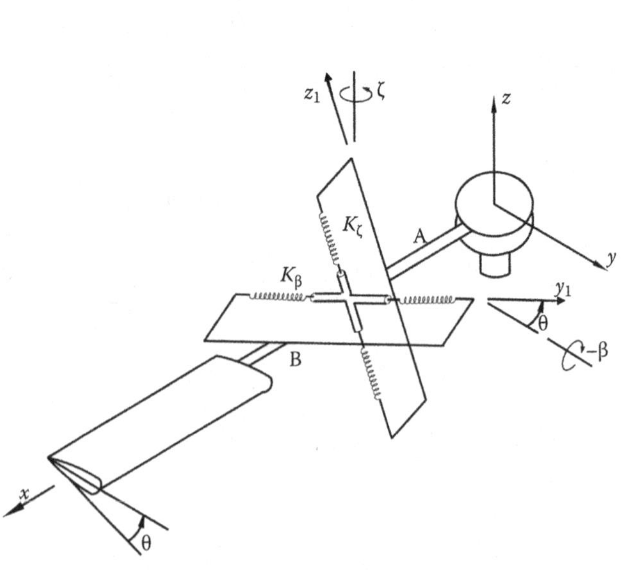

FIGURE 6.9
Flap and lag deformations.

Resolving these moments along the y and z axes system, we have

$$-M_y = -M_{y1}\cos\theta - M_{z1}\sin\theta$$

$$M_z = M_{z1}\cos\theta - M_{y1}\sin\theta$$

Substituting the respective moments, the relation between the flap and lag moments and the flap and lag deformations can be written in matrix form as

$$\begin{Bmatrix} M_y \\ M_z \end{Bmatrix} = \begin{bmatrix} K_\beta\cos^2\theta + K_\zeta\sin^2\theta & -(K_\beta - K_\zeta)\sin\theta\cos\theta \\ -(K_\beta - K_\zeta)\sin\theta\cos\theta & (K_\beta\sin^2\theta + K_\zeta\cos^2\theta) \end{bmatrix} \begin{Bmatrix} \beta \\ \zeta \end{Bmatrix} \tag{6.43}$$

It may be noted that, in Equation 6.43, flap-up is positive (i.e., the positive value of β represents flap-up condition) and lead-forward is positive (the positive value of ζ represents lead-forward condition).

If the pitch bearing is out-board of the flap and lag spring assembly, the change in pitch angle will not rotate the spring assembly. In this case, the moment deformation relation can be written as

$$\begin{aligned} M_y &= K_\beta\beta \\ M_z &= K_\varsigma\varsigma \end{aligned} \tag{6.44}$$

Combining Equations 6.43 and 6.44, the moment–deformation relationship can be written in a general form as

$$\begin{Bmatrix} M_y \\ M_z \end{Bmatrix} = \begin{bmatrix} K_\beta\cos^2 R\theta + K_\zeta\sin^2 R\theta & -(K_\beta - K_\zeta)\sin R\theta\cos R\theta \\ -(K_\beta - K_\zeta)\sin R\theta\cos R\theta & (K_\beta\sin^2 R\theta + K_\zeta\cos^2 R\theta) \end{bmatrix} \begin{Bmatrix} \beta \\ \zeta \end{Bmatrix} \tag{6.45}$$

From the above relationship, it can be seen that, because of the coupling term, a flap motion introduces a lag moment and a lag motion produces a flap moment. This type of coupling is called "structural flap–lag coupling." For $\theta = 0$, the coupling is 0. Also, when $(K_\beta - K_\varsigma)$ is equal to 0, the coupling is 0 for all values of θ. This condition is known as "matched stiffness" blade configuration. In general, practical blades do not have matched stiffness (chordwise bending stiffness is ~20–40 times greater than flap bending stiffness). However, it is possible to design a blade to have matched stiffness in flap and lag; thereby, one can eliminate structural flap–lag coupling.

These various coupling factors significantly influence the aeroelastic behavior of the rotor blade and, hence, must be treated in a proper manner while performing rotor blade aeroelastic analysis.

Isolated Blade Lag Dynamics in Uncoupled Mode

Consider a rigid blade with a root offset and a root spring undergoing only lead–lag motion, as shown in Figure 6.10.

The rotor blade is hinged at a distance e from the center of the hub, and a spring representing the lead–lag stiffness is situated at the hinge. The spring constant is taken as K_ς. The rotor blade is undergoing a lag deformation ζ about the hinge, as shown in Figure 6.10. During lag deformation, the root spring K_ς and a component of centrifugal force provide restoring moments to lag deformation ς. The equation of the motion of the blade can be obtained by balancing the moment about the root hinge. Assume that the mass of the blade is uniformly distributed along the blade reference axis.

The reference coordinate systems are given as

$\hat{e}_{x1}, \hat{e}_{y1}, \hat{e}_{z1}$ are the unit vectors of the hub-fixed rotating coordinate system, with origin at the hub center.

$\hat{e}_{x2}, \hat{e}_{y2}, \hat{e}_{z2}$ are blade-fixed rotating system in the deformed state of the blade, with origin at the hinge offset location. Note that unit vectors $\hat{e}_{z1}$ and $\hat{e}_{z2}$ are parallel and are normal to the plane of Figure 6.10.

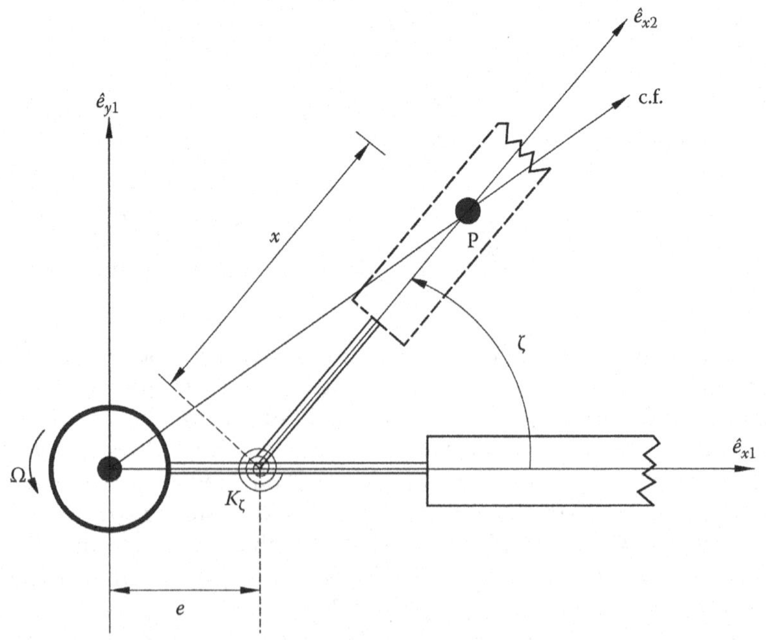

FIGURE 6.10
Idealized offset hinged spring restrained the rigid blade model for lead–lag motion.

The position vector of any point P on the deformed blade can be written as

$$\bar{r}_{\text{P}} = e\hat{e}_{x1} + r\hat{e}_{x2} = e\hat{e}_{x1} + r\cos\zeta\hat{e}_{x1} + r\sin\zeta\hat{e}_{y1} \qquad (6.46)$$

Assuming the constant angular velocity (Ω) of the blade, the absolute velocity and absolute acceleration of point P can be obtained.

The absolute velocity of point P can be obtained as (noting that the position vector is defined in the rotating coordinate system)

$$\vec{v}_P = \left\{\dot{\vec{r}}_P\right\}_{\text{rel}} + \Omega\hat{e}_{z1} \;\; X\vec{r}_P \qquad (6.47)$$

Differentiating Equation 6.46 and substituting the respective quantities, the absolute velocity of point P can be written as

$$\vec{v}_{\text{P}} = -r\sin\zeta\,\frac{d\zeta}{dt}\hat{e}_{x1} + r\cos\zeta\frac{d\zeta}{dt}\hat{e}_{y1} + \Omega\hat{e}_{z1}X\vec{r}_{\text{P}} \qquad (6.48)$$

Substituting for the position vector and taking the cross-product, the velocity vector can be expanded as

$$\vec{v}_{\text{P}} = -r\sin\zeta\frac{d\zeta}{dt}\hat{e}_{x1} + r\cos\zeta\frac{d\zeta}{dt}\hat{e}_{y1} + \Omega(e + r\cos\zeta)\hat{e}_{y1} - \Omega r\sin\zeta\hat{e}_{x1} \qquad (6.49)$$

The absolute acceleration of the mass point P can be obtained from the relationship

$$\vec{a}_{\text{P}} = \left(\frac{d^2\vec{r}_p}{dt^2}\right)_{\text{rel}} + \frac{d\bar{w}}{dt}X\vec{r}_{\text{p}} + 2\bar{w}X\left(\frac{d\vec{r}_p}{dt}\right)_{\text{rel}} + \bar{w}X\left(\bar{w}X\vec{r}_{\text{p}}\right) \qquad (6.50)$$

(Note: $\bar{\omega} = \Omega\hat{e}_{z1}$)

Differentiating Equation 6.46 and substituting various quantities, the absolute acceleration of mass point P can be expressed as

$$\vec{a}_p = \left(-r\cos\zeta\left(\frac{d\zeta}{dt}\right)^2 - r\sin\zeta\frac{d^2\zeta}{dt^2} - 2\Omega r\cos\zeta\frac{d\zeta}{dt} - \Omega^2 e - \Omega^2 r\cos\zeta\right)\hat{e}_{x1}$$
$$+\left(-r\sin\zeta\left(\frac{d\zeta}{dt}\right)^2 + r\cos\zeta\frac{d^2\zeta}{dt^2} - 2\Omega r\sin\zeta\frac{d\zeta}{dt} - \Omega^2 r\sin\zeta\right)\hat{e}_{y1} \qquad (6.51)$$

The inertia moment about the lag hinge is given by

$$Q_{\text{I}} = \int_0^l (r\cos\zeta\hat{e}_{x1} + r\sin\zeta\hat{e}_{y1})X(-m\vec{a}_{\text{p}})\,dr \qquad (6.52)$$

where $l = R - e$ is the length of the blade, and m is the mass per unit length of the blade.

Substituting for $\vec{a}_p$ from Equation 6.51 in Equation 6.52 and simplifying, the inertia moment in lag mode, which is directed along the $\hat{e}_{z1}$ axis, can be written as

$$Q_I = \int_0^l -\left(r^2 \frac{d^2\zeta}{dt^2} + \Omega^2 er \sin\zeta\right) m\,dr \tag{6.53}$$

Invoking small angle assumption (i.e., sin $\zeta \simeq \zeta$), the inertia moment in the lag mode can be written as

$$Q_I = -\left(I_b \frac{d^2\zeta}{dt^2} + \Omega^2 MX_{c.g.} e\zeta\right) \tag{6.54}$$

where the mass moment of inertia of the blade about the lag hinge is $I_b = \int_0^l r^2 m\,dx$, the static mass moment about the lag hinge is $MX_{c.g.} = \int_0^l rm\,dx$, and the mass of the blade is $M = \int_0^l m\,dx$.

For a blade with a uniform mass distribution, $I_b = \frac{Ml^2}{3}$, $MX_{c.g.} = \frac{Ml}{2}$, and $M = m\,l$.

Assume that the aerodynamic drag force acting on the blade gives rise to an aerodynamic moment Q_A about the lag hinge. Combining the inertia, aerodynamic and elastic root moments, the equation of the motion in lag mode can be written in symbolic form as,

$$Q_A + Q_I = K_\zeta \zeta \tag{6.55}$$

Rearranging and substituting for Q_I from Equation 6.54,

$$I_b \frac{d^2\zeta}{dt^2} + (K_\zeta + \Omega^2 MX_{c.g.} e)\zeta = Q_A \tag{6.56}$$

Assuming uniform blade, the lead–lag dynamic equation can be simplified as

$$\frac{Ml^2}{3}\frac{d^2\zeta}{dt^2} + \left(K_\zeta + \frac{\Omega^2 M\,l\,e}{2}\right)\zeta = Q_A \tag{6.57}$$

From Equation 6.57, the nondimensional rotating natural frequency in the lag mode can be written as

$$\bar{\omega}_{\mathrm{RL}} = \frac{\omega_{\mathrm{RL}}}{\Omega} = \sqrt{\frac{K_\zeta}{I_b\Omega^2} + \frac{3}{2}\frac{e}{l}} \tag{6.58}$$

where $\frac{K_\zeta}{I_b\Omega^2}$ represents the nondimensional nonrotating natural frequency in the lag mode. Equation 6.58 can be written in a modified form as

$$\bar{\omega}_{\mathrm{RL}} = \sqrt{\bar{\omega}_{N\mathrm{RL}}^2 + \frac{3}{2}\frac{e}{R-e}} \tag{6.59}$$

(Note: $l = R - e$ and $\bar{\omega}_{N\mathrm{RL}}$ is the nondimensional nonrotating natural frequency in lag mode.)

For articulated rotors, the root spring $K_\zeta = 0$. Therefore, the rotating natural frequency in the lag mode is directly proportional to the hinge offset e. Typically, in articulated rotors, the rotating natural frequency in the lag mode is around $\bar{\omega}_{\mathrm{RL}} = 0.25$ to 0.3, which corresponds to a hinge offset of $\frac{e}{R} \approx 0.04$ to 0.06 or (4% ~ 6%) of the rotor radius. For hingeless rotors, the natural frequency in the lag mode is of the order of $\bar{\omega}_{\mathrm{RL}} \approx 0.7$, which corresponds to an equivalent hinge offset of about 25%. When the rotating nondimensional natural frequency in the lag mode $\bar{\omega}_{\mathrm{RL}} > 1$, then the rotor blade is said to be a stiff-in plane rotor blade, and when $\bar{\omega}_{\mathrm{RL}} < 1$, the rotor blade is denoted as a soft-in plane rotor.

The aerodynamic moment in the lag mode (Q_A) is very small in magnitude compared to the aerodynamic moment in the flap mode. In the flap mode, the aerodynamic moment is due to the lift forces acting on the blade, whereas in the lag mode, the moment is due to aerodynamic drag force, which is about 1% of the lift force. Hence, the damping in the lag mode is usually very small. It may be noted that isolated lag mode is always stable. However, the lag mode can become unstable under certain operating conditions. There are several types of instabilities involving the lag mode, namely, flap–lag aeroelastic instability and coupled rotor–fuselage aeromechanical instability (ground resonance or air resonance). Hence, to avoid these instabilities, external lag dampers are provided in the rotor blades. The various aeroelastic instabilities will be discussed in later chapters.

Isolated Blade Torsional Dynamics in Uncoupled Mode

Now, let us consider the dynamics of the rotor blade undergoing only torsion motion. The rotor blade is assumed to be a rigid blade that is free to rotate about the pitch bearing. The torsional stiffness of the blade and the control system stiffness are combined together. This combined stiffness is represented as K_c, and it is assumed to be located at the root of the blade, with an offset a from the axis of the blade (bearing axis), as shown in Figure 6.11. The effective torsional stiffness of the blade is $K_\phi = K_c a^2$.

The reference coordinate systems are given as

$\hat{e}_{x1}$, $\hat{e}_{y1}$, $\hat{e}_{z1}$ are the unit vectors of the hub-fixed rotating coordinate system, with origin at the hub center.

$\hat{e}_{x2}$, $\hat{e}_{\eta}$, $\hat{e}_{\xi}$ are the unit vectors of the cross-sectional coordinate system fixed to the blade, with origin at the cross section of the blade pitch axis. Note that unit vectors $\hat{e}_{x1}$ and $\hat{e}_{x2}$ are parallel, and they represent the axis of torsion rotation.

The blade pitch angle consists of two parts. They are θ_0 representing the initial pitch angle of the blade due to the control input from the pilot, and ϕ is due to the elastic deformation of the root spring.

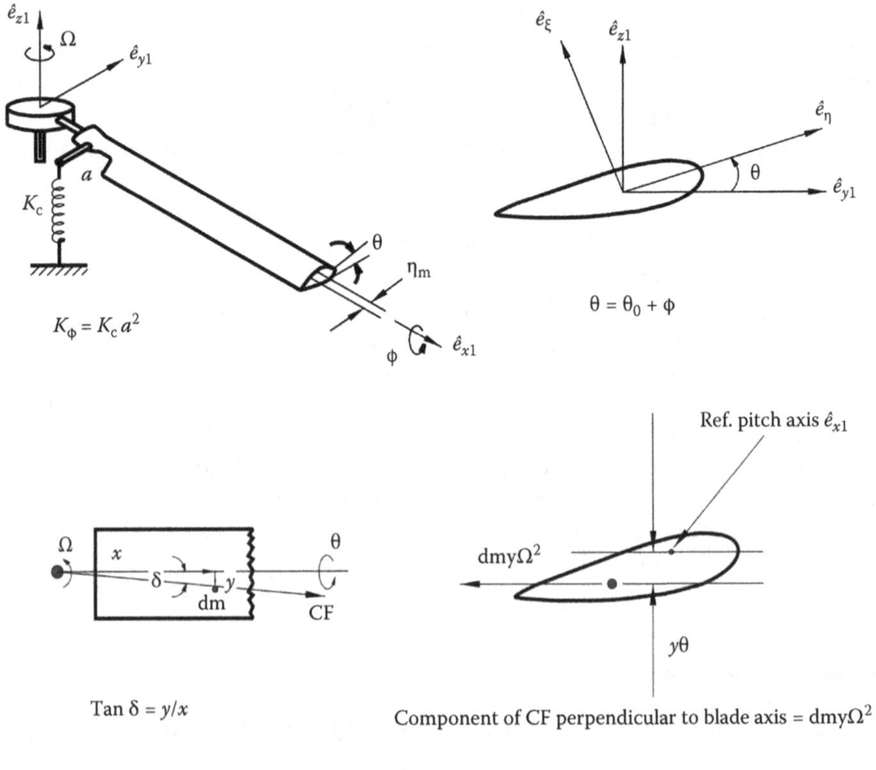

FIGURE 6.11
Idealized blade for torsion dynamics.

The transformation relationship between the unit vectors along the hub-fixed rotating coordinate system and the cross-sectional coordinate system can be written as

$$\begin{Bmatrix} \hat{e}_{x1} \\ \hat{e}_{y1} \\ \hat{e}_{z1} \end{Bmatrix} = \begin{bmatrix} 1 & 0 & 0 \\ 0 & \cos(\theta_0+\phi) & -\sin(\theta_0+\phi) \\ 0 & \sin(\theta_0+\phi) & \cos(\theta_0+\phi) \end{bmatrix} \begin{Bmatrix} \hat{e}_{x2} \\ \hat{e}_{\eta} \\ \hat{e}_{\xi} \end{Bmatrix} \tag{6.60}$$

The coordinates η, ξ represent the cross-sectional position of any arbitrary point P from the origin at the reference pitch axis. The position vector of any mass point P in the cross section is given as

$$\vec{r}_P = r\hat{e}_{x1} + \eta\cos(\theta_0+\phi)\hat{e}_{y1} + \eta\sin(\theta_0+\phi)\hat{e}_{z1} - \xi\sin(\theta_0+\phi)\hat{e}_{y1} + \xi\cos(\theta_0+\phi)\hat{e}_{z1} \tag{6.61}$$

Assuming that the rotor angular rate Ω is a constant, the absolute velocity of the mass point P can be obtained from the relationship

$$\vec{v}_P = \left\{\dot{\vec{r}}_P\right\}_{\text{rel}} + \Omega\hat{e}_{z1} X \vec{r}_P \tag{6.62}$$

Differentiating Equation 6.61 and substituting the respective quantities, the absolute velocity of point P can be written as

$$\begin{aligned} \vec{v}_P &= \{-\eta\sin(\theta_0+\phi) - \xi\cos(\theta_0+\phi)\}\frac{d\phi}{dt}\hat{e}_{y1} \\ &+ \{\eta\cos(\theta_0+\phi) + \xi\sin(\theta_0+\phi)\}\frac{d\phi}{dt}\hat{e}_{z1} + \Omega\hat{e}_{z1} X \vec{r}_P \end{aligned} \tag{6.63}$$

Substituting for the position vector from Equation 6.61 and taking the cross-product, the velocity vector can be expanded as

$$\begin{aligned} \vec{v}_P = {} & \{-\eta\sin(\theta_0+\phi) - \xi\cos(\theta_0+\phi)\}\frac{d\phi}{dt}\hat{e}_{y1} + \{\eta\cos(\theta_0+\phi) + \xi\sin(\theta_0+\phi)\}\frac{d\phi}{dt}\hat{e}_{z1} \\ & \Omega r\hat{e}_{y1} - \Omega\{\eta\cos(\theta_0+\phi) - \xi\sin(\theta_0+\phi)\}\hat{e}_{x1} \end{aligned} \tag{6.64}$$

The absolute acceleration of mass point P can be obtained from the relationship

$$\vec{a}_p = \left(\frac{d^2\vec{r}_p}{dt^2}\right)_{\text{rel}} + \frac{d\bar{w}}{dt} X\vec{r}_p + 2\bar{w} X \left(\frac{d\vec{r}_p}{dt}\right)_{\text{rel}} + \bar{w} X \left(\bar{w} X \vec{r}_p\right) \tag{6.65}$$

(Note: $\bar{\omega} = \Omega\hat{e}_{z1}$)

The absolute acceleration of point P can be written as

$$\vec{a}_p =$$

$$-2\Omega\left\{-\eta\sin(\theta_0+\phi)\frac{d\phi}{dt}-\xi\cos(\theta_0+\phi)\frac{d\phi}{dt}\right\}\hat{e}_{x1}-r\Omega^2\hat{e}_{x1}$$

$$+\left\{\begin{aligned}&-\eta\sin(\theta_0+\phi)\frac{d^2\phi}{dt^2}-\eta\cos(\theta_0+\phi)\left(\frac{d\phi}{dt}\right)^2-\xi\cos(\theta_0+\phi)\frac{d^2\phi}{dt^2}\\&+\xi\sin(\theta_0+\phi)\left(\frac{d\phi}{dt}\right)^2-\Omega^2\eta\cos(\theta_0+\phi)+\Omega^2\xi\sin(\theta_0+\phi)\end{aligned}\right\}\hat{e}_{y1}$$

$$+\left\{\eta\cos(\theta_0+\phi)\frac{d^2\phi}{dt^2}-\eta\sin(\theta_0+\phi)\left(\frac{d\phi}{dt}\right)^2-\xi\sin(\theta_0+\phi)\frac{d^2\phi}{dt^2}-\xi\cos(\theta_0+\phi)\left(\frac{d\phi}{dt}\right)^2\right\}\hat{e}_{z1} \tag{6.66}$$

The inertia moment about the blade root can be obtained by integrating over the volume of the blade as

$$\vec{M}_\mathrm{I} = \int_{\mathrm{vol}} \vec{r}_\mathrm{p} X(-\rho_\mathrm{b}\vec{a}_\mathrm{p})\,dv \tag{6.67}$$

where ρ_b is the mass per unit volume of the blade.

Since we are interested in the torsional motion, let us consider only the moment about the reference blade axis ($\hat{e}_{x1}$ component). Substituting the position vector and the acceleration vector in Equation 6.67, taking the vector cross-product, and performing the integration over the cross section, the torsional inertia moment can be written as

$$\begin{aligned}(M_\phi)_\mathrm{I} = \int_0^l \Big\{&-(I_{m\xi\xi}+I_{m\eta\eta})\frac{d^2\phi}{dt^2}-\Omega^2(I_{m\xi\xi}-I_{m\eta\eta})\sin(\theta_0+\phi)\cos(\theta_0+\phi)\\&-\Omega^2 I_{m\eta\xi}(\cos^2(\theta_0+\phi)-\sin^2(\theta_0+\phi))\Big\}dr\end{aligned} \tag{6.68}$$

where the cross-sectional integrals are defined as

$$\iint \rho_\mathrm{b}\eta^2\,dA = I_{m\xi\xi};\quad \iint \rho_\mathrm{b}\xi^2\,dA = I_{m\eta\eta};\quad \iint \rho_\mathrm{b}\eta\xi\,dA = I_{m\eta\xi} \tag{6.69}$$

In Equation 6.68, the integration is performed over the length l of the blade.

Assuming a uniform cross section along the length of the blade, Equation 6.68 can be integrated to obtain the total torsional inertia moment of the blade. The torsional equation of motion can be written, by balancing the inertia, aerodynamic (M_A), and elastic torsional moments about the reference axis, as

$$-l(I_{m\xi\xi}+I_{m\eta\eta})\frac{d^2\phi}{dt^2}-l\Omega^2(I_{m\xi\xi}-I_{m\eta\eta})\sin(\theta_0+\phi)\cos(\theta_0+\phi)$$
$$-\Omega^2 l\,\mathrm{I}_{m\eta\xi}(\cos^2(\theta_0+\phi)-\sin^2(\theta_0+\phi))+M_A-K_\phi\phi=0 \tag{6.70}$$

where K_ϕ is the torsional spring constant at the root.

Assuming a small angle for torsional deformation ϕ,

$$\sin(\theta_0+\phi)\approx\sin\theta_0+\phi\cos\theta_0 \tag{6.71}$$

$$\cos(\theta_0+\phi)\approx\cos\theta_0-\phi\sin\theta_0 \tag{6.72}$$

Substituting Equations 6.71 and 6.72 in Equation 6.70 and neglecting higher-order terms, the torsional equation of motion can be written as

$$l(I_{m\xi\xi}+I_{m\eta\eta})\frac{d^2\phi}{dt^2}+l\Omega^2\left[(I_{m\xi\xi}-I_{m\eta\eta})(\cos^2\theta_0-\sin^2\theta_0)+\mathrm{K}_\phi/l\Omega^2-I_{m\eta\xi}4\sin\theta_0\cos\theta_0\right]\phi$$
$$=M_A-l\Omega^2\left(I_{m\xi\xi}-I_{m\eta\eta}\right)\sin\theta_0\cos\theta_0-l\Omega^2 I_{m\eta\xi}(\cos^2\theta_0-\sin^2\theta_0) \tag{6.73}$$

It can be seen from Equation 6.73 that the natural frequency in torsion is dependent on the pitch angle θ_0. If we assume that θ_0 is equal to 0 and consider only free vibration in torsional motion, the nondimensional rotating natural frequency in torsional mode can be written as

$$\bar{\omega}_{\mathrm{RT}}=\frac{\omega_{\mathrm{RT}}}{\Omega}=\sqrt{\frac{K_\phi}{(I_{m\xi\xi}+I_{m\eta\eta})l\Omega^2}+\frac{(I_{m\xi\xi}-I_{m\eta\eta})}{(I_{m\xi\xi}+I_{m\eta\eta})}} \tag{6.74}$$

Generally, for rotor blades, $\mathrm{I}_{m\eta\eta}\ll I_{m\xi\xi}$; therefore, the rotating natural frequency in torsion can be written as

$$\bar{\omega}_{RT}\simeq\sqrt{\bar{\omega}^2_{N\mathrm{RT}}+1} \tag{6.75}$$

where $\bar{\omega}_{N\mathrm{RT}}=\sqrt{\dfrac{K_\phi}{(\mathrm{I}_{m\xi\xi}+\mathrm{I}_{m\eta\eta})l\Omega^2}}$ is the nondimensional nonrotating torsion natural frequency of the blade.

Equation 6.75 indicates that the angular rate Ω of the blade adds to the stiffness in the torsion. In general the rotating natural frequency in torsion is above 3.5/rev or more than 4/rev. When $K_{\phi} = 0$, the natural frequency in the torsion of a rotating blade is equal to 1/rev. This stiffness (or the restoring moment) arises due to the effect of the centrifugal force acting on an element of mass, which is away from the axis of rotation, as shown in Figure 6.11. This effect is known as the "tennis racquet effect."

Damping in the torsional mode is provided by the aerodynamic moment due to the unsteady pitch motion of the blade. It may be noted that Theodorsen two-dimensional unsteady aerodynamic theory for an oscillating airfoil can be used for obtaining the torsional aerodynamic moment on the oscillating blade. In isolated torsional mode, the aerodynamic damping is reasonably high, and it is found to be in the order of 10% of the critical damping.

7

Rotor Blade Aeroelastic Stability: Coupled Mode Dynamics

In the previous chapter, the dynamics of an isolated idealized rotor blade in various uncoupled modes was considered. In this chapter, let us consider the dynamics of coupled motion. The coupled dynamics leads to various aeroelastic instabilities in the rotor blade. The analysis of coupled flap–lag–torsion dynamics of a rotor blade is rather complicated. However, one can bring out the essential characteristics of aeroelastic stability of the blade by considering two separate problems. They are (1) the coupled flap–lag dynamics and (2) the coupled flap–torsion dynamics of an isolated rotor blade. The fundamental understanding of these two simplified cases is a precursor to the complicated study of coupled flag–lag–torsion and axial dynamics of an elastic rotor blade, which is essential for the analysis of rotor blade aeroelastic stability, response, loads, and vibration.

Coupled Flap–Lag Dynamics

Consider a rigid blade having a hinge offset and root springs simulating the flexibility of the blade in the flap and lag modes, as shown in Figure 7.1. Lag and flap deformations are shown in Figure 7.2 for clarity. The formulation of the coupled flap–lag dynamic equations requires elastic, inertia, and aerodynamic operators. Since these operators depend on blade motion, the kinematic description of blade motion forms the first step in the formulation of equations of motion. This requires the choice and definition of several hub-fixed and blade-fixed, deformed and undeformed state coordinate systems and the transformation relationships between them. Using the kinematic description of blade motion, the inertia operator is formulated. Knowing the blade motion and the relative air velocity components, the aerodynamic operator is formulated. The procedure essentially follows the one outlined in the previous chapter on isolated blade dynamics in uncoupled modes. The systematic formulation of the equations of motion of coupled flap–lag dynamics is presented in the following. The equations of motion for the system shown in Figure 7.1 can be written in symbolic form as

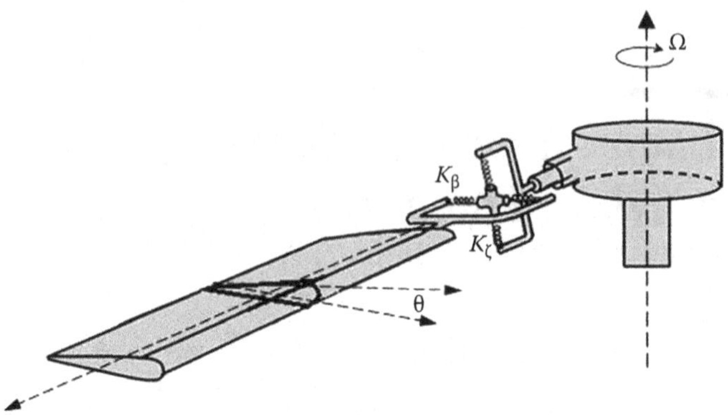

FIGURE 7.1
Idealization of rotor blade for coupled flap–lag dynamics.

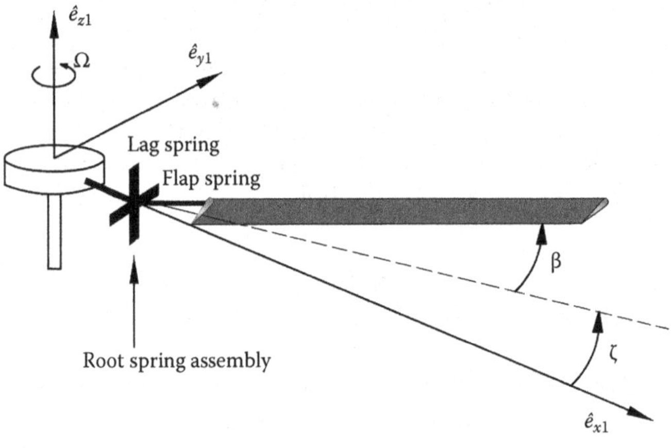

FIGURE 7.2
Flap–lag deformations of the rigid blade.

Flap equation

$$M_\beta + Q_{Iy} + Q_{Ay} = 0 \tag{7.1}$$

where M_β is the moment due to root spring restraint, Q_{Iy} is the inertia moment in the flap mode, and Q_{Ay} is the aerodynamic moment in the flap mode.

Lag equation

$$M_\zeta + Q_{Iz} + Q_{Az} = 0 \tag{7.2}$$

where M_ζ is the moment due to the root spring, Q_{Iz} is the inertia moment in the lag mode, and Q_{Az} is the aerodynamic moment in the lag mode.

In the following, a systematic approach to the derivation of the coupled flap–lag dynamics of the isolated rotor blade is presented.

Ordering Scheme

When deriving equations of motion for the coupled flap–lag dynamics of a rotor blade, a large number of higher-order terms have to be considered. Research has clearly indicated that many higher-order terms can be neglected systematically by using an ordering scheme.

The ordering scheme is based on defining a small dimensionless parameter ε, which represents typical slopes due to the elastic deflections of the rotor blade. It is known that, for helicopter blades, ε is in the range $0.1 \le \varepsilon \le 0.15$.

The ordering scheme is based on the assumption that

$$1 + O(\varepsilon^2) \cong 1$$

that is, the terms of order $O(\varepsilon^2)$ are neglected in comparison with unity. The orders of magnitude for various parameters governing this problem are given in the following.

$$\cos\psi, \sin\psi, \frac{\rho_a abR}{m}, \frac{x_t}{R}, \frac{y_t}{R}, \frac{z_t}{R} = O(1)$$

$$\frac{1}{\Omega}\frac{\partial}{\partial t}(\ ) = \frac{\partial}{\partial\psi}(\ ) = O(1)$$

$$\theta_k = O\left(\varepsilon^{1/2}\right)$$

$$\beta_k(t), \zeta_k(t), \psi_k(t), \frac{e}{R}, \frac{b}{R}, \lambda, \dot{\theta}_k, \ddot{\theta}_k = O(\varepsilon)$$

$$\frac{C_{d0}}{a} = O\left(\varepsilon^{3/2}\right)$$

$$\frac{X_A}{R} = O(\varepsilon^2)$$

Coordinate Transformations

In the derivation of equations of motion of the helicopter, various reference coordinate systems are used. The transformation relationship between quantities referred in various inertial and non-inertial coordinate systems has to be established before deriving the equations of motion. The relationship between two orthogonal coordinate systems with axes X_i, Y_i, Z_i and X_j, Y_j, Z_j with $\hat{e}_{xi}$, $\hat{e}_{yi}$, $\hat{e}_{zi}$ and $\hat{e}_{xj}$, $\hat{e}_{yj}$, $\hat{e}_{zj}$ as unit vectors along the respective axes is given as

$$\begin{bmatrix} e_{xi} \\ e_{yi} \\ e_{zi} \end{bmatrix} = [T_{ij}] \begin{bmatrix} e_{xj} \\ e_{yj} \\ e_{zj} \end{bmatrix} \tag{7.3}$$

where $[T_{ij}]$, the transformation matrix, can be found using the Euler angles required to rotate the jth system so as to make it parallel to the ith system.

Summary of Coordinate Systems for Flap–Lag Motion

The complete set of coordinate systems used in the development of the flap–lag dynamics of the rotor blade is described in the following for convenience.

H: Inertial hub-fixed nonrotating system, with origin at the center of the hub. The unit vectors are $\hat{e}_{xH}$, $\hat{e}_{yH}$, $\hat{e}_{zH}$.

1k: Hub-fixed rotating system, which rotates with the kth blade, with origin at the center of the hub. The unit vectors are $\hat{e}_{x1}$, $\hat{e}_{y1}$, $\hat{e}_{z1}$.

2k: The origin is at the hinge offset, and it is parallel to the 1k system. The 2k system rotates with the kth blade. The unit vectors are $\hat{e}_{x2}$, $\hat{e}_{y2}$, $\hat{e}_{z2}$.

3k: Undeformed coordinate system of the blade after precone, with origin at the kth-blade hinge offset. The unit vectors are $\hat{e}_{x3}$, $\hat{e}_{y3}$, $\hat{e}_{z3}$.

4k: Deformed coordinate system after undergoing flap angle β_k and lag angle ζ_k with respect to the 3k system, with origin at the kth-blade hinge offset. The unit vectors are $\hat{e}_{x4}$, $\hat{e}_{y4}$, $\hat{e}_{z4}$.

Figure 7.3 shows the H and 1k coordinate systems. The H system is a hub-fixed nonrotating coordinate system whose origin is at the center of the rotor hub. The 1k system is a coordinate system with origin at the center of the hub and rotates with the kth blade. This 1k system is rotated from the H system by an azimuth angle ψ_k about the Z_1 axis, as shown in Figure 7.3.

The azimuth angle ψ_k is measured from $\hat{e}_{X_H}$, and counterclockwise rotation is taken as positive. The transformation matrix between the H and the 1k system is given by

$$\left[T_{1H}\right] = \begin{bmatrix} \cos\psi_k & \sin\psi_k & 0 \\ -\sin\psi_k & \cos\psi_k & 0 \\ 0 & 0 & 1 \end{bmatrix} \tag{7.4}$$

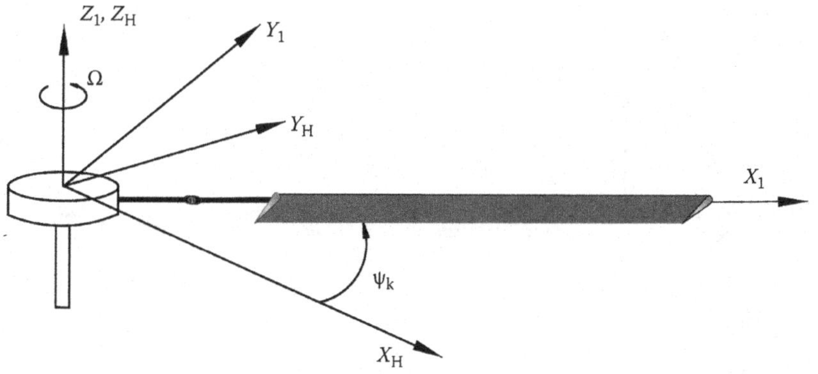

FIGURE 7.3
Hub-fixed nonrotating and rotating coordinate systems.

where ψ_k is the azimuth position of the kth blade. It is defined as $\psi_k = \psi + \frac{2\pi}{N}(k-1)$, and $\psi = \Omega t$ represents the nondimensional time (or the azimuth location of the first blade).

N is the number of blades in the rotor system.

The $2k$ system is also a rotating system, with its origin at the hinge offset location, as shown in Figure 7.4. The $2k$ system and the $1k$ system are parallel, and the transformation matrix is given by

$$[T_{21}] = \begin{bmatrix} 1 & 0 & 0 \\ 0 & 1 & 0 \\ 0 & 0 & 1 \end{bmatrix} \tag{7.5}$$

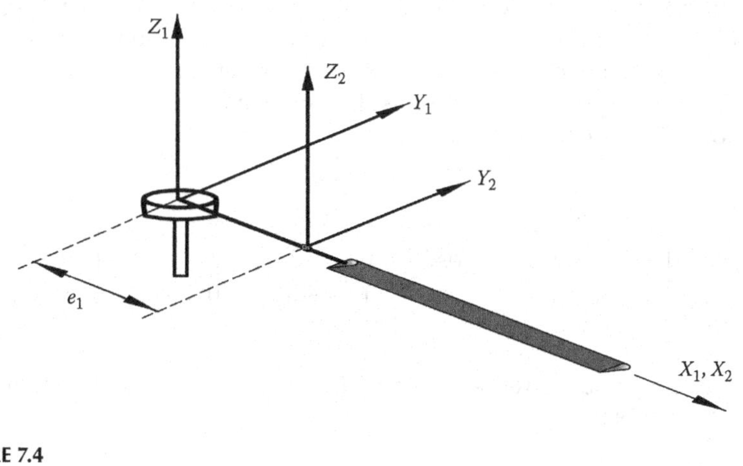

FIGURE 7.4
Blade root–fixed rotating axis system, with origin at the hinge offset.

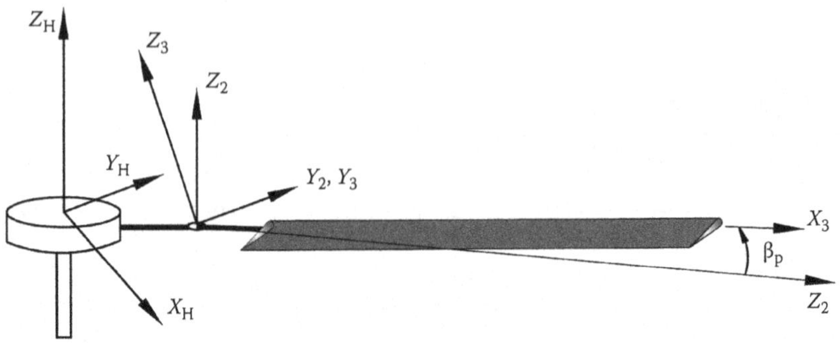

FIGURE 7.5
Undeformed reference state of the *k*th blade with precone.

Rotating the $2k$ system by an angle β_p (precone angle) about the Y_2 axis, the $3k$ system is obtained, which is shown in Figure 7.5. Assuming that β_p is very small, one can make the small assumption such that $\sin \beta_p \approx \beta_p$ and $\cos \beta_p \approx 1$. The transformation matrix is given by

$$[T_{32}] = \begin{bmatrix} 1 & 0 & \beta_p \\ 0 & 1 & 0 \\ -\beta_p & 0 & 1 \end{bmatrix} \tag{7.6}$$

It may be noted that, in the undeformed state, the reference axis of the *k*th blade is along the X_{3k} axis.

The $4k$ system is the blade-fixed deformed coordinate system. Considering flapping and lagging motion, the $3k$ system is rotated first by flap angle β_k and followed by lag angle ζ_k (i.e., first, a clockwise rotation about $\hat{e}_{y3k}$, followed by a counterclockwise rotation about the rotated $\hat{e}_{z3k}$ axis), as shown in Figure 7.6. The sequence of rotation is flap followed by lag. The transformation matrix $[T_{43}]$ can be written as

$$[T_{43}] = \begin{bmatrix} \cos\zeta_k & \sin\zeta_k & 0 \\ -\sin\zeta_k & \cos\zeta_k & 0 \\ 0 & 0 & 1 \end{bmatrix} \begin{bmatrix} \cos\beta_k & 0 & \sin\beta_k \\ 0 & 1 & 0 \\ -\sin\beta_k & 0 & \cos\beta_k \end{bmatrix} \tag{7.7}$$

(Note: X_{3i}, Y_{3i}, and Z_{3i} represent the intermediate coordinate system after executing flap rotation.)

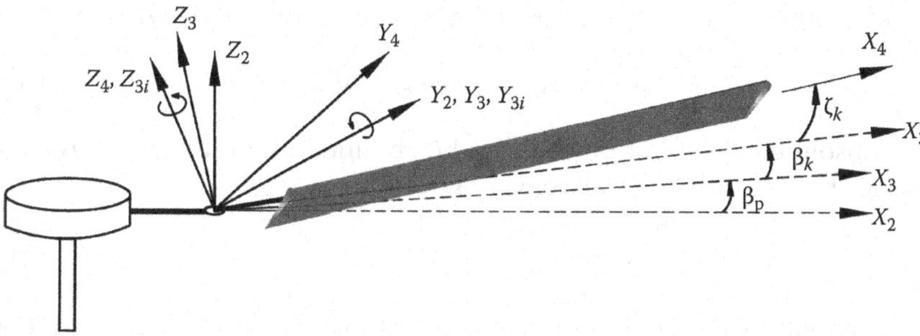

FIGURE 7.6
Deformed state of the blade after flap and lag rotations.

Assuming small angles $[\zeta_k, \beta_k \approx O(\varepsilon)]$, the transformation matrix can simplified as

$$[T_{43}] = \begin{bmatrix} 1 & \zeta_k & \beta_k \\ -\zeta_k & 1 & -\zeta_k\beta_k \\ -\beta_k & 0 & 1 \end{bmatrix} \tag{7.8}$$

Acceleration of a Point "P" on the *k*th Blade

Considering the rigid rotor blade to be a straight line, the position vector (from the hub center) of a point "P" on the *k*th blade in the deformed state is given as (using the inverse of the transformation given in Equation 7.7)

$$\begin{aligned} \vec{r}_P &= e\hat{e}_{x1} + x\hat{e}_{x4} = e\hat{e}_{x1} + x\cos\zeta_k \cos\beta_k \hat{e}_{x3} \\ &\quad + x\sin\zeta_k \hat{e}_{y3} + x\cos\zeta_k \sin\beta_k \hat{e}_{z3} \end{aligned} \tag{7.9}$$

(Note that the position vector of point P consists of two parts, namely position from the hub center to the hinge and position from the hinge to point P.)

Writing the position vector in the 1*k* system (using Equations 7.5–7.7), we have

$$\begin{aligned} \vec{r}_P &= (e + x\cos\zeta_k \cos\beta_k - x\cos\zeta_k \sin\beta_k \beta_p)\hat{e}_{x1} + x\sin\zeta_k \hat{e}_{y1} \\ &\quad + (x\cos\zeta_k \cos\beta_k \beta_p + x\cos\zeta_k \sin\beta_k)\hat{e}_{z1} \end{aligned} \tag{7.10}$$

Making small angle assumption, the position vector can be written as

$$\vec{r}_{\mathrm{P}} = (e + x - x\beta_k\beta_{\mathrm{p}})\hat{e}_{x1} + x\zeta_k\hat{e}_{y1} + (x\beta_{\mathrm{P}} + x\beta_k)\hat{e}_{z1} \tag{7.11}$$

The absolute velocity of point P can be obtained as (note that the rotor is rotating at a constant angular velocity $\Omega\hat{e}_{z1}$)

$$\vec{v}_{\mathrm{P}} = \left\{\dot{\vec{r}}_{\mathrm{P}}\right\}_{\mathrm{rel}} + \Omega\hat{e}_{z1} X \vec{r}_{\mathrm{P}} \tag{7.12}$$

Differentiating Equation 7.10 and substituting the respective quantities, the absolute velocity of point P can be written as (after nondimensionalizing the time derivative with rotor angular velocity Ω) (note that a small angle assumption is used in making the approximation and, also, higher-order terms are neglected)

$$\vec{v}_{\mathrm{P}} = \Omega\left\{\begin{array}{c} (-x\zeta_k\dot{\zeta}_k - x\beta_k\dot{\beta}_k - x\beta_{\mathrm{p}}\dot{\beta}_k)\hat{e}_{x1} + x\dot{\zeta}_k\hat{e}_{y1} + x\dot{\beta}_k\hat{e}_{z1} \\ +\left[(e + x - x\beta_k\beta_{\mathrm{p}})\hat{e}_{y1} - x\zeta_k\hat{e}_{x1}\right] \end{array}\right\} \tag{7.13}$$

The absolute acceleration of mass point P can be obtained from the expression

$$\vec{a}_p = \left(\frac{d^2\vec{r}_p}{dt^2}\right)_{rel} + \frac{d\bar{\omega}}{dt} X\,\vec{r}_p + 2\bar{\omega}\,X\left(\frac{d\vec{r}_p}{dt}\right)_{rel} + \bar{\omega}\,X(\bar{\omega}\,X\,\vec{r}_p) \tag{7.14}$$

Note: $\bar{\omega} = \Omega\hat{e}_{z1}$

Differentiating Equation 7.10, and substituting various quantities in Equation 7.14, the absolute acceleration of mass point P can be obtained. Making a small angle assumption and neglecting higher-order terms, the acceleration at point P can be expressed as

$$\vec{a}_{\mathrm{P}} = \Omega^2\left\{\begin{array}{c} (-x\zeta_k\ddot{\zeta}_k - x\dot{\zeta}_k^2 - x\dot{\beta}_k^2 - x\beta_k\ddot{\beta}_k - x\beta_{\mathrm{p}}\ddot{\beta}_k)\hat{e}_{x1} + x\ddot{\zeta}_k\hat{e}_{y1} + x\ddot{\beta}_k\hat{e}_{z1} \\ +2(-x\zeta_k\dot{\zeta}_k - x\beta_k\dot{\beta}_k - x\beta_{\mathrm{p}}\dot{\beta}_k)\hat{e}_{y1} - 2x\dot{\zeta}_k\hat{e}_{x1} \\ +\left[-(e + x - x\beta_k\beta_{\mathrm{p}})\hat{e}_{x1} - x\zeta_k\hat{e}_{y1}\right] \end{array}\right\} \tag{7.15}$$

Rearranging the terms, the acceleration of point P on the kth blade is expressed as

$$\vec{a}_{\mathrm{P}} = \Omega^2\left\{\begin{array}{c} \left[-x\zeta_k\ddot{\zeta}_k - x\dot{\zeta}_k^2 - x\dot{\beta}_k^2 - x\beta_k\ddot{\beta}_k - x\beta_{\mathrm{p}}\ddot{\beta}_k - (e + x - x\beta_k\beta_{\mathrm{p}}) - 2x\dot{\zeta}_k\right]\hat{e}_{x1} \\ +\left[2(-x\zeta_k\dot{\zeta}_k - x\beta_k\dot{\beta}_k - x\beta_{\mathrm{p}}\dot{\beta}_k) + x\ddot{\zeta}_k - x\zeta_k\right]\hat{e}_{y1} \\ +x\ddot{\beta}_k\hat{e}_{z1} \end{array}\right\} \tag{7.16}$$

Using coordinate transformation matrices, the acceleration of point P on the blade is expressed in the 3*k* system as

$$\vec{a}_p = \Omega^2 \left\{ \begin{array}{c} \left[-x\zeta_k\ddot{\zeta}_k - x\dot{\zeta}_k^2 - x\dot{\beta}_k^2 - x\beta_k\ddot{\beta}_k - (e + x - x\beta_k\beta_p) - 2x\dot{\zeta}_k\right]\hat{e}_{x3} \\ +\left[2(-x\zeta_k\dot{\zeta}_k - x\beta_k\dot{\beta}_k - x\beta_p\dot{\beta}_k) + x\ddot{\zeta}_k - x\zeta_k\right]\hat{e}_{y3} \\ +\left[x\ddot{\beta}_k + \beta_p(e+x) + \beta_p 2x\dot{\zeta}_k\right]\hat{e}_{z3} \end{array} \right\} \tag{7.17}$$

Distributed Inertia Force

The distributed inertia force per unit length of the *k*th blade is obtained from d'Alembert's principle. The distributed inertia force acting on the *k*th blade is expressed as

$$p_I = \iint -\rho \vec{a}_p \, dA \tag{7.18}$$

where ρ is the mass density of the blade, and the integral is taken over the cross section of the blade.

Substituting the various components of acceleration from Equation 7.17 in Equation 7.18, the distributed inertia forces can be obtained in the blade-fixed 3*k* system. The components of these distributed inertia loads are given by

$$p_{Ix3k} = m\Omega^2 \left\{ x\zeta_k\ddot{\zeta}_k + x\dot{\zeta}_k^2 + x\dot{\beta}_k^2 + x\beta_k\ddot{\beta}_k + (e + x - x\beta_k\beta_p) + 2x\dot{\zeta}_k \right\} \tag{7.19}$$

$$p_{Iy3k} = m\Omega^2 \left\{ 2(x\zeta_k\dot{\zeta}_k + x\beta_k\dot{\beta}_k + x\beta_p\dot{\beta}_k) + x\ddot{\zeta}_k + x\zeta_k \right\} \tag{7.20}$$

$$p_{Iz3k} = m\Omega^2 \left\{ -x\ddot{\beta}_k - \beta_p(e+k) - \beta_p 2x\dot{\zeta}_k \right\} \tag{7.21}$$

where the mass per unit length of the blade is given by $m = \iint \rho dA$.

Root Inertia Moment

The inertia moment about the root hinge of the blade can be obtained by using d'Alembert's principle by evaluating the integral of the vector product, which is given as

$$Q_I = \iiint -\rho\left\{ (x\cos\zeta_k \cos\beta_k \hat{e}_{x3} + x\sin\zeta_k \hat{e}_{y3} + x\cos\zeta_k \sin\beta_k \hat{e}_{z3}) X\vec{a}_p dA \right\} dx \tag{7.22}$$

Substituting for acceleration from Equation 7.17 and making a small angle assumption after taking the vector cross-product and integrating over the cross section, the inertia moment in the component form can be written as (after neglecting higher-order terms)

$$Q_{Ix3k} = \int_0^l q_{Ix3k}\, dx, \quad Q_{Iy3k} = \int_0^l q_{Iy3k}\, dx, \quad Q_{Iz3k} = \int_0^l q_{Iz3k}\, dx$$

where $l = (R-e)$ is the length of the blade from the hinge offset, and

$$q_{Ix3k} = m\Omega^2 \left\{ x^2 \zeta_k + xe\zeta_k + 2\Omega x^2 \zeta_k \dot{\zeta}_k + x^2 \beta_k \ddot{\zeta}_k - x^2 \beta_k \zeta_k \right\} \qquad (7.23)$$

$$q_{Iy3k} = m\Omega^2 \left\{ x^2 \ddot{\beta}_k + (x^2 + xe)(\beta_p + \beta_k) + 2x^2 \dot{\zeta}_k (\beta_p + \beta_k) \right\} \qquad (7.24)$$

$$q_{Iz3k} = m\Omega^2 \left\{ 2x^2 \dot{\beta}_k (\beta_p + \beta_k) - x^2 \ddot{\zeta}_k - xe\zeta_k \right\} \qquad (7.25)$$

Integrating over the length of the blade and simplifying, the inertia moment about the root can be obtained. Since we are interested in only the flap and lead–lag moments, expressions for these two components are provided in the following.

Inertia moments in flap and lag motion:

$$Q_{Iy3k} \cong I_b \Omega^2 \ddot{\beta}_k + 2I_b \Omega^2 \dot{\zeta}_k (\beta_p + \beta_k) + (I_b \Omega^2 + eMX_{c.g.}\Omega^2)(\beta_p + \beta_k) \qquad (7.26)$$

$$Q_{Iz3k} = -I_b \Omega^2 \ddot{\zeta}_k + 2I_b \Omega^2 \dot{\beta}_k (\beta_p + \beta_k) - \Omega^2 eMX_{c.g.}\zeta_k \qquad (7.27)$$

where $I_b = \displaystyle\int_0^{R-e} mx^2\, dx$ is the mass moment of inertia of the blade about the hinge, $MX_{c.g.} = \displaystyle\int_0^{R-e} mx\, dx$ is the first moment of mass of the blade about the hinge offset, and $M = \displaystyle\int_0^{R-e} m\, dx$ is the mass of the blade.

Aerodynamic Loads

The aerodynamic loads for coupled flap–lag dynamics are obtained by using a quasi-static aerodynamic model. The blade is assumed to be a straight blade having a zero twist. A uniform inflow model based on the momentum

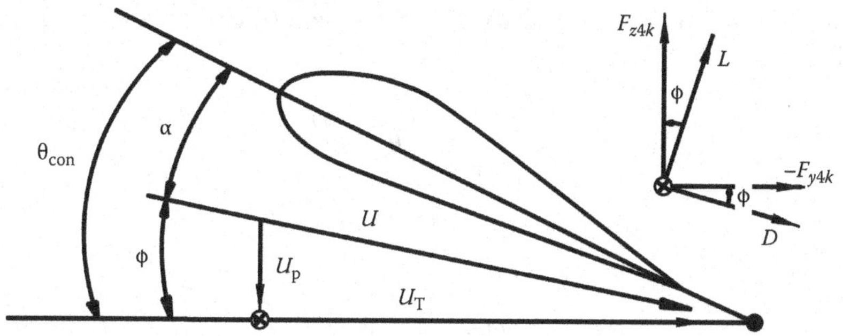

FIGURE 7.7
Relative air velocities and aerodynamic loads on a blade element.

theory is used. Figure 7.7 shows the velocity components and the resultant aerodynamic forces acting on a typical cross section of the rotor blade. The blade is set at an initial pitch angle θ_{con}. The oncoming air velocities U_T and U_P represent the tangential and perpendicular components, respectively.

The resultant velocity U and the inflow angle are given by

$$U = \sqrt{U_T^2 + U_P^2} \tag{7.28}$$

$$\tan\phi = \frac{U_P}{U_T} \tag{7.29}$$

The effective angle of attack of the blade section is given as

$$\alpha = \theta_{con} - \phi \tag{7.30}$$

The sectional lift and drag forces can be written, respectively, as

$$L = \frac{1}{2}\rho U^2 c C_l \tag{7.31}$$

and

$$D = \frac{1}{2}\rho U^2 c C_{d0} \tag{7.32}$$

where ρ is the density of air, c is the blade chord, and C_l and C_{d0} the aerodynamic lift and drag coefficients, respectively, which are functions of the angle of attack and the Mach number. Note that the sectional aerodynamic moment is taken as 0.

Assuming $U_P \ll U_T$, one can make the approximation, $U \cong U_T$ and $\tan\phi \cong \phi$.

The expressions for lift and drag per unit length can be written, respectively, as

$$L = \frac{1}{2}\rho a U_T^2 c\left(\theta_{\text{con}} - \frac{U_P}{U_T}\right) \tag{7.33}$$

and

$$D = \frac{1}{2}\rho U_T^2 c C_{d0} \tag{7.34}$$

Relative Air Velocity Components at a Typical Cross Section of the *k*th Blade

The cross-sectional aerodynamic loads are obtained in the deformed state of the blade. Hence, the components of velocity have to be defined in the blade-fixed ($\hat{e}_{x4k}, \hat{e}_{y4k}, \hat{e}_{z4k}$) rotating coordinate system.

The net relative velocity of airflow has two components. They are (1) the forward speed of the vehicle ($\mu\Omega R$) and the induced flow ($\lambda\Omega R$), and (2) the velocity due to blade motion.

Let $\vec{V}_F$ be the free-stream velocity of air defined in the hub plane. It is given as

$$\vec{V}_F = \Omega R\mu\hat{e}_{xH} - \Omega R\lambda\hat{e}_{zH} = \Omega R\mu\cos\psi_k\hat{e}_{x1} - \Omega R\mu\sin\psi_k\hat{e}_{y1} - \Omega R\lambda\hat{e}_{z1} \tag{7.35}$$

Velocity at a point "P" on the reference axis due to blade motion in the 1*k* system is given as (from Equation 7.13)

$$\vec{v}_P = \Omega\begin{Bmatrix} (-x\zeta_k\dot{\zeta}_k - x\beta_k\dot{\beta}_k - x\beta_p\dot{\beta}_k - x\zeta_k)\hat{e}_{x1} \\ + \left[(e + x - x\beta_k\beta_p + x\dot{\zeta}_k)\hat{e}_{y1}\right] + x\dot{\beta}_k\hat{e}_{z1} \end{Bmatrix} \tag{7.36}$$

The net relative air velocity vector at the typical cross section of the blade can be written as

$$\vec{V}_{\text{net-air}1k} = \vec{V}_F - \vec{v}_P \tag{7.37}$$

Using the transformation relationships given in Equations 7.5 to 7.8, the net air velocity in the 4*k* system can be written as

$$\vec{V}_{\text{net-air}4k} = [T_{43}][T_{32}][T_{21}]\vec{V}_{\text{net-air}1k} \tag{7.38}$$

The net air velocity can be written in symbolic form as

$$\vec{V}_{net-air4k} = U_R\hat{e}_{x4} - U_T\hat{e}_{y4} - U_P\hat{e}_{z4} \tag{7.39}$$

Using Equations 7.35 to 7.38, the radial, tangential, and perpendicular velocity components are obtained and are given as (for convenience, the length quantities are nondimensionalized with respect to the rotor radius R, and the time derivatives are nondimensionalized with the rotor angular velocity Ω).

$$U_R = \Omega R\left\{(1-\beta_p\beta_k)\mu\cos\psi_k - \zeta_k\mu\sin\psi_k - (\beta_p + \beta_k)\lambda - e\zeta_k\right\} \tag{7.40}$$

$$U_T = \Omega R\left\{\zeta_k\mu\cos\psi_k + \mu\sin\psi_k - \zeta_k(\beta_p + \beta_k)\lambda + x\dot{\zeta}_k + x + e - x\beta_k\beta_p - x\zeta_k^2\right\} \tag{7.41}$$

$$U_P = \Omega R\left\{(\beta_p + \beta_k)\mu\cos\psi_k + (1-\beta_p\beta_k)\lambda + x\dot{\beta}_k + x\zeta_k(\beta_p + \beta_k)\right\} \tag{7.42}$$

Note that the velocity components U_T and U_P represent the oncoming air velocity in the directions, as shown in Figure 7.7. The velocity components can be simplified by neglecting higher-order terms with respect to the leading term. The final expressions for the air velocity components are given as

$$U_T \cong \Omega R\left\{\zeta_k\mu\cos\psi_k + \mu\sin\psi_k + x\dot{\zeta}_k + x + e\right\} \tag{7.43}$$

$$U_P = \Omega R\left\{(\beta_p + \beta_k)\mu\cos\psi_k + \lambda + x\dot{\beta}_k + x\zeta_k(\beta_p + \beta_k)\right\} \tag{7.44}$$

Aerodynamic Loads and Moments at the Blade Root

The aerodynamics lift and drag forces acting on a typical cross section are resolved along the blade cross-sectional coordinate system, as shown in Figure 7.7. They are given as

$$F_{y4k} = -L\sin\phi - D\cos\phi \tag{7.45}$$

$$F_{z4k} = L\cos\phi - D\sin\phi \tag{7.46}$$

Assuming that the drag is very small in comparison to the lift, and that the induced angle ϕ is very small, the sectional aerodynamic loads can be approximated as

$$F_{y4k} \cong -L\phi - D \tag{7.47}$$

$$F_{z4k} \cong L \tag{7.48}$$

Substituting from Equations 7.33 and 7.34, the sectional aerodynamic loads can be written as

$$F_{y4k} = -\frac{1}{2}\rho ac\left(U_P U_T \theta_{con} - U_P^2 + \frac{U_T^2 C_{d0}}{a}\right) \tag{7.49}$$

$$F_{z4k} = \frac{1}{2}\rho ac\left(U_T^2 \theta_{con} - U_P U_T\right) \tag{7.50}$$

Note that the aerodynamic force along the x_{4k} axis is taken as 0. The aerodynamic moment about the hinge at the root can be obtained by using the two forces in the 4k system, which is written as

$$Q_A = \int_0^l x\hat{e}_{x4k} X (F_{y4k}\hat{e}_{y4k} + F_{z4k}\hat{e}_{z4k})\, dx \tag{7.51}$$

where $l = (R - e)$ is the length of the blade. Substituting the aerodynamic forces Equations 7.49 and 7.50, and using Equations 7.43 and 7.44, the aerodynamic moment about the hinge can be obtained. The components of the aerodynamic moment are given as

$$Q_{Ay4k} = -\int_0^l xF_{z4k}\, dx \tag{7.52}$$

$$Q_{Az4k} = \int_0^l xF_{y4k}\, dx \tag{7.53}$$

Using the transformation relationship given in Equation 7.8, the aerodynamic moments can be transformed to the components in the 3k system. Neglecting higher-order terms, it may be expressed as

$$Q_{Ay3k} = Q_{Ay4k} \tag{7.54}$$

$$Q_{Az3k} \cong Q_{Az4k} - \beta_k \zeta_k Q_{Ay4k} \tag{7.55}$$

The detailed expressions for these aerodynamic moments in flap and lead–lag motion are given in the following.

Aerodynamic Loads

The aerodynamic loads are obtained using a quasi-steady aerodynamic model. The blade is assumed to be a straight blade having a zero twist. A uniform inflow model based on the momentum theory is used. Stall, compressibility, and reverse flow regions have not been considered.

$$
\begin{aligned}
Q_{Ay3k} \cong{} & \frac{\rho a C\Omega^2 R^4}{2}\left\{-\theta_{\text{con}}\left\{\frac{\bar{l}^4}{4}+2\bar{e}\frac{\bar{l}^3}{3}+\bar{e}^2\frac{\bar{l}^2}{2}+2\left(\frac{\bar{l}^3}{3}+\bar{e}\frac{\bar{l}^2}{2}\right)\mu\sin\psi_k\right.\right.\\
&+\frac{\bar{l}^2}{2}(\mu\sin\psi_k)^2+2\dot{\zeta}_k\left(\frac{\bar{l}^4}{4}+\bar{e}\frac{\bar{l}^3}{3}+\frac{\bar{l}^3}{3}\mu\sin\psi_k\right)\\
&\left.+2\zeta_k\mu\cos\psi_k\left(\frac{\bar{l}^3}{3}+\bar{e}\frac{\bar{l}^2}{2}+\frac{\bar{l}^2}{2}\mu\sin\psi_k\right)\right\}\\
&+\dot{\beta}_k\left(\frac{\bar{l}^4}{4}+\bar{e}\frac{\bar{l}^3}{3}+\frac{\bar{l}^3}{3}\mu\sin\psi_k\right)+(\beta_p+\beta_k)\mu\cos\psi_k\left(\frac{\bar{l}^3}{3}+\bar{e}\frac{\bar{l}^2}{2}+\frac{\bar{l}^2}{2}\mu\sin\psi_k\right)\\
&\left.+\lambda\left(\frac{\bar{l}^3}{3}+\bar{e}\frac{\bar{l}^2}{2}+\frac{\bar{l}^2}{2}\mu\sin\psi_k\right)+\lambda\frac{\bar{l}^3}{3}\dot{\zeta}_k+\lambda\zeta_k\frac{\bar{l}^2}{2}\mu\cos\psi_k\right\}\\
&+\frac{\rho a C\,\Omega^2 R^4}{2}\left\{-\theta_{\text{con}}\left\{2\zeta_k\dot{\zeta}_k\frac{\bar{l}^3}{3}\mu\cos\psi_k+\zeta_k^2\frac{\bar{l}^2}{2}(\mu\cos\psi_k)^2\right\}\right.\\
&+\zeta_k(\beta_p+\beta_k)\left(\frac{\bar{l}^4}{4}+\bar{e}\frac{\bar{l}^3}{3}+\frac{\bar{l}^3}{3}\mu\sin\psi_k\right)+\dot{\zeta}_k(\beta_p+\beta_k)\frac{\bar{l}^3}{3}\mu\cos\psi_k\\
&\left.+(\beta_p+\beta_k)\zeta_k\frac{\bar{l}^2}{2}(\mu\cos\psi_k)^2+\zeta_k\dot{\beta}_k\frac{\bar{l}^3}{3}\mu\cos\psi_k\right\}
\end{aligned}
\tag{7.56}
$$

The moment expression can be written in two parts as

$$
Q_{Ay3k} = Q_{Ay3k}^{\text{L}} + Q_{Ay3k}^{\text{NL}} \tag{7.57}
$$

where the superscript "L" represents the linear term and NL represents the nonlinear term.

$$
\begin{aligned}
Q_{Az3k} ={} & \frac{\rho a C\Omega^2 R^4}{2}\left\{-\frac{C_{d0}}{\alpha}\left\{\frac{\bar{l}^4}{4}+2\bar{e}\frac{\bar{l}^3}{3}+\bar{e}^2\frac{\bar{l}^2}{2}\right.\right.\\
&+2\left(\frac{\bar{l}^3}{3}+\bar{e}\frac{\bar{l}^2}{2}\right)(\mu\sin\psi_k)+\frac{\bar{l}^2}{2}(\mu\sin\psi_k)^2\\
&+2\dot{\zeta}_k\left(\frac{\bar{l}^4}{4}+\bar{e}\frac{\bar{l}^3}{3}+\frac{\bar{l}^2}{2}\mu\sin\psi_k\right)
\end{aligned}
$$

$$+2\zeta_k \mu \cos \psi_k \left(\frac{\bar{l}^3}{3} + \bar{e}\frac{\bar{l}^2}{2} + \frac{\bar{l}^2}{2}\mu \sin \psi_k \right) \Bigg\}$$

$$+\dot{\beta}_k 2\lambda \frac{\bar{l}^3}{3} + \mu \cos \psi_k (\beta_p + \beta_k)\frac{\bar{l}^2}{2} 2\lambda + \frac{\bar{l}^2}{2}\lambda^2$$

$$-\theta_{con}\left\{ \dot{\beta}_k \left(\frac{\bar{l}^4}{4} + \bar{e}\frac{\bar{l}^3}{3} + \frac{\bar{l}^3}{3}\mu \sin \psi_k \right) + (\beta_p + \beta_k)\mu \cos \psi_k \left(\frac{\bar{l}^3}{3} + \bar{e}\frac{\bar{l}^2}{2} + \frac{\bar{l}^2}{2}\mu \sin \psi_k \right) \right.$$

$$\left. + \lambda \left(\frac{\bar{l}^3}{3} + \bar{e}\frac{\bar{l}^2}{2} + \frac{\bar{l}^2}{2}\mu \sin \psi_k \right) + \lambda \dot{\zeta}_k \frac{\bar{l}^3}{3} + \lambda \dot{\zeta}_k \frac{\bar{l}^2}{2}\mu \cos \psi_k \right\} \Bigg\}$$

$$+ \frac{\rho a C \ \Omega^2 R^4}{2}\left\{ -\frac{C_{d0}}{\alpha}\left\{ 2\dot{\zeta}_k\zeta_k \frac{\bar{l}^3}{3}\mu \cos \psi_k + \zeta_k^2 \frac{\bar{l}^2}{2}(\mu \cos \psi_k)^2 \right\} \right.$$

$$+ \dot{\beta}_k(\beta_p + \beta_k)\frac{\bar{l}^3}{3} 2\mu \cos \psi_k + \zeta_k(\beta_p + \beta_k)\frac{\bar{l}^3}{3} 2\lambda + (\beta_p + \beta_k)^2 \frac{\bar{l}^2}{2}(\mu \cos \psi_k)^2 \Bigg\}$$

$$- \theta \left\{ \zeta_k(\beta_p + \beta_k)\left(\frac{\bar{l}^4}{4} + \bar{e}\frac{\bar{l}^3}{3} + \frac{\bar{l}^2}{2}\mu \sin \psi_k \right) \right.$$

$$+ \dot{\zeta}_k(\beta_p + \beta_k)\frac{\bar{l}^3}{3}\mu \sin \psi_k + \zeta_k(\beta_p + \beta_k)\frac{\bar{l}^2}{2}(\mu \sin \psi_k)^2$$

$$\left. + \dot{\beta}_k \zeta_k \frac{\bar{l}^3}{3}\mu \sin \psi_k \right\} \Bigg\}$$

$$+ \frac{\rho a C \Omega^2 R^4}{2}\beta_k\zeta_k \left\{ -\dot{\beta}_k\left(\frac{\bar{l}^4}{4} + \bar{e}\frac{\bar{l}^3}{3} + \frac{\bar{l}^2}{2}\mu \sin \psi_k \right) \right.$$

$$+ (\beta_p + \beta_k)\mu \cos \psi_k \left(\frac{\bar{l}^3}{3} + \bar{e}\frac{\bar{l}^2}{2} + \frac{\bar{l}^2}{2}\mu \sin \psi_k \right)$$

$$+ \lambda \left(\frac{\bar{l}^3}{3} + \bar{e}\frac{\bar{l}^2}{2} + \frac{\bar{l}^2}{2}\mu \sin \psi_k \right)$$

$$+ \theta_{con}\left\{ \frac{\bar{l}^4}{4} + 2\bar{e}\frac{\bar{l}^3}{3} + \bar{e}^2\frac{\bar{l}^2}{2} + 2\left(\frac{\bar{l}^3}{3} + \bar{e}\frac{\bar{l}^2}{2} \right)\mu \sin \psi_k + \frac{\bar{l}^2}{2}(\mu \sin \psi_k)^2 \right.$$

$$+ 2\dot{\zeta}_k\left(\frac{\bar{l}^3}{3} + \bar{e}\frac{\bar{l}^2}{2} + \frac{\bar{l}^2}{2}\mu \sin \psi_k \right)$$

$$\left. + 2\zeta_k\mu \cos \psi_k \left(\frac{\bar{l}^3}{3} + \bar{e}\frac{\bar{l}^2}{2} + \frac{\bar{l}^2}{2}\mu \sin \psi_k \right) \right\} \Bigg\} \Bigg\}$$

$$(7.58)$$

The aerodynamic moment in the lag can be written in two parts as

$$Q_{z3k} = Q_{Az3k}^{L} + Q_{Az3k}^{NL} \qquad (7.59)$$

where the superscript "L" represents the linear term and NL represents the nonlinear term.

Root Spring Moment

The restoring moments due to root spring assembly can be written as (from Equation 6.45)

$$M_\beta = (K_\beta \cos^2 R\theta + K_\zeta \sin^2 R\theta)\beta_k - (K_\beta - K_\zeta) \sin R\theta \cos R\theta\, \zeta_k \tag{7.60}$$

and

$$M_\zeta = -(K_\beta \sin^2 R\theta + K_\zeta \cos^2 R\theta)\zeta_k + (K_\beta - K_\zeta) \sin R\theta \cos R\theta\, \beta_k \tag{7.61}$$

Subscript "k" refers to the kth blade in the rotor system.

Substituting the various load expressions in Equations 7.1 and 7.2, the coupled flap and lag equations of motion can be expressed in an expanded form. These equations are coupled nonlinear ordinary differential equations. The stability of the blade is analyzed using linearized perturbation analysis about a nonlinear equilibrium position. This analysis is usually denoted as "linearized aeroelastic stability analysis." The procedure for the linearized stability analysis is described in the following.

1. Assume that

$$\begin{aligned} \theta &= \theta_0 + \tilde{\theta}(t) \\ \beta_k &= \beta_0 + \tilde{\beta}(t) \\ \zeta_k &= \zeta_0 + \tilde{\zeta}(t) \\ \lambda &= \lambda_0 + \tilde{\lambda}(t) \end{aligned} \tag{7.62}$$

   where θ_0, β_0, ζ_0, and λ_0 represent the equilibrium quantities. In forward flight condition, because of periodic loading, the equilibrium quantities will also be time varying. However, in the specialized case of hover, the equilibrium quantities are constants. $\tilde{\theta}(t)$, $\bar{\beta}(t)$, $\bar{\zeta}(t)$, and $\bar{\lambda}(t)$ are the perturbational quantities about the equilibrium (trim) state of the blade.

2. Substituting the assumed form of the four parameters (Equation 7.62) in the flap–lag equations and collecting all the terms corresponding to the equilibrium state and the perturbational quantities and equating them separately to 0, we obtain two sets of equations. The first set of equations, containing only trim quantities (β_0, ζ_0, θ_0, λ_0), are called "trim or equilibrium state equations." In forming the other set of equations containing the perturbational quantities, the product

of perturbational quantities is neglected due to their small order. This second set of equations containing the perturbation quantities $\left(\bar{\beta}(t),\ \bar{\zeta}(t),\ \tilde{\theta}(t),\ \bar{\lambda}(t)\right)$ are called "linearized stability equations." For example, the approximation used for converting a nonlinear term into linearized perturbation terms is shown in the following.

$$\beta_k\zeta_k = (\beta_0 + \tilde{\beta}(t))(\zeta_0 + \tilde{\zeta}(t)) \cong \beta_0\zeta_0 + \beta_0\tilde{\zeta}(t) + \zeta_0\tilde{\beta}(t) \tag{7.63}$$

It may be noted that the product of perturbation quantities are neglected while forming the linearized perturbation equations.

3. In forward flight, the equilibrium (or trim state) equations are nonlinear differential equations with time-varying coefficients. These equations are solved by time integration or by harmonic balance approach to obtain the steady-state response (equilibrium response) of the blade. In the case of hover, the equilibrium state equations are nonlinear algebraic equations that can be solved by the Newton–Raphson method.
4. The linearized perturbation equations are linear differential equations. For forward flight, they contain time-varying periodic coefficients. The stability analysis in forward flight has to be performed by using the Floquet–Lyapunov theory or the approximate method of multiblade coordinate transformation. In the case of hover, the equations are linear differential equations with constant coefficients. Following the standard procedure of eigenvalue analysis, the stability of the system in hover can be analyzed. The eigenvalues appear as $(s_j = \sigma_j \pm i\omega_j)$ complex quantities, where σ_j represents the damping and ω_j represents the frequency of the jth aeroelastic mode. If σ_j is negative, the mode is stable, and if σ_j is positive, the mode is unstable.

The whole exercise of aeroelastic stability analysis is carried out to identify whether a mode is stable or not and how the stability of the blade is affected by various system parameters and operating conditions.

Example

Let us consider the simplest case of hovering flight. Assume that $\mu = 0$, $\beta_p = 0$ and include pitch–flap and pitch–lag couplings. Let us represent the perturbation as

$$\begin{aligned} \beta_k &= \beta_0 + \tilde{\beta}(t) \\ \zeta_k &= \zeta_0 + \tilde{\zeta}(t) \\ \theta &= \theta_0 + \tilde{\theta}(t) \\ \lambda &= \lambda_0 \text{ (a constant)} \end{aligned} \tag{7.64}$$

The assumption of constant inflow simplifies the problem. On the other hand, if we include the time variation in inflow $\left(\tilde{\lambda}(t)\right)$, one has to formulate an additional equation for $\tilde{\lambda}$ using the perturbation momentum theory, which is an extension of the momentum theory. This extension of formulating an additional equation for $\lambda(t)$ is the fundamental basis for the dynamic inflow model, which is not covered in this basic book.

The inclusion of lag–pitch coupling (δ_1) and pitch–flap coupling (δ_3) modifies the pitch angle as

$$\theta = (\theta_{con} - K_{P\beta}\beta_0 + K_{P\zeta}\zeta_0) + (-K_{P\beta}\tilde{\beta} + K_{P\zeta}\tilde{\zeta}) \tag{7.65}$$

It may be noted from Equation 7.63 that

$$\theta_0 = (\theta_{con} - K_{P\beta}\beta_0 + K_{P\zeta}\zeta_0) \text{ and } \tilde{\theta}(t) = (-K_{P\beta}\tilde{\beta} + K_{P\zeta}\tilde{\zeta})$$

where θ_{con} is the control pitch input given at the blade root.

Substituting these expressions (Equations 7.64 and 7.65) in the flap–lag equations and separating the equilibrium and perturbation equations yield the following (for the sake of simplicity, the nonlinear terms in aerodynamic moments are neglected in formulating the equations given in the following).

Equilibrium Equations

The equilibrium state equations in the flap and lag modes are given as follows:

Flap mode:

$$\begin{aligned}
&(K_\beta \cos^2 R\theta_{con} + K_\zeta \sin^2 R\theta_{con})\beta_0 - (K_\beta - K_\zeta)\sin R\theta_{con} \cos R\theta_{con}\zeta_0 \\
&+ I_b\Omega^2\beta_0 + eMX_{c.g.}\Omega^2\beta_0 \\
&+ \frac{\rho a C R^4 \Omega^2}{2}\left\{\begin{aligned}
&-\theta_{con}\left\{\frac{\bar{l}^4}{4} + 2\bar{e}\frac{\bar{l}^3}{3} + \bar{e}^2\frac{\bar{l}^2}{2}\right\} + \lambda_0\left\{\frac{\bar{l}^3}{3} + \bar{e}\frac{\bar{l}^2}{2}\right\} \\
&+\left\{\frac{\bar{l}^4}{4} + 2\bar{e}\frac{\bar{l}^3}{3} + \bar{e}^2\frac{\bar{l}^2}{2}\right\}K_{P\beta}\beta_0 \\
&-\left\{\frac{\bar{l}^4}{4} + 2\bar{e}\frac{\bar{l}^3}{3} + \bar{e}^2\frac{\bar{l}^2}{2}\right\}K_{P\zeta}\zeta_0
\end{aligned}\right\} = 0
\end{aligned} \tag{7.66}$$

Lag mode:

$$
\begin{aligned}
&-(K_\beta \sin^2 R\theta_{\text{con}} + K_\zeta \cos^2 R\theta_{\text{con}})\zeta_0 + (K_\beta - K_\zeta)\sin R\theta_{\text{con}} \cos R\theta_{\text{con}}\beta_0 \\
&- eMX_{\text{c.g.}}\Omega^2\zeta_0 \\
&+ \frac{\rho a C R^4 \Omega^2}{2}\left\{
\begin{array}{c}
-\frac{C_{d0}}{a}\left\{\frac{\bar{l}^4}{4} + 2\bar{e}\frac{\bar{l}^3}{3} + \bar{e}^2\frac{\bar{l}^2}{2}\right\} + \lambda_0^2\frac{\bar{l}^2}{2} \\
-\theta_{\text{con}}\lambda_0\left\{\frac{\bar{l}^3}{3} + \bar{e}\frac{\bar{l}^2}{2}\right\} \\
+\left\{\frac{\bar{l}^3}{3} + \bar{e}\frac{\bar{l}^2}{2}\right\}K_{P\beta}\beta_0\lambda_0 \\
-\left\{\frac{\bar{l}^3}{3} + \bar{e}\frac{\bar{l}^2}{2}\right\}K_{P\zeta}\zeta_0\lambda_0
\end{array}
\right\} = 0
\end{aligned}
\tag{7.67}
$$

Dividing by $I_b\Omega^2$ and collecting terms, the equilibrium equations in nondimensional form can be written as follows:

Equilibrium equation in the flap:

$$
\begin{aligned}
&\left(\bar{\omega}_{\text{NRF}}^2 \cos^2 R\theta_{\text{con}} + \bar{\omega}_{\text{NRL}}^2 \sin^2 R\theta_{\text{con}}\right)\beta_0 \\
&-\left(\bar{\omega}_{\text{NRF}}^2 - \bar{\omega}_{\text{NRL}}^2\right)\sin R\theta_{\text{con}} \cos R\theta_{\text{con}}\zeta_0 \\
&+\beta_0 + \frac{eMX_{\text{c.g.}}}{I_b}\beta_0 + \frac{\gamma}{2}\left\{\left\{\frac{\bar{l}^4}{4} + 2\bar{e}\frac{\bar{l}^3}{3} + \bar{e}^2\frac{\bar{l}^2}{2}\right\}K_{P\beta}\beta_0\right. \\
&\left.-\left\{\frac{\bar{l}^4}{4} + 2\bar{e}\frac{\bar{l}^3}{3} + \bar{e}^2\frac{\bar{l}^2}{2}\right\}K_{P\zeta}\zeta_0\right\} \\
&= \frac{\gamma}{2}\left\{\theta_{\text{con}}\left\{\frac{\bar{l}^4}{4} + 2\bar{e}\frac{\bar{l}^3}{3} + \bar{e}^2\frac{\bar{l}^2}{2}\right\} - \lambda_0\left\{\frac{\bar{l}^3}{3} + \bar{e}\frac{\bar{l}^2}{2}\right\}\right\}
\end{aligned}
\tag{7.68}
$$

Equilibrium equation in the lag:

$$
\begin{aligned}
&-\left(\bar{\omega}_{\text{NRF}}^2 \sin^2 R\theta_{\text{con}} + \bar{\omega}_{\text{NRL}}^2 \cos^2 R\theta_{\text{con}}\right)\zeta_0 \\
&+\left(\bar{\omega}_{\text{NRF}}^2 - \bar{\omega}_{\text{NRL}}^2\right)\sin R\theta_{\text{con}} \cos R\theta_{\text{con}}\beta_0 \\
&-\frac{eMX_{\text{c.g.}}}{I_b}\zeta_0 + \frac{\gamma}{2}\left\{\left\{\frac{\bar{l}^3}{3} + \bar{e}\frac{\bar{l}^2}{2}\right\}K_{P\beta}\beta_0\lambda_0 - \left\{\frac{\bar{l}^3}{3} + \bar{e}\frac{\bar{l}^2}{2}\right\}K_{P\zeta}\zeta_0\lambda_0\right\} \\
&= \frac{\gamma}{2}\left\{\frac{C_{d0}}{a}\left\{\frac{\bar{l}^4}{4} + 2\bar{e}\frac{\bar{l}^3}{3} + \bar{e}^2\frac{\bar{l}^2}{2}\right\} - \lambda_0^2\frac{\bar{l}^2}{2} + \theta_{\text{con}}\lambda_0\left\{\frac{\bar{l}^3}{3} + \bar{e}\frac{\bar{l}^2}{2}\right\}\right\}
\end{aligned}
\tag{7.69}
$$

where

$$\bar{\omega}_{NRF}^2 = \frac{K_\beta}{I_b\Omega^2},\ \bar{\omega}_{NRL}^2 = \frac{K_\zeta}{I_b\Omega^2} \text{ and } \gamma = \frac{\rho a C R^4}{I_b}$$

Linearized Stability Equations (Perturbation Equations)

The linearized perturbation equations can be written as follows:

Perturbation equation in the flap mode:

$$\begin{aligned}&(K_\beta \cos^2 R\theta_{con} + K_\zeta \sin^2 R\theta_{con})\tilde{\beta} - (K_\beta - K_\zeta)\sin R\theta_{con} \cos R\theta_{con}\tilde{\zeta}\\&+ I_b\Omega^2\ddot{\tilde{\beta}} + 2I_b\Omega^2\beta_0\dot{\tilde{\zeta}} + (I_b\Omega^2 + eMX_{c.g.}\Omega^2)\tilde{\beta}\\&+\frac{\rho a C R^4\Omega^2}{2}\left\{\begin{array}{c}-\tilde{\theta}\left\{\dfrac{\bar{l}^4}{4} + 2\bar{e}\dfrac{\bar{l}^3}{3} + \bar{e}^2\dfrac{\bar{l}^2}{2}\right\} - \theta_0 2\dot{\tilde{\zeta}}\left\{\dfrac{\bar{l}^4}{4} + \bar{e}\dfrac{\bar{l}^3}{3}\right\}\\ +\dot{\tilde{\beta}}\left\{\dfrac{\bar{l}^4}{4} + \bar{e}\dfrac{\bar{l}^3}{3}\right\} + \lambda_0\dfrac{\bar{l}^3}{3}\dot{\tilde{\zeta}}\end{array}\right\} = 0\end{aligned} \tag{7.70}$$

Perturbation equation in the lag mode:

$$\begin{aligned}&-(K_\beta \sin^2 R\theta_{con} + K_\zeta \cos^2 R\theta_{con})\tilde{\zeta} + (K_\beta - K_\zeta)\sin R\theta_{con} \cos R\theta_{con}\tilde{\beta}\\&- I_b\Omega^2\ddot{\tilde{\zeta}} + 2I_b\Omega^2\beta_0\dot{\tilde{\beta}} - eMX_{c.g.}\Omega^2\tilde{\zeta}\\&+\frac{\rho a C R^4\Omega^2}{2}\left\{\begin{array}{c}-\dfrac{C_{d0}}{a}\left(2\dot{\tilde{\zeta}}\left[\dfrac{\bar{l}^4}{4} + \bar{e}\dfrac{\bar{l}^3}{3}\right]\right) + \dot{\tilde{\beta}}2\lambda_0\dfrac{\bar{l}^3}{3}\\ -\theta_0\left\{\dot{\tilde{\beta}}\left\{\dfrac{\bar{l}^4}{4} + \bar{e}\dfrac{\bar{l}^3}{3}\right\} + \lambda_0\dfrac{\bar{l}^3}{3}\dot{\tilde{\zeta}}\right\}\\ -\tilde{\theta}\left\{\lambda_0\dfrac{\bar{l}^3}{3} + \bar{e}\dfrac{\bar{l}^2}{2}\right\}\end{array}\right\} = 0\end{aligned} \tag{7.71}$$

It should be noted that, in Equations 7.70 and 7.71, $\theta_0 = \theta_{con} - K_{P\beta}\beta_0 + K_{P\zeta}\zeta_0$ and $\tilde{\theta} = -K_{P\beta}\tilde{\beta} + K_{P\zeta}\tilde{\zeta}$.

The inflow λ_0 can be obtained from either of the two following equations:

The global momentum theory provides the constant inflow as

$$\lambda_0 = \sqrt{\frac{C_T}{2}} \tag{7.72}$$

The local momentum theory relates inflow at 75% of the radius to the blade pitch angle:

$$\lambda_0 = \frac{\sigma a}{16}\left[\sqrt{1+\frac{32}{\sigma\alpha}\theta_0\left(\frac{3}{4}\right)}-1\right] \tag{7.73}$$

where $\theta_0 = \theta_{con} - K_{P\beta}\beta_0 + K_{P\zeta}\zeta_0$.

The set of equilibrium, perturbation, and inflow equations can be solved for two different sets of conditions. They are as follows: (1) given the thrust coefficient C_T, evaluate the equilibrium and the stability of the system; and (2) given the pitch angle θ_{con}, evaluate the equilibrium and the stability of the blade.

Problem 1

Given the thrust coefficient C_T, solve for the equilibrium and stability conditions. In this approach, knowing C_T, a preliminary estimate of λ_0 can be obtained as $\lambda_0 = \sqrt{\frac{C_T}{2}}$.

Using the relationship,

$$C_T = \frac{\sigma a}{2}\left[\frac{\theta_0}{3} - \frac{\lambda_0}{2}\right] \tag{7.74}$$

where $\theta_0 = \theta_{con} - K_{P\beta}\beta_0 + K_{P\zeta}\zeta_0$.

Assuming β_0 and ζ_0 to be 0, an initial estimate of the pitch angle θ_{con} can be obtained from Equation 7.74. Solving the four equations (two equilibrium equations [Equations 7.68 and 7.69], the differential inflow equation [Equation 7.73], and the thrust coefficient equation [Equation 7.74]) iteratively, the equilibrium quantities θ_{con}, β_0, ζ_0, and λ_0 can be obtained.

Problem 2

Given the pitch angle θ_{con}, evaluate the inflow from the differential inflow equation (Equation 7.73) (by assuming that, in the first iteration, β_0, and ζ_0 is 0). Using the three equations (two equilibrium equations [Equations 7.68 and 7.69] and the differential inflow equation [Equation 7.73]), solve iteratively for β_0, ζ_0, and λ_0. The resulting thrust can be obtained from the thrust equation (Equation 7.74).

Perturbational Stability Equations

The linearized perturbation equations (Equations 7.70 and 7.71) can be written in matrix form as (after dividing by $I_b\Omega^2$)

$$[M]\begin{Bmatrix}\ddot{\tilde{\beta}}\\ \ddot{\tilde{\zeta}}\end{Bmatrix}+[C]\begin{Bmatrix}\dot{\tilde{\beta}}\\ \dot{\tilde{\zeta}}\end{Bmatrix}+[K]\begin{Bmatrix}\tilde{\beta}\\ \tilde{\zeta}\end{Bmatrix}=0 \tag{7.75}$$

where

$$[M]\begin{bmatrix}1 & 0\\ 0 & -1\end{bmatrix}$$

$$[C]\begin{bmatrix}C_{11} & C_{12}\\ C_{21} & C_{22}\end{bmatrix}$$

$$[K]\begin{bmatrix}K_{11} & K_{12}\\ K_{21} & K_{22}\end{bmatrix}.$$

The elements of the damping and stiffness matrices are given as

$$C_{11}=\frac{\gamma}{2}\left\{\frac{\bar{l}^4}{4}+\bar{e}\frac{\bar{l}^3}{3}\right\}$$

$$C_{12}=2\beta_0-\left\{\frac{\bar{l}^4}{4}+\bar{e}\frac{\bar{l}^3}{3}\right\}\gamma(\theta_{\text{con}}-K_{P\beta}\beta_0+K_{P\zeta}\zeta_0)+\gamma\frac{\bar{l}^3}{6}\lambda_0$$

$$C_{21}=2\beta_0-\frac{\gamma}{2}\left\{\frac{\bar{l}^4}{4}+\bar{e}\frac{\bar{l}^3}{3}\right\}(\theta_{\text{con}}-K_{P\beta}\beta_0+K_{P\zeta}\zeta_0)+\gamma\frac{\bar{l}^3}{3}\lambda_0$$

$$C_{22}=-\gamma\frac{C_{d0}}{a}\left\{\frac{\bar{l}^4}{4}+\bar{e}\frac{\bar{l}^3}{3}\right\}-\gamma\frac{\bar{l}^3}{6}(\theta_{con}-K_{P\beta}\beta_0+K_{P\zeta}\zeta_0)\lambda_0$$

$$K_{11}=\left(\bar{\omega}_{\text{NRF}}^2\cos^2 R\theta_{\text{con}}+\bar{\omega}_{\text{NRL}}^2\sin^2 R\theta_{\text{con}}+1+\frac{eMX_{\text{c.g.}}}{I_b}\right)+\frac{\gamma}{2}\left\{\frac{\bar{l}^4}{4}+2\bar{e}\frac{\bar{l}^3}{3}+\bar{e}^2\frac{\bar{l}^2}{2}\right\}K_{P\beta}$$

$$K_{12}=-\left(\bar{\omega}_{\text{NRF}}^2-\bar{\omega}_{\text{NRL}}^2\right)\sin R\theta_{\text{con}}\cos R\theta_{\text{con}}-\frac{\gamma}{2}\left\{\frac{\bar{l}^4}{4}+2\bar{e}\frac{\bar{l}^3}{3}+\bar{e}^2\frac{\bar{l}^2}{2}\right\}K_{P\zeta}$$

$$K_{21}=\left(\bar{\omega}_{\text{NRF}}^2-\bar{\omega}_{\text{NRL}}^2\right)\sin R\theta_{\text{con}}\cos R\theta_{\text{con}}+\frac{\gamma}{2}\left\{\frac{\bar{l}^3}{3}+\bar{e}\frac{\bar{l}^2}{2}\right\}\lambda_0 K_{P\beta}$$

$$K_{22}=-\left(\bar{\omega}_{\text{NRF}}^2\sin^2 R\theta_{\text{con}}+\bar{\omega}_{\text{NRL}}^2\cos^2 R\theta_{\text{con}}+\frac{eMX_{\text{c.g.}}}{I_b}\right)-\frac{\gamma}{2}\left\{\frac{\bar{l}^3}{3}+\bar{e}\frac{\bar{l}^2}{2}\right\}\lambda_0 K_{P\zeta}$$

It is evident from these equations that the damping matrix is a function of equilibrium position. Therefore, depending on the operating condition and system parameters, the elements of the damping matrix will vary. For certain combinations of operating condition and system parameters, the coupled flap–lag dynamic system can become unstable. The stability analysis is carried out as follows:

Assume a solution of the form

$$\begin{Bmatrix} \tilde{\beta} \\ \tilde{\zeta} \end{Bmatrix} = \begin{Bmatrix} \bar{\beta} \\ \bar{\zeta} \end{Bmatrix} e^{st} \tag{7.76}$$

Substituting Equation 7.76 in the stability equation (Equation 7.75) yields

$$\begin{bmatrix} M_{11}s^2 + C_{11}s + K_{11} & C_{12}s + K_{12} \\ C_{21}s + K_{21} & M_{22}s^2 + C_{22}s + K_{22} \end{bmatrix} \begin{Bmatrix} \bar{\beta} \\ \bar{\zeta} \end{Bmatrix} = 0 \tag{7.77}$$

The characteristics polynomial can be written as

$$(M_{11}s^2 + C_{11}s + K_{11})(M_{22}s^2 + C_{22}s + K_{22}) - (C_{12}s + K_{12})(C_{21}s + K_{21}) = 0 \tag{7.78}$$

Solving for all the roots, one can obtain the information about the stability of the system. The roots appear as complex conjugate pairs as $s_j = \sigma_j \pm i\omega_j$, where σ_j represents the modal damping and ω_j represents the modal frequency. If σ_j is positive, the mode is unstable, and if σ_j is negative, the mode is stable.

A parametric study can be performed to obtain the effect of each parameter on the stability of the rotor blade. In the open literature, several studies have been performed to analyze the flap–lag stability of the rotor blade in hover. Sample results are shown in Figure 7.8, which have been evaluated using the data given in Table 7.1.

The effect of the rotating natural frequencies of the blade in the flap and lag modes on the flap–lag stability is shown in Figure 7.8. The nondimensional flap frequency is varied from 1.05 to 1.5, whereas the nondimensional lag frequency is varied from 0.8 to 1.8. Five different values of the blade pitch angle are considered from 0.2 to 0.6 rad in steps of 0.1 rad. The nondimensional hinge offset of the blade is taken as 0.1. It may be noted that, for a given pitch angle, the region inside loop is an unstable zone and the region outside loop is a stable zone. The unstable region increases with an increase in blade pitch angle. An important point to note is that, unless the nondimensional lag frequency is greater than 0.95/rev, flap–lag instability does not occur. The region of instability depends on the pitch angle of the blade. The lowest value of pitch angle at which the blade can go unstable is around 0.197 rad. The results shown in Figure 7.8 indicate that flap–lag instability is more likely to be a problem of stiff-in plane rotors. Usually, main rotor blades are

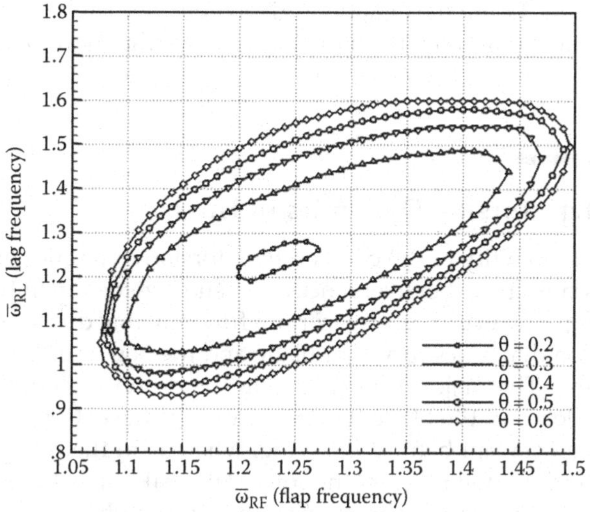

FIGURE 7.8
Stability boundary of flap–lag dynamics with coupling parameter $R = 0$.

TABLE 7.1
Data for the Flap–Lag Stability Analysis of the Main Rotor Blade

Variable	Quantity
Number of blades, N	4
Air density at sea level, ρ (kg/m^3)	1.224
Coefficient of drag, c_{d0}	0.008
Weight of the helicopter, W (N)	45,000
Radius of the main rotor blade, R (m)	6.6
Chord of the main rotor blade, c (m)	0.5
Main rotor rotating speed, Ω (rpm)	300
Mass of the main rotor blade, m_0 (kg/m)	11.24
Lift curve slope, a	2π
Precone, β_p (°)	0
Hinge offset, e (m)	0.66
Flap–pitch coupling, $K_{p\beta}$	0
Lag–pitch coupling, $K_{p\zeta}$	0
Structural flap–lag coupling parameter, R	0

soft-in-plane rotors (i.e., $\bar{\omega}_{RL} < 1.0$); hence, flap-lag instability is unlikely to occur. On the other hand, for stiff-in-plane tail rotors, flap–lag instability is a possibility at high thrust conditions.

It is shown in the literature that structural flap–lag coupling, pitch–flap, and pitch–lag couplings have a significant influence on the flap–lag stability of the rotor blade. Using the equations given here, one can study the effect of various individual parameters on the flap–lag stability of rotor blade.

Coupled Flap–Torsion Dynamics in Hover

The aeroelastic stability of the coupled flap–torsion dynamics of an isolated rotor blade is similar to the divergence and bending–torsion flutter problems of a fixed wing. However, there are certain fundamental differences between the flap–torsion dynamics of a rotor blade and the bending–torsion problem of a fixed wing. The major difference is in the description of the unsteady aerodynamic load on the blade. In the fixed wing case, the unsteady wake from the wing is swept behind the wing, whereas in the rotary wing case, the wake is pushed down below the rotor. The wake structures are entirely different for the two cases. The prediction of unsteady aerodynamic loads on a rotor blade is very difficult, and it is still a topic of research. Therefore, for the purpose of understanding, a simple aerodynamic model, based on the fixed wing theory, will be used; however, certain modifications will be incorporated to make the model suitable for rotary wing aeroelastic analysis. Such a model will provide a good understanding of the fundamental aspects of the coupled flap–torsion dynamics. Another difference between the rotary wing and the fixed wing aeroelastic problem is the effect of rotation, which introduces a coupling between flap and torsion, when the blade sectional center of mass is offset from the feathering (pitching) axis.

Figure 7.9 shows a uniform blade with two springs at the root. The spring K_β represents the flap spring, and the linear spring K_c represents the combined

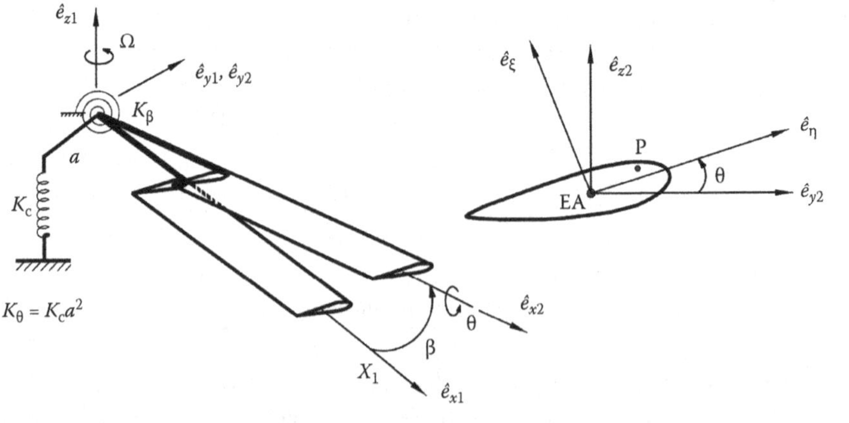

FIGURE 7.9
Idealization of the rotor blade for coupled flap–torsion dynamic analysis and the coordinate systems.

effect of control system stiffness and blade torsional stiffness. This spring is attached at the root with an offset a. The effective torsional stiffness of the blade is given as $K_\theta = K_c a^2$.

The rotor blade undergoes flap deformation β, followed by a torsional deformation θ, as shown in Figure 7.9.

$X_1 - Y_1 - Z_1$ represents the hub-fixed rotating coordinate system, which is also referred to as the "undeformed blade coordinate system," with unit vectors $\hat{e}_{x1}, \hat{e}_{y1}, \hat{e}_{z1}$.

$X_2 - Y_2 - Z_2$ represents the deformed blade coordinate system, with unit vectors $\hat{e}_{x2}, \hat{e}_{y2}, \hat{e}_{z2}$.

$\eta - \xi$ represents the blade cross-sectional coordinate system, whose origin is at the reference elastic axis (EA) at a distance r from the hub center. The unit vectors along the cross-sectional coordinate system are $\hat{e}_\eta, \hat{e}_\xi$.

The transformation relationship between the X_1, Y_1, Z_1 and the X_2, Y_2, Z_2 axes system is given by

$$\begin{Bmatrix} \hat{e}_{x1} \\ \hat{e}_{y1} \\ \hat{e}_{z1} \end{Bmatrix} = \begin{bmatrix} \cos\beta & 0 & -\sin\beta \\ 0 & 1 & 0 \\ \sin\beta & 0 & \cos\beta \end{bmatrix} \begin{Bmatrix} \hat{e}_{x2} \\ \hat{e}_{y2} \\ \hat{e}_{z2} \end{Bmatrix} \tag{7.79}$$

The transformation between the X_2, η, ξ and the X_2, Y_2, Z_2 coordinate systems can be given as

$$\begin{Bmatrix} \hat{e}_{x2} \\ \hat{e}_{y2} \\ \hat{e}_{z2} \end{Bmatrix} = \begin{bmatrix} 1 & 0 & 0 \\ 0 & \cos\theta & -\sin\theta \\ 0 & \sin\theta & \cos\theta \end{bmatrix} \begin{Bmatrix} \hat{e}_{x2} \\ \hat{e}_{\eta} \\ \hat{e}_{\xi} \end{Bmatrix} \tag{7.80}$$

The position vector of any arbitrary point P in the cross section can be written as

$$\vec{r}_P = r\hat{e}_{x2} + \eta\hat{e}_\eta + \xi\hat{e}_\xi \tag{7.81}$$

Using these transformation relationships given in Equations 7.78 and 7.79, the position vector can be written as

$$\begin{aligned} \vec{r}_P = {} & r\cos\beta\hat{e}_{x1} + r\sin\beta\hat{e}_{z1} + (\eta\cos\theta - \xi\sin\theta)\hat{e}_{y1} - \sin\beta(\eta\sin\theta + \xi\cos\theta)\hat{e}_{x1} \\ & + \cos\beta(\eta\sin\theta + \xi\cos\theta)\hat{e}_{z1} \end{aligned} \tag{7.82}$$

Rewriting the position vector using a small angle assumption for flap deflection β, the position vector can be written as

$$\vec{r}_P = [r - \beta(\eta \sin\theta + \xi \cos\theta)]\hat{e}_{x1} + (\eta \cos\theta - \xi \sin\theta)\hat{e}_{y1} + (r\beta + \eta \sin\theta + \xi \cos\theta)\hat{e}_{z1} \tag{7.83}$$

Since the cross-sectional dimension of the blade is very small compared to the length, one can neglect the term $\beta(\eta \sin\theta + \xi \cos\theta)$ in comparison to r. Neglecting this term, the position vector can be simplified as

$$\vec{r}_P \cong r\hat{e}_{x1} + (\eta \cos\theta - \xi \sin\theta)\hat{e}_{y1} + (r\beta + \eta \sin\theta + \xi \cos\theta)\hat{e}_{z1} \tag{7.84}$$

The absolute velocity of point P is

$$\vec{V}_P = \left(\dot{\vec{r}}_P\right)_{rel} + \vec{\omega} X \vec{r}_P \tag{7.85}$$

where $\vec{\omega} = \Omega\hat{e}_{z1}$ is the angular velocity of the rotor system. Expanding Equation 7.84, the absolute velocity of point P can be written as

$$\vec{V}_P = \Omega\Big[-(\eta \cos\theta - \xi \sin\theta)\hat{e}_{x1} + \left\{r - (\eta \sin\theta + \xi \cos\theta)\dot{\theta}\right\}\hat{e}_{y1} + \left\{r\dot{\beta} + \left(\eta \cos\theta - \xi \sin\theta\right)\dot{\theta}\right\}\hat{e}_{z1}\Big] \tag{7.86}$$

Note that the time derivatives are taken w.r.t. nondimensional time $\psi = \Omega t$. The velocity at the reference EA at location x is given as

$$\vec{V}_{EA} = \Omega[r\hat{e}_{y1} + r\dot{\beta}\hat{e}_{z1}] \tag{7.87}$$

The absolute acceleration of point P is

$$\vec{a}_P = \ddot{\vec{r}}_P + \dot{\vec{\omega}} X \vec{r}_P + 2\vec{\omega} X \vec{r}_P + \vec{\omega} X (\vec{\omega} X \vec{r}_P) \tag{7.88}$$

Substituting various terms and noting that Ω is a constant, the acceleration at point P in expanded form can be written as

$$\begin{aligned}\vec{a}_P = \Omega^2\Big[&\left\{-r + 2\dot{\theta}(\eta \sin\theta + \xi \cos\theta)\right\}\hat{e}_{x1}\\ &+ \left\{-(\eta \cos\theta - \xi \sin\theta) - (\eta \sin\theta + \xi \cos\theta)\ddot{\theta} - (\eta \cos\theta - \xi \sin\theta)\dot{\theta}^2\right\}\hat{e}_{y1}\\ &+ \left\{r\ddot{\beta} + (\eta \cos\theta - \xi \sin\theta)\ddot{\theta} - (\eta \sin\theta + \xi \cos\theta)\dot{\theta}^2\right\}\hat{e}_{z1}\Big]\end{aligned} \tag{7.89}$$

The acceleration can be approximated by neglecting higher-order terms, such as $\dot{\theta}^2$ and $2\dot{\theta}(\eta\sin\theta+\xi\cos\theta)$ in comparison to other terms. The simplified expression can be written as

$$\vec{a}_{\rm P} = \Omega^2\Big[-r\hat{e}_{x1} + \big\{-(\eta\cos\theta-\xi\sin\theta)-(\eta\sin\theta+\xi\cos\theta)\ddot{\theta}\big\}\hat{e}_{y1} + \big\{r\ddot{\beta}+(\eta\cos\theta-\xi\sin\theta)\ddot{\theta}\big\}\hat{e}_{z1}\Big] \tag{7.90}$$

The cross-sectional inertia force and inertia moment about the reference EA can be obtained by evaluating the following integrals over the cross section.

$$\vec{p}_{\rm I} = \iint_A -\rho_b\vec{a}_{\rm P}\,dA \tag{7.91}$$

$$\tilde{q}_{\rm I} = \iint_A [\eta\hat{e}_\eta + \xi\hat{e}_\xi]X(-\rho_b\vec{a}_{\rm P})\,dA \tag{7.92}$$

where ρ_b is the mass per unit volume of the blade.

Equation 7.92 can be written as

$$\tilde{q}_{\rm I} = \iint_A \Big[(\eta\cos\theta-\xi\sin\theta)\hat{e}_{y1}+(\eta\sin\theta+\xi\cos\theta)\hat{e}_{z1}\Big]\,X(-\rho_b\vec{a}_{\rm P})\,dA \tag{7.93}$$

Defining the cross-sectional integrals as follows:

Mass per unit length of the blade:

$$m = \iint \rho_b\,dA$$

Static mass moment per unit length:

$$m\eta_m = \iint \rho_b\eta\,dA$$
$$m\xi_m = \iint \rho_b\xi\,dA$$

Mass moment of inertia per unit length:

$$\begin{aligned} I_{\xi\xi} &= \iint \rho_b\eta^2\,dA \\ I_{\eta\eta} &= \iint \rho_b\xi^2\,dA \\ I_{\eta\xi} &= \iint \rho_b\eta\xi\,dA \end{aligned} \tag{7.94}$$

Using the integrals given in Equation 7.94, the distributed inertia force per unit length of the blade can be written as

$$p_{Ix1} = m\Omega^2 r \tag{7.95}$$

$$p_{Iy1} = \Omega^2 \left\{(m\eta_m \cos\theta - m\xi_m \sin\theta) + (m\eta_m \sin\theta + m\xi_m \cos\theta)\ddot{\theta}\right\} \tag{7.96}$$

$$p_{Iz1} = \Omega^2 \left\{(-mr\ddot{\beta} - (m\eta_m \cos\theta - m\xi_m \sin\theta)\ddot{\theta}\right\} \tag{7.97}$$

The distributed inertia moment per unit length can be written as

$$\begin{aligned} q_{Ix1} = \Omega^2 \iint -\rho_b \Big[&(\eta\cos\theta - \xi\sin\theta)\left\{r\ddot{\beta} + (\eta\cos\theta - \xi\sin\theta)\ddot{\theta}\right\} \\ &- (\eta\sin\theta + \xi\cos\theta)\left\{-(\eta\cos\theta - \xi\sin\theta) - (\eta\sin\theta + \xi\cos\theta)\ddot{\theta}\right\}\Big] dA \end{aligned} \tag{7.98}$$

$$q_{Iy1} = \Omega^2 \iint -\rho_b \left[(\eta\sin\theta + \xi\cos\theta)\{-r\}\right] dA \tag{7.99}$$

$$q_{Iz1} = \Omega^2 \iint -\rho_b \left[-(\eta\cos\theta - \xi\sin\theta)\{-r\}\right] dA \tag{7.100}$$

Integrating over the cross section and simplifying, the distributed inertia moment per unit span can be written as

$$\begin{aligned} q_{Ix1} = -\Omega^2 \Big[&(m\eta_m \cos\theta - m\xi_m \sin\theta) r\ddot{\beta} + (I_{\xi\xi} + I_{\eta\eta})\ddot{\theta} \\ &+ (I_{\xi\xi} - I_{\eta\eta})\sin\theta\cos\theta + I_{\eta\xi}(\cos^2\theta - \sin^2\theta)\Big] \end{aligned} \tag{7.101}$$

$$q_{Iy1} = -\Omega^2 \left[-(m\eta_m \sin\theta + m\xi_m \cos\theta) r\right] \tag{7.102}$$

$$q_{Iz1} = -\Omega^2 \left[+(m\eta_m \cos\theta - m\xi_m \sin\theta) r\right] \tag{7.103}$$

The important sectional inertia forces and moments are shown in Figure 7.10.

The inertia moment about the root of the rotor blade, assuming a zero hinge offset, can be obtained from the expression

$$\vec{\bar{Q}}_I = \int_0^R \left\{\vec{q}_I + (r\hat{e}_{x1} + r\beta\hat{e}_{z1}) X \vec{p}_I\right\} dr \tag{7.104}$$

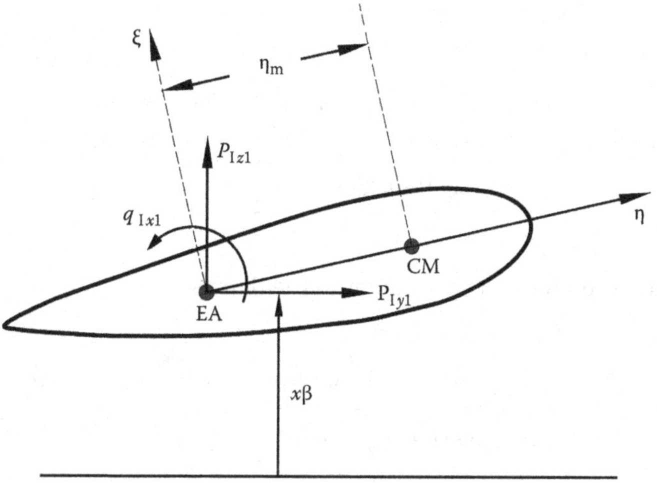

FIGURE 7.10
Sectional inertia loads.

Identifying the relevant quantities from the above integral, the torsional and flap moments at the blade root can be written as follows:

Torsional inertia moment at the root:

$$Q_{Ix1} = \int_0^R \left\{ q_{Ix1} + \{-r\beta p_{Iy1}\} \right\} dr \tag{7.105}$$

Assuming uniform properties along the span of the blade, Equation 7.105 can be written in expanded form as

$$\begin{aligned} Q_{Ix1} = -\Omega^2 \Bigg[& (m\eta_m \cos - m\xi_m \sin\theta)\frac{R^2}{2}\ddot{\beta} + (I_{\xi\xi} + I_{\eta\eta})R\ddot{\theta} \\ & + \left\{ (I_{\xi\xi} - I_{\eta\eta})\sin\theta\cos\theta + I_{\eta\xi}(\cos^2\theta - \sin^2\theta) \right\} R \\ & + \beta \left\{ (m\eta_m \cos\theta - m\xi_m \sin\theta)\frac{R^2}{2} + (m\eta_m \sin\theta + m\xi_m \cos\theta)\frac{R^2}{2}\ddot{\theta} \right\} \Bigg] \end{aligned} \tag{7.106}$$

Flap inertia moment at the root:

$$Q_{Iy1} = \int_0^R \left\{ q_{Iy1} + r\beta p_{Ix1} - r p_{Iz1} \right\} dr \tag{7.107}$$

Equation 7.107 can be expanded as

$$Q_{Iy1} = \Omega^2 \left[(m\eta_m \sin + m\xi_m \cos\theta)\frac{R^2}{2} + m\frac{R^3}{3}\beta + m\frac{R^3}{3}\ddot{\beta} \right.$$
$$\left. + (m\eta_m \cos\theta - m\xi_m \sin\theta)\frac{R^2}{2}\ddot{\theta} \right]$$

Rewriting the root inertia moment in the flap as

$$Q_{Iy1} = m\Omega^2 \frac{R^3}{3}[\ddot{\beta} + \beta] + \frac{R^2}{2}\Omega^2 \left\{ (m\eta_m \cos\theta - m\xi_m \sin\theta)\ddot{\theta} + (m\eta_m \sin\theta + m\xi_m \cos\theta) \right\} \quad (7.108)$$

The coupled flap–torsion equations of motion can be obtained by moment balance at the root of the blade. These equations can be symbolically written as follows:

Flap equation:

$$Q_{Iy1} + K_\beta \beta + Q_{Ay1} = 0 \quad (7.109)$$

Torsion equation:

$$Q_{Ix1} + K_\theta \theta + Q_{Ax1} = 0 \quad (7.110)$$

where Q_{Ax1} and Q_{Ay1} are the torsion and flap root moments, respectively, due to the aerodynamic loads acting on the blade.

In evaluating the aerodynamic loads on a typical cross section of the rotor blade undergoing flap (plunging) and torsion (pitching) motions, care must be exercised. The unsteady aerodynamic loads acting on the rotor blade is very difficult to be evaluated because of the complex rotor wake piling up beneath the rotor. The first unsteady aerodynamic model was developed by Theodorsen (1935) for an oscillating thin airfoil undergoing pitching and plunging motions in an incompressible flow. This theory is not directly applicable to the rotor blade because of the difference in the wake geometry. This theory was later modified by Greenberg (1947) to include the effect of pulsating oncoming flow and constant angle of attack. However, the wake geometry was similar to the Theodorsen model. Later, in 1957, Loewy (1957) considered a hovering model with wake layers beneath the rotor blade. This is the first two-dimensional unsteady aerodynamic model applicable for rotor blades. The lift and moment expressions obtained in the Loewy theory are similar to those obtained in the Theodorsen theory, except that the lift deficiency function has a different form.

It is important to recognize that the Loewy theory is difficult to implement in a rotor blade analysis. Hence, the Greenberg or the Theodorsen model is used in evaluating rotor blade cross-sectional aerodynamic loads. However, in practical situations, the static airfoil data are dynamically corrected and used as a table look-up for calculating aerodynamic loads. Currently, efforts are underway to calculate unsteady aerodynamic loads using computational fluid dynamics (CFD) techniques.

For a highly simplified analysis, a quasi-steady aerodynamic approximation of the Theodorsen theory will be used in the following. This model is reasonable to bring out the essential features of coupled flap–torsion dynamic aeroelastic stability. It should be pointed out that the quasi-static aerodynamic model based on the instantaneous effective angle attack used in the previous chapters and section (while considering flap dynamics and flap–lag dynamics) cannot be applied for coupled flap–torsion dynamics because this quasi-static aerodynamic model does not include the effects of the rate of change of the pitch angle of the blade. Hence, this model is inadequate to be used in coupled flap–torsion dynamics.

Let us now develop the aerodynamic loads acting on the rotor blade undergoing flap–torsion dynamics using a quasi-steady approximation of the Theodorsen unsteady aerodynamic theory. Consider an airfoil executing

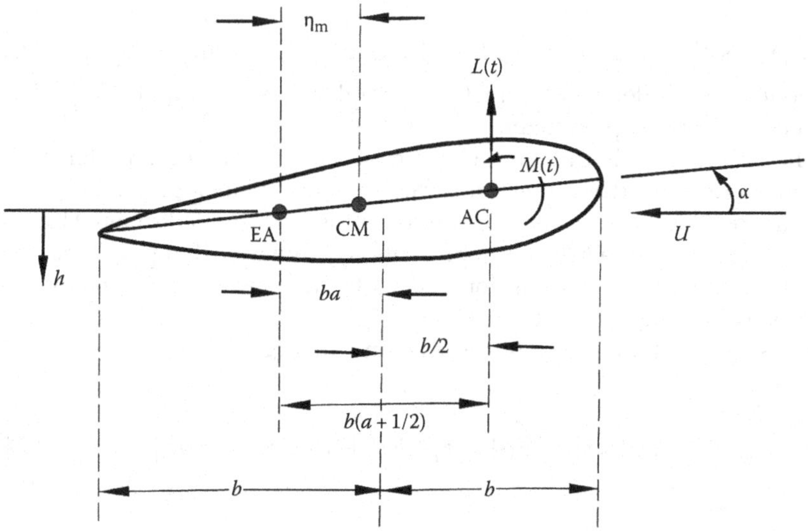

FIGURE 7.11
Typical cross section of a rotor blade undergoing plunging and pitching motions.

a simple harmonic motion in pitching and plunging motions, as shown in Figure 7.11.

In Figure 7.11, b represents the blade semi-chord, EA represents the elastic axis, AC represents the aerodynamic center (at quarter-chord location), and CM represents the center of mass location of the cross section of the blade. The plunging motion is represented by the displacement of the point at EA by h, and the pitching motion is represented by α, measured with respect to the oncoming flow U.

The expressions for the unsteady lift and moment acting at the reference EA are given by the Theodorsen theory as

$$L(t) = \pi\rho b^2[\ddot{h} + U\dot{\alpha} - ba\ddot{\alpha}] + 2\pi\rho UbC(k)\left[\dot{h} + U\alpha + b\left(\frac{1}{2} - a\right)\dot{\alpha}\right] \tag{7.111}$$

$$M_{EA}(t) = M(t) + Lb\left(a + \frac{1}{2}\right) \tag{7.112}$$

$$\begin{aligned} M_{EA}(t) = {} & \pi\rho b^2\left[ba\ddot{h} - Ub\left(\frac{1}{2} - a\right)\dot{\alpha} - b^2\left(\frac{1}{8} + a^2\right)\ddot{\alpha}\right] \\ & + 2\pi\rho Ub^2\left(\frac{1}{2} + a\right)C(k)\left[\dot{h} + U\alpha + b\left(\frac{1}{2} - a\right)\dot{\alpha}\right] \end{aligned} \tag{7.113}$$

where ρ is the density of air, $b = C/2$ is the blade semi-chord, $C(k)$ is the Theodorsen lift deficiency function (complex quantity), and $k = \frac{\omega b}{U}$ is the reduced frequency parameter.

The first term in both L and M_{EA} is called the "apparent mass term." The second-order time derivative terms can be neglected in apparent mass terms because of their order of smallness in comparison to the blade mass. However, the $\dot{\alpha}$ term is retained since it represents aerodynamic damping. The quasi-steady aerodynamic assumption relates to neglecting the unsteady wake effects by making $C(k) = 1.0$.

The simplified quasi-steady aerodynamic model can be written as

$$L(t) = \pi\rho b^2[U\dot{\alpha}] + 2\pi\rho Ub\left[\dot{h} + U\alpha + b\left(\frac{1}{2} - a\right)\dot{\alpha}\right] \tag{7.114}$$

$$M_{EA}(t) = \pi\rho b^2\left[-Ub\left(\frac{1}{2} - a\right)\dot{\alpha}\right] + 2\pi\rho Ub^2\left(\frac{1}{2} + a\right)\left[\dot{h} + U\alpha + b\left(\frac{1}{2} - a\right)\dot{\alpha}\right] \tag{7.115}$$

Relative Air Velocity Components at a Typical Cross Section

The cross-sectional aerodynamic loads are obtained in the deformed state of the blade. Hence, the components of velocity have to be defined in the blade-fixed deformed ($\hat{e}_{x2}$, $\hat{e}_{y2}$, $\hat{e}_{z2}$) coordinate system.

The net relative velocity of airflow has two components. They are (1) the forward speed of the vehicle ($\mu\Omega R$) and the induced flow ($\lambda\Omega R$), and (2) the velocity due to blade motion.

Let $\vec{V}_F$ be the free-stream velocity of air defined in the hub plane. It is given as (from Equation 7.35)

$$\vec{V}_F = \Omega R\mu\cos\psi_k\hat{e}_{x1} - \Omega R\mu\sin\psi_k\hat{e}_{y1} - \Omega R\lambda\hat{e}_{z1} \quad (7.116)$$

The velocity at the reference EA at location r is given as (Equation 7.87)

$$\vec{V}_{EA} = \Omega[r\hat{e}_{y1} + r\dot{\beta}\hat{e}_{z1}]$$

The net relative air velocity vector at the typical cross section of the blade can be written as

$$\vec{V}_{\text{net-air}1k} = \vec{V}_F - \vec{V}_{EA} \quad (7.117)$$

Using the transformation relationships given in Equation 7.78, net air velocity in the $2k$ system can be written as

$$\vec{V}_{\text{net-air}2k} = [T_{21}]\vec{V}_{\text{net-air}1k} \quad (7.118)$$

The net air velocity can be written in symbolic form as

$$\vec{V}_{\text{net-air}2k} = U_R\hat{e}_{x2} - U_T\hat{e}_{y2} - U_P\hat{e}_{z2} \quad (7.119)$$

Using Equation 7.119, the radial, tangential, and perpendicular velocity components are written as (for convenience, the length quantities are nondimensionalized with respect to R and the time derivatives are nondimensionalized with rotor angular speed Ω)

$$U_R = \Omega\left\{R\mu\cos\psi_k\cos\beta - (R\lambda + r\dot{\beta})\sin\beta\right\} \quad (7.120)$$

$$U_T = \Omega\{+R\mu\sin\psi_k + r\} \quad (7.121)$$

$$U_P = \Omega\left\{\sin\beta\, R\mu\cos\psi_k + (R\lambda + r\dot{\beta})\cos\beta\right\} \tag{7.122}$$

Neglecting the radial velocity component, and assuming a small angle for flap deflection β, the velocity components can be written as

$$U_T = \Omega\{R\mu \sin\psi_k + r\} \tag{7.123}$$

$$U_P = \Omega\left\{\beta R\mu\cos\psi_k + R\lambda + r\dot{\beta}\right\} \tag{7.124}$$

Before applying these velocity components for aerodynamic load evaluation, one must identify the velocity components in relation to Figure 7.11. For the sake of simplicity, let us assume that the rotor is operating under hovering condition (μ = 0). Using Equations 7.123 and 7.124, the velocity components, angle of attack, and its rate can be identified as

$$\begin{aligned} U &= +\Omega r \\ \dot{h} &= -r\Omega\dot{\beta} - \lambda\Omega R \\ \alpha &= \theta \end{aligned} \tag{7.125}$$

$$\dot{\alpha} = \Omega\dot{\theta}$$

$$ba = x_A - \frac{b}{2}$$

(Note that x_A represents the distance from the EA to the aerodynamic center in the blade cross section; Figure 7.11.)

Substituting the expressions given in Equation 7.125 in the unsteady lift and moment expressions given in Equations 7.114 and 7.115 yields the following:

Lift per unit span:

$$L(t) = \pi\rho b^2\{\Omega r\}(\dot{\theta}\Omega) + 2\pi\rho\{\Omega r\}b\left[-r\Omega\dot{\beta} - \lambda\Omega R + \Omega r\theta + (b - x_A)\Omega\dot{\theta}\right] \tag{7.126}$$

Moment about the EA per unit span:

$$\begin{aligned} M_{EA}(t) = {} & \pi\rho b^2\left[-\Omega r(b - x_A)\Omega\dot{\theta}\right] \\ & + 2\pi\rho\{\Omega r\}bx_A\left[-r\Omega\dot{\beta} - \lambda\Omega R + \Omega r\theta + (b - x_A)\Omega\dot{\theta}\right] \end{aligned} \tag{7.127}$$

It is to be noted that lift acts normal to the resultant of the oncoming flow. For a small angle assumption for flap angle, the lift $L(t)$ can be assumed to act along the z_1 axis.

Aerodynamic Flap Moment

The aerodynamic moment in the flap can be obtained from integrating the moment due to lift over the entire span of the blade, and it is given by

$$Q_{Ay1} = -\int_0^R L(t) r\, dr \tag{7.128}$$

Substituting for the lift expression from Equation 7.126, the expanded expression for aerodynamic flap moment can be written as

$$Q_{Ay1} = -\int_0^R \pi\rho\Omega^2(2b)\left[-r^3\dot{\beta} - \lambda R r^2 + r^3\theta + (b - x_A)r^2\dot{\theta} + \frac{b}{2}r^2\dot{\theta}\right]dr \tag{7.129}$$

Assuming uniform properties along the span of the blade, one can integrate Equation 7.129 and obtain an expression for the aerodynamic flap moment about the root flap hinge.

$$Q_{Ay1} = \pi\rho\Omega^2(2b)R^4\left[-\frac{\theta}{4} + \frac{\dot{\beta}}{4} + \frac{\lambda}{3} - \left(\frac{b}{R} - \frac{x_A}{R}\right)\frac{\dot{\theta}}{3} - \frac{b}{2R}\frac{\dot{\theta}}{3}\right] \tag{7.130}$$

Noting that the lift curve slope of airfoil can be taken as $a = 2\pi$, Equation 7.130 can be modified as

$$Q_{Ay1} = \rho ab\Omega^2R^4\left[-\frac{\theta}{4} + \frac{\dot{\beta}}{4} + \frac{\lambda}{3} - \left(\frac{b}{R} - \frac{x_A}{R}\right)\frac{\dot{\theta}}{3} - \frac{b}{2R}\frac{\dot{\theta}}{3}\right] \tag{7.131}$$

Aerodynamic Torsional Moment

The aerodynamic torsion moment about the EA, acting on the entire blade, can be obtained by integrating the sectional aerodynamic moment over the length of the blade. It is given as

$$Q_{Ax1} = \int_0^R M_{EA}\, dr \tag{7.132}$$

Substituting Equation 7.127 in Equation 7.132, one obtains

$$\begin{aligned} Q_{Ax1} = \int_0^R &\Big\{\pi\rho b^2\left[-\Omega^2 r(b - x_A)\dot{\theta}\right] \\ &+ 2\pi\rho\{\Omega r\}bx_A\left[-r\Omega\dot{\beta} - \lambda\Omega R + \Omega r\theta + (b - x_A)\Omega\dot{\theta}\right]\Big\}dr \end{aligned} \tag{7.133}$$

Assuming uniform properties along the span of the blade, Equation 7.133 can be integrated to obtain the aerodynamic pitch moment acting on the rotor blade about the EA.

$$Q_{\mathrm{A}x1} = 2\pi\rho b R^4\Omega^2 \frac{x_{\mathrm{A}}}{R}\left[\frac{\theta}{3} - \frac{\lambda}{2} - \frac{1}{3}\dot{\beta} + \left(\frac{b}{R} - \frac{x_{\mathrm{A}}}{R}\right)\frac{1}{2}\dot{\theta}\right] - \pi\rho b\Omega^2 R^4 \frac{b}{R}\left(\frac{b}{R} - \frac{x_{\mathrm{A}}}{R}\right)\frac{1}{2}\dot{\theta} \tag{7.134}$$

Noting that the lift curve slope of airfoil can be taken as $a = 2\pi$, Equation 7.134 can be modified as

$$Q_{\mathrm{A}x1} = a\rho b R^4\Omega^2 \frac{x_{\mathrm{A}}}{R}\left[\frac{\theta}{3} - \frac{\lambda}{2} - \frac{1}{3}\dot{\beta} + \left(\frac{b}{R} - \frac{x_{\mathrm{A}}}{R}\right)\frac{1}{2}\dot{\theta}\right] - \frac{a}{2}\rho b\Omega^2 R^4 \frac{b}{R}\left(\frac{b}{R} - \frac{x_{\mathrm{A}}}{R}\right)\frac{1}{2}\dot{\theta} \tag{7.135}$$

Collecting the inertia, elastic, and aerodynamic loads acting on the blade, the coupled flap–torsion equations can be written as follows:

Flap equation:

$$\begin{aligned} m\Omega^2 \frac{R^3}{3}\left[\ddot{\beta} + \beta\right] &+ \frac{R^2}{2}\Omega^2\left\{\left(m\eta_{\mathrm{m}}\cos\theta - m\xi_{\mathrm{m}}\sin\theta\right)\ddot{\theta} + \left(m\eta_{\mathrm{m}}\sin\theta + m\xi_{\mathrm{m}}\cos\theta\right)\right\} \\ &+ \rho a b\Omega^2 R^4\left[-\frac{\theta}{4} + \frac{\dot{\beta}}{4} + \frac{\lambda}{3} - \left(\frac{b}{R} - \frac{x_{\mathrm{A}}}{R}\right)\frac{\dot{\theta}}{3} - \frac{b}{2R}\frac{\dot{\theta}}{3}\right] + K_\beta\beta = 0 \end{aligned} \tag{7.136}$$

Torsion equation:

$$\begin{aligned} -\Omega^2\Bigg[&(m\eta_{\mathrm{m}}\cos - m\xi_{\mathrm{m}}\sin\theta)\frac{R^2}{2}\ddot{\beta} + (I_{\xi\xi} + I_{\eta\eta})R\ddot{\theta} \\ &+\left\{(I_{\xi\xi} - I_{\eta\eta})\sin\theta\cos\theta + I_{\eta\xi}(cos^2\theta - sin^2\theta)\right\}R \\ &+\beta\left\{(m\eta_{\mathrm{m}}cos\theta - m\xi_{\mathrm{m}}\sin\theta)\frac{R^2}{2} + (m\eta_{\mathrm{m}}\sin\theta + m\xi_{\mathrm{m}}\cos\theta)\frac{R^2}{2}\ddot{\theta}\right\}\Bigg] \\ &+ a\rho b R^4\Omega^2\frac{x_A}{R}\left[\frac{\theta}{3} - \frac{\lambda}{2} - \frac{1}{3}\dot{\beta} + \left(\frac{b}{R} - \frac{x_A}{R}\right)\frac{1}{2}\dot{\theta}\right] - \frac{a}{2}\rho b\Omega^2 R^4\frac{b}{R}\left(\frac{b}{R} - \frac{x_A}{R}\right)\frac{1}{2}\dot{\theta} - K_\theta\theta = 0 \end{aligned} \tag{7.137}$$

Equations 7.136 and 7.137 are coupled nonlinear differential equations. These nonlinear equations can be further simplified by making a small angle assumption as

$$\cos\theta \simeq 1 \; \sin\theta \simeq \theta$$

In addition, assume that the cross section of the blade is symmetric (i.e., $\xi_m = 0$ and $I_{\eta\xi} = 0$). Using all these approximations, the flap–torsion equations of motion can be simplified as follows:

Flap equation:

$$\ddot{\beta} + \frac{3}{2}\frac{\eta_m}{R}\{\ddot{\theta} + \theta\} - \frac{\gamma}{2}\left[\frac{\theta}{4} - \frac{\beta}{4} - \frac{\lambda}{3} + \left(\frac{b - x_A}{R} + \frac{b}{2R}\right)\frac{\dot{\theta}}{3}\right] + \left(1 + \bar{\omega}_{NRF}^2\right)\beta = 0 \tag{7.138}$$

Torsion equation:

$$\begin{aligned} &-\frac{3}{2}\frac{\eta_m}{R}(\ddot{\beta} + \beta) - \left(\bar{I}_{\xi\xi} + \bar{I}_{\eta\eta}\right)\ddot{\theta} - \left(\bar{I}_{\xi\xi} - \bar{I}_{\eta\eta}\right)\theta \\ &+ \frac{\gamma}{2}\left\{\frac{x_A}{R}\left[\frac{\theta}{3} - \frac{\dot{\beta}}{3} - \frac{\lambda}{2}\right] + \left(\frac{x_A}{R} - \frac{b}{2R}\right)\left(\frac{b - x_A}{R}\right)\frac{\dot{\theta}}{2}\right\} - \bar{\omega}_T^2\left(\bar{I}_{\xi\xi} + \bar{I}_{\eta\eta}\right)\phi = 0 \end{aligned} \tag{7.139}$$

where

$$I_b = \frac{mR^3}{3}$$

$$\bar{I}_{\xi\xi} + \bar{I}_{\eta\eta} = \frac{(I_{\xi\xi} + I_{\eta\eta})R}{I_b}$$

$$\bar{I}_{\xi\xi} = \frac{I_{\xi\xi}R}{I_b}$$

$$\bar{I}_{\eta\eta} = \frac{I_{\eta\eta}R}{I_b}$$

$$\bar{\omega}_T^2 = \frac{K_\theta}{(I_{\xi\xi} + I_{\eta\eta})R}$$

$$\gamma = \frac{\rho a 2 b R^4}{I_b}$$

The aeroelastic stability analysis requires the equilibrium equations and the perturbation equations about the equilibrium state.

Assuming a solution of the form,

$$\theta = \theta_0 + \phi_0 + \tilde{\phi}(t)$$

$$\beta = \beta_0 + \tilde{\beta}(t)$$

where ϕ_0 and β_0 are constant quantities representing the equilibrium deformation of the blade and $\tilde{\phi}$ and $\tilde{\beta}$ are time-dependent perturbational quantities. θ_0 represents the pitch input given to the blade. Substituting these expressions, and collecting the constant and the time-varying terms and equating then separately to 0, the equilibrium equations and linearized stability equations are obtained.

Equilibrium equations:

$$\left(1+\bar{\omega}_{\mathrm{NRF}}^2\right)\beta_0 + \frac{3}{2}\frac{\eta_m}{R}(\theta_0+\phi_0) - \gamma\left[\frac{\theta_0}{8}+\frac{\phi_0}{8}-\frac{\lambda}{6}\right] = 0$$

$$-\frac{3}{2}\frac{\eta_m}{R}\beta_0 - \left(\bar{I}_{\xi\xi}-\bar{I}_{\eta\eta}\right)\theta_0 - \left(\bar{I}_{\xi\xi}-\bar{I}_{\eta\eta}\right)\phi_0 + \gamma\left\{\frac{x_A}{R}\left[\frac{\theta_0}{6}-\frac{\lambda}{4}\right]\right\} \tag{7.140}$$

$$+\gamma\frac{x_A}{R}\frac{\phi_0}{6} - \bar{\omega}_T^2\left(\bar{I}_{\xi\xi}+\bar{I}_{\eta\eta}\right)\phi_0 = 0 \tag{7.141}$$

Writing in matrix form,

$$\begin{bmatrix} \left(1+\bar{\omega}_{\mathrm{NRF}}^2\right) & -\frac{\gamma}{8}+\frac{3}{2}\frac{\eta_m}{R} \\ -\frac{3}{2}\frac{\eta_m}{R} & -\left(\bar{I}_{\xi\xi}-\bar{I}_{\eta\eta}\right) - \bar{\omega}_T^2\left(\bar{I}_{\xi\xi}+\bar{I}_{\eta\eta}\right) + \frac{\gamma}{6}\frac{x_A}{R} \end{bmatrix} \begin{Bmatrix} \beta_0 \\ \phi_0 \end{Bmatrix} = \begin{Bmatrix} \Upsilon_1 \\ \Upsilon_2 \end{Bmatrix} \tag{7.142}$$

where

$$\Upsilon_1 = \gamma\left[\frac{\theta_0}{8}-\frac{\lambda}{6}\right] - \frac{3}{2}\frac{\eta_m}{R}\theta_0$$

$$\Upsilon_2 = \left(\bar{I}_{\xi\xi}-\bar{I}_{\eta\eta}\right)\theta_0 - \gamma\frac{x_A}{R}\left[\frac{\theta_0}{6}-\frac{\lambda}{4}\right]$$

The equilibrium values of β_0 and ϕ_0 can be obtained by inverting the square matrix and multiplying the column vector in the right-hand side. Divergence is a static instability phenomenon, indicating that the equilibrium values of β_0 and ϕ_0 are indeterminate (or infinity). Such a situation is possible, if the determinant of the square matrix is equal to 0. This condition provides the divergence limit of the blade. It is given as

$$\left(1+\bar{\omega}_{\mathrm{NRF}}^2\right)\left(\bar{\omega}_{\mathrm{T}}^2\left(\bar{I}_{\xi\xi}+\bar{I}_{\eta\eta}\right)+\left(\bar{I}_{\xi\xi}-\bar{I}_{\eta\eta}\right)-\frac{\gamma}{6}\frac{x_{\mathrm{A}}}{R}\right)-\left(\frac{\eta_{\mathrm{m}}}{R}\frac{3}{2}\left\{-\frac{\gamma}{8}+\frac{3}{2}\frac{\eta_{\mathrm{m}}}{R}\right\}\right)=0 \quad (7.143)$$

This equation indicates that rotor blade divergence depends on flap frequency, torsional frequency, blade torsional inertia, the offset between the aerodynamic center and the elastic center, the offset between the mass center and the elastic center, and the Lock number. If we make an assumption that the blade can be treated as a thin plate, then $\bar{I}_{\eta\eta}=0$, and the divergence condition can be written as

$$\left(1+\bar{\omega}_{\mathrm{NRF}}^2\right)\left\{\left(\bar{\omega}_{\mathrm{T}}^2+1\right)\bar{I}_{\xi\xi}-\frac{\gamma}{6}\frac{x_{\mathrm{A}}}{R}\right\}-\frac{3}{2}\frac{\eta_{\mathrm{m}}}{R}\left(\frac{3}{2}\frac{\eta_{\mathrm{m}}}{R}-\frac{\gamma}{8}\right)=0 \quad (7.144)$$

Using the above equation, one can formulate different conditions for avoiding blade divergence. (Note: The determinant must be greater than 0 to avoid divergence instability.) From Equation 7.144, the condition for avoiding divergence instability can be expressed as

Condition A:

$$\bar{\omega}_{\mathrm{T}}^2>\left[\frac{1}{\left(1+\bar{\omega}_{\mathrm{NRF}}^2\right)}\left[\frac{3}{2}\frac{\eta_{\mathrm{m}}}{R}\left(\frac{3}{2}\frac{\eta_{\mathrm{m}}}{R}-\frac{\gamma}{8}\right)\right]+\frac{\gamma}{6}\frac{x_{\mathrm{A}}}{R}\right]\frac{1}{\bar{I}_{\xi\xi}}-1 \quad (7.145)$$

Condition B: Neglecting $\frac{3}{2}\frac{\eta_m}{R}$ in comparison to $\frac{\gamma}{8}$ in Equation 7.143, a modified form of the condition can be written

$$\frac{3}{2}\frac{\eta_{\mathrm{m}}}{R}>-\frac{1}{\gamma/8}\left[\left(1+\bar{\omega}_{\mathrm{NRF}}^2\right)\left\{\left(\bar{\omega}_{\mathrm{T}}^2+1\right)\bar{I}_{\xi\xi}-\frac{\gamma}{6}\frac{x_{\mathrm{A}}}{R}\right\}\right] \quad (7.146)$$

Flutter or dynamic stability equations: The perturbation equations can be written as (to avoid confusion, different notations have been used for the damping and the stiffness matrices)

$$\begin{bmatrix} M_{11} & M_{12} \\ M_{21} & M_{22} \end{bmatrix}\begin{Bmatrix} \ddot{\beta} \\ \ddot{\phi} \end{Bmatrix}+\begin{bmatrix} P_{11} & P_{12} \\ P_{21} & P_{22} \end{bmatrix}\begin{Bmatrix} \dot{\beta} \\ \dot{\phi} \end{Bmatrix}+\begin{bmatrix} S_{11} & S_{12} \\ S_{21} & S_{22} \end{bmatrix}\begin{Bmatrix} \beta \\ \phi \end{Bmatrix}=0 \quad (7.147)$$

$$M_{11} = 1$$

$$M_{12} = \frac{3}{2}\frac{\eta_m}{R}$$

$$M_{21} = -\frac{3}{2}\frac{\eta_m}{R}$$

$$M_{22} = -\left(\bar{I}_{\xi\xi} + \bar{I}_{\eta\eta}\right)$$

$$P_{11} = \frac{\gamma}{8}$$

$$P_{12} = -\frac{\gamma}{6}\left(\frac{b - x_A}{R} + \frac{b}{2R}\right)$$

$$P_{21} = -\frac{\gamma}{6}\frac{x_A}{R}$$

$$P_{22} = \frac{\gamma}{4}\left(\frac{x_A - b/2}{R}\right)\left(\frac{b - x_A}{R}\right)$$

$$S_{11} = 1 + \bar{\omega}_{NRF}^2$$

$$S_{12} = -\frac{\gamma}{8} + \frac{3}{2}\frac{\eta_m}{R}$$

$$S_{21} = -\frac{3}{2}\frac{\eta_m}{R}$$

$$S_{22} = -\left(\bar{I}_{\xi\xi} - \bar{I}_{\eta\eta}\right) + \frac{\gamma}{6}\frac{x_A}{R} - \bar{\omega}_T^2\left(\bar{I}_{\xi\xi} + \bar{I}_{\eta\eta}\right)$$

Assuming a solution of the form $\tilde{\beta} = \bar{\beta}e^{st}$ and $\tilde{\phi} = \bar{\phi}e^{st}$ and substituting in Equation 7.147, the characteristic equation is formed. The four roots of the characteristic equation have to be evaluated numerically. The roots appear as complex conjugates. The real part represents the modal damping, and the imaginary part represents the modal frequency. The stability of the system has been studied by several researchers using quasi-steady aerodynamics or the

TABLE 7.2

Data for Flap–Torsion Stability Analysis

Variable	Quantity
Number of blades, N	4
Air density at sea level, r (kg/m^3)	1.224
Coefficient of drag, c_d	0.008
Radius of the main rotor blade, R (m)	6.6
Chord of the main rotor blade, C (m)	0.5
Main rotor rotating speed, Ω (rpm)	300
Mass of the main rotor blade, m_0 (kg/m)	11.24
Flap frequency, $\bar{\omega}_{RF}$ (rad/s)	1.09
Lift curve slope, a	2π
Offset between the EA and the aerodynamic center, x_A	0
Mass moment of inertia per unit length	
$\bar{I}_{\eta\eta}$	0
$\bar{I}_{\xi\xi}$	0.004
$\bar{I}_{\eta\xi}$	0

Theodorsen unsteady aerodynamic theory. The significant parameters affecting the stability of the system are the natural frequency in torsion $\bar{\omega}_T$, and the offset parameters x_A and η_m. Generally, to avoid high oscillatory loads in the control system during flight, the rotor blade is designed in such a manner to make x_A equal to 0 or to a very small value. Using the data given in Table 7.2, typical results showing flutter and divergence boundaries are obtained.

The nondimensional torsional frequency is varied from 0 to 4.8. The nondimensional offset distance between sectional c.g. and the EA ($\eta_m/2b$) is varied from 0 to −0.08 in a step of 0.01. The negative sign indicates that the mass center is behind the EA. The results of the divergence and stability boundaries are shown in Figure 7.12.

If the torsional frequency is fixed at the value corresponding to point A (2.0) and the chordwise c.g. position is varied, the line AB is traced. The blade is completely stable until the line intersects the divergence boundary at point B. Increasing η_m to a higher value, the flutter boundary is reached at point C. Between points B and C, the stability roots are found to be complex pairs with a negative real part, implying dynamic stability. To the right of point C, the stability roots become a complex pair, with a positive real part indicating instability. It can be noted that, for low values of torsional frequency, the divergence occurs first followed by flutter. However, at higher values of torsional frequency, the flutter occurs first, followed by the divergence. It is observed that the region of stability of the blade increases with increasing torsional frequency. At high torsional frequencies, a large value of negative $\eta_m/2b$ is required to destabilize the blade. It is noted that moving the c.g.

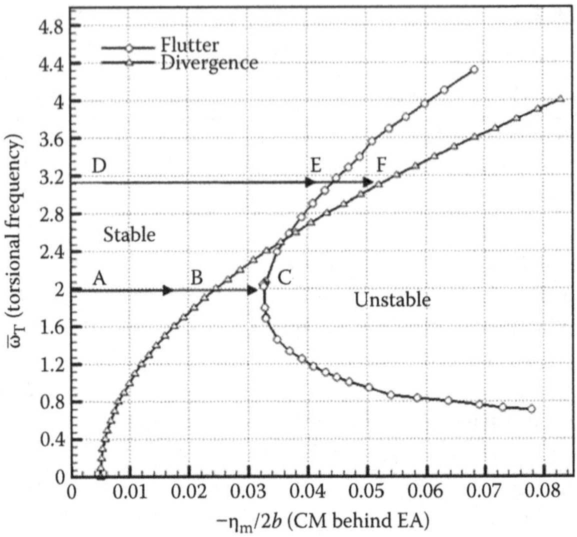

FIGURE 7.12
Stability boundaries for flap–torsional motion.

forward toward the leading edge of the rotor blade stabilizes the blade. In practice, the leading edge mass is added to bring the CM as close as possible to the EA, which is usually designed to be close to the aerodynamic center.

Increasing torsional frequency improves both the static stability (divergence) and the dynamic stability (flutter) of the blade. Hence, most of the blades are designed to have a torsional frequency in the range of 4.0/rev to 5.0/rev.

During forward flight, because of the periodic nature of the aerodynamic loads, the equations of motion will have time-varying periodic coefficients. The stability of time-varying periodic systems has to be treated in a different manner. Several researchers have studied the aeroelastic stability problem in forward flight using the Floquet–Liapunov theory. Another approach is to modify the periodic coefficient equations into a set of constant coefficient equations (in an approximate manner) using multiblade coordinate or Fourier coordinate transformation. The stability analysis is then performed using the constant coefficient equations. It may be noted that this procedure of transforming periodic coefficient equations to constant coefficient equations is approximate and is applicable only for low forward speeds. It is worthwhile pointing out that it was only during the late 1970s and the early 1980s that the aeroelastic stability analysis of the coupled flag–lag–torsional dynamics of isotropic rotor blades was solved. The composite blade aeroelastic analysis is the topic of research during the early and mid-1990s. During the last decade, research effort is focused on developing smart rotors with a trailing edge flap control and tip shape modifications to reduce rotor loads and rotor noise.

8

Rotor Modes: Multi-Blade Coordinate Transformation

The dynamics of a coupled rotor–fuselage helicopter system is a very complex problem. A detailed model should include the blade modes in axial, flap, lag, and torsion; the rigid body motion of the fuselage in both translation and rotation; and the flexible modes of the fuselage. Figure 8.1 shows a schematic of the interactions in a rotor—fuselage system.

Even if it is assumed that the fuselage is rigid and the blade is idealized as a rigid blade with a root spring, the total number of degrees of freedom of the dynamic model becomes (6 fuselage rigid body modes + $3N$ flap, lag, and torsion modes of all blades), where N is the number of blades in the main rotor system. For example, say for $N = 4$, the total number of degrees of freedom is 18. In addition, if one includes the fuselage flexible modes, the problem will become a formidable one. Furthermore, inclusion of rotor aerodynamics will make the problem highly complex. Now, the question is, how do we go about solving this complex problem? The approach adopted in the simplification of this problem, based on the certain fundamental characteristics of the rotor–fuselage dynamics, is quite interesting.

There are three types of problems associated with rotor–fuselage dynamics. They are (1) vehicle dynamics related to flying and handling characteristics; (2) aeromechanical instabilities, such as ground resonance and air resonance; and (3) vibration problems. These classifications are based on the frequency range of interest for each problem. In addition, the physics of the problem is also different for each one of them, and hence, different assumptions are used in solving these problems.

The range of the frequency of motion is below 0.5–1.0 Hz for vehicle dynamic problems, which involve analyzing the stability and control characteristics of the helicopter. These characteristics describe the handling qualities and/or flying qualities of the helicopter. A successful vehicle should have very good handling qualities meeting the MIL-H-8501A, FAR, DEFSTAN, or ADS-33 requirements. Ground and air resonance problem involves the coupling of rotor lead–lag modes with body modes in pitch and roll, causing instability. When the helicopter is on ground, the instability is denoted as ground resonance, which is generally very dangerous. This instability is avoided by providing adequate damping in the lead–lag mode of the rotor blade and in the landing gear of the helicopter. These instabilities occur in the frequency range of 2–5 Hz depending on the blade lag frequency and the

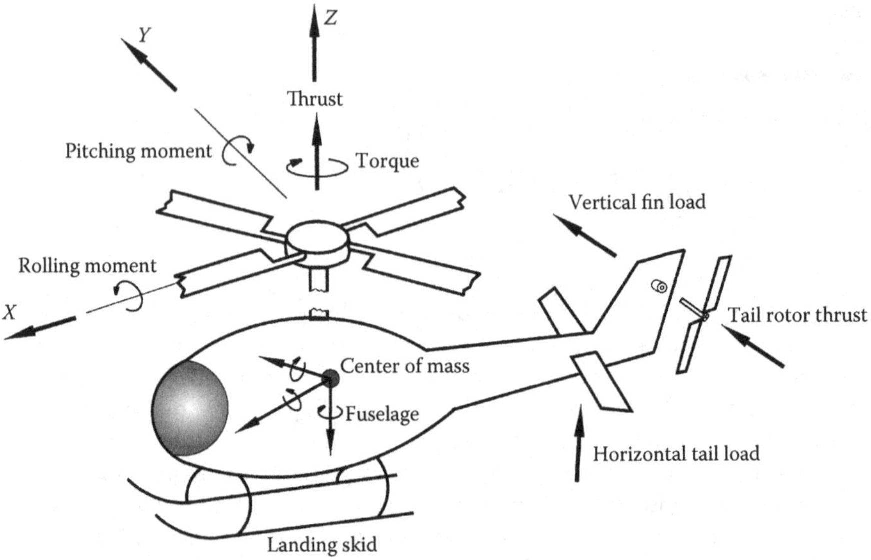

FIGURE 8.1
Schematic of a coupled rotor–fuselage interaction.

fuselage frequencies in pitch and roll. The problem of vibration in helicopters involves the estimation of fuselage response to rotor vibratory loads. The frequency of interest in vibration problems falls in the range of about 5Hz and above. Hence response problem requires the inclusion of a large number of blade and fuselage degrees of freedom.

Multiblade Coordinate Transformation

When dealing with coupled rotor–fuselage dynamics, it must be recognized that the fuselage motion is influenced by the combined effect of the hub loads due to all the blades. Therefore, one can address the problem by analyzing the dynamics of the rotor system rather than by analyzing the dynamics of the individual rotor blades due to perturbation in hub motion. Please note that the dynamics of the individual blades contributes to the dynamics of the rotor system. The response of individual blades is analyzed in the rotating frame, whereas the dynamics of the rotor system is studied from a nonrotating frame. The mathematical approach to convert the degrees of freedom of individual blades into rotor degrees of freedom is known as "multiblade coordinate transformation." Note that this transformation can be applied only to linear differential equations. In the following, a clear description of the multiblade coordinate transformation is presented, and it will be applied to the flap dynamics of the rotor blade to highlight the essential features of the flap dynamics as observed by a person in the nonrotating frame.

To have a physical understanding of the rotor modes, let us consider a four-bladed rotor system undergoing flap motion $\beta_i (i = 1, \ldots 4)$. Since there are four blades, there are four degrees of freedom in the rotating frame. These degrees of freedom can be converted into four rotor degrees of freedom describing the motion of the rotor tip-path plane. These rotor modes are described as collective, cyclic, and differential (or alternating) modes.

Figure 8.2 shows all these rotor modes for both the flap and lag motions of the blade. In the collective mode, all blades move through the same angle

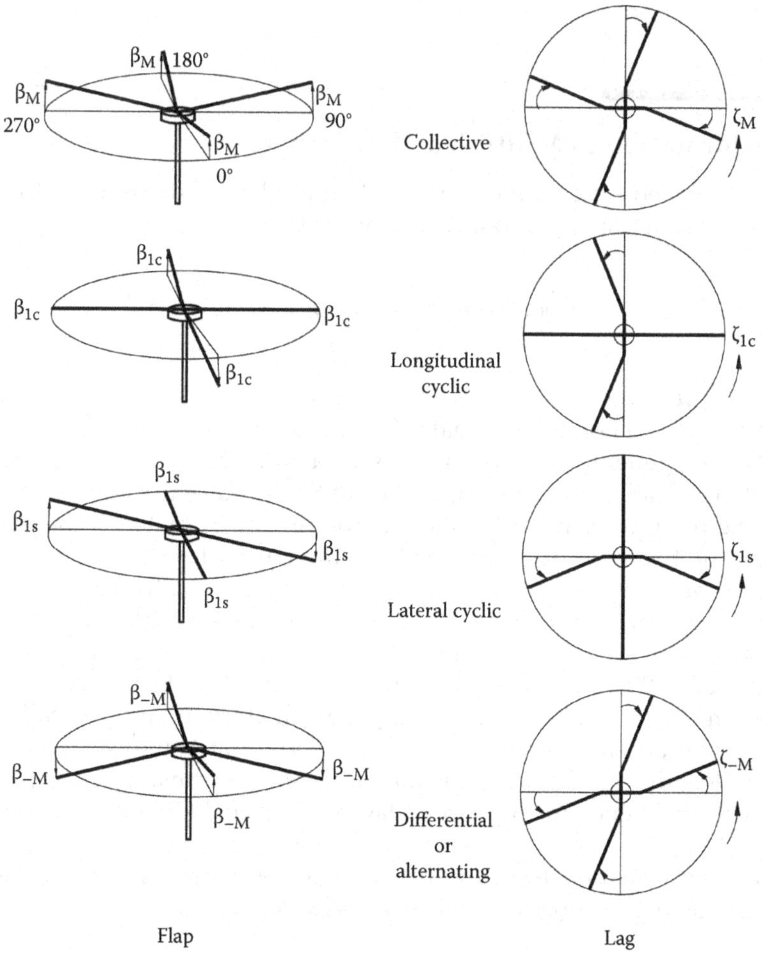

FIGURE 8.2
Rotor degrees of freedom in flap and lag motions.

irrespective of the azimuth. In the longitudinal and lateral cyclic modes, the tip-path plane tilts in longitudinal and lateral directions. In the differential mode, alternate blades move through the same angle but in opposite directions.

It can be seen that, in the collective flap mode, the c.g. of the rotor system moves up and down; in cyclic modes, it moves in the plane of rotation; and in the alternating mode, it does not move. In collective and differential lag modes, the c.g. of the rotor system remains at the hub center, but in cyclic lag modes, the c.g. moves in the plane of rotation.

In the following, a description of the transformation from blade degrees of freedom to rotor degrees of freedom is provided. This transformation is known as "multiblade coordinate transformation."

Transformation to Multiblade Coordinates

For an *N*-bladed rotor with blades evenly spaced around, the azimuth angle for the *k*th blade, at any instant, can be written as

$$\psi_K = \psi + \frac{2\pi}{N}(k-1) \quad \kappa = 1, \ldots N \tag{8.1}$$

where $\psi = \Omega t$ is the azimuth location of the first blade, and the variable ψ can also be used as a nondimensional time variable.

Let α_k be a generalized coordinate associated with any degree of freedom (say, flap or lead–lag or torsion) of the *k*th blade. Since this α_k is associated with the rotating blade, it is called a "rotating coordinate." If there are *N* blades, the behavior of all the blade in the particular degree of freedom can be represented by *N* rotating coordinates $\alpha_1 \ldots \alpha_N$. By suitably choosing a transformation, these *N* rotating coordinates can be transformed to another set of *N* coordinates, each of which is associated with a specific variation of all the α_k's as viewed from a nonrotating frame. This type of transformation is called the "multiblade transformation." Basically, this transformation transforms the rotating coordinates into a nonrotating frame. Usually, the physical explanation about this transformation is given only with reference to the flap or lag motion of the blade, as shown in Figure 8.2.

The transformation from the rotating to the nonrotating coordinate is obtained by performing the following operations.

$$\alpha_M = \frac{1}{N}\sum_{k=1}^{N}\alpha_k$$

$$\alpha_{-M} = \frac{1}{N}\sum_{k=1}^{N}(-1)^k\alpha_k \quad \text{(for even } N \text{ only)}$$

$$\alpha_{nc} = \frac{2}{N}\sum_{k=1}^{N}\cos(n\psi_k)\alpha_k \tag{8.2}$$

$$\alpha_{ns} = \frac{2}{N}\sum_{k=1}^{N}\sin(n\psi_k)\alpha_k$$

where $n = 1,\ldots L$, and

$$L = \frac{N-1}{2} \text{ for odd } N$$

and

$$L = \frac{N-2}{2} \text{ for even } N.$$

The inverse transformation is given by

$$\alpha_k = \alpha_M + \sum_{n=1}^{L}(\alpha_{nc}\cos(n\psi_k) + \alpha_{ns}\sin(n\psi_k) + (-1)^k\alpha_{-M} \tag{8.3}$$

The last term will appear only for even N.

The transformation, given in Equation 8.3, appears like a truncated Fourier series, except for the last term. The major difference between this transformation and the usual Fourier transformation is that, here, the coefficients α_M, α_{nc}, α_{ns}, α_{-M} are all functions of time, whereas in the Fourier series, the coefficients are constants. That is why, sometimes, these multiblade coordinates are also referred to as "Fourier coordinates."

Differentiating Equation 8.2 with respect to nondimensional time $\psi = \Omega t$ (where Ω is a constant), one obtains

$$\dot{\alpha}_M = \frac{1}{N}\sum_{k=1}^{N}\dot{\alpha}_k$$

$$\dot{\alpha}_{-M} = \frac{1}{N}\sum_{k=1}^{N}(-1)^k\dot{\alpha}_k$$

$$\dot{\alpha}_{nc} + n\alpha_{ns} = \frac{2}{N}\sum_{k=1}^{N}\cos(n\psi_k)\,\dot{\alpha}_k \tag{8.4}$$

$$\dot{\alpha}_{ns} - n\alpha_{nc} = \frac{2}{N}\sum_{k=1}^{N}\sin(n\psi_k)\dot{\alpha}_k$$

Differentiating once again with respect to ψ,

$$\ddot{\alpha}_M = \frac{1}{N}\sum_{k=1}^{N}\ddot{\alpha}_k$$

$$\ddot{\alpha}_{-M} = \frac{1}{N}\sum_{k=1}^{N}(-1)^k\ddot{\alpha}_k$$

$$\ddot{\alpha}_{nc} + 2n\dot{\alpha}_{ns} - n^2\alpha_{nc} = \frac{2}{N}\sum_{k=1}^{N}\cos(n\psi_k)\ddot{\alpha}_k \tag{8.5}$$

$$\ddot{\alpha}_{ns} - 2n\dot{\alpha}_{nc} - n^2\alpha_{ns} = \frac{2}{N}\sum_{k=1}^{N}\sin(n\psi_k)\ddot{\alpha}_k$$

It can be seen in Equation 8.5 that the transformation of the second derivative terms of the rotating degrees of freedom introduces Coriolis and centrifugal terms in the nonrotating frame.

The transformation from the rotating frame to the nonrotating frame is accomplished by applying the following N operators to the complete set of linear perturbation equations of the blade dynamics in the rotating frame.

They are

$$
\begin{aligned}
&\frac{1}{N}\sum_{k=1}^{N}(\ldots) \quad \text{collective operator}\\
&\frac{1}{N}\sum_{k=1}^{N}(-1)^k(\ldots) \quad \text{alternating operator (for even } N \text{ only)}\\
&\frac{2}{N}\sum_{k=1}^{N}\cos(n\psi_k)(\ldots) \quad n-\text{cosine operator}\\
&\frac{2}{N}\sum_{k=1}^{N}\sin\,(n\psi_k)(\ldots) \quad n-\text{sine operator}
\end{aligned}
\tag{8.6}
$$

The application of these four operators to the blade equation of motion transforms the blade degrees of freedom to rotor degrees of freedom. The resulting equations will have the multiblade coordinates as the new set of generalized coordinates. These equations represent the dynamics of the rotor as a whole as viewed from the nonrotating frame.

To understand the physics of the transformation from blade degrees of freedom to rotor degrees of freedom, let us consider the dynamics of a centrally hinged, spring-restrained rigid blade undergoing only flap motion in hover. Let β_i be the degree of freedom of the ith blade. The flap equation can be written as simplified after setting $e = 0$ and $\beta_p = 0$ in Equation 6.32

$$\ddot{\beta}_i + \frac{\gamma}{8}\dot{\beta}_i + \left(1 + \bar{\omega}_{NRF}^2\right)\beta_i = \frac{\gamma}{8}\theta_0 - \frac{\gamma}{6}\lambda_0 \quad (i = 1\ldots N) \tag{8.7}$$

Note that $\left(1 + \bar{\omega}_{NRF}^2 = \bar{\omega}_{RF}^2\right)$

Let us assume that there are four blades ($N = 4$). Under hovering conditions, the collective pitch angle θ_0 and the inflow ratio λ_0 remain the same for all blades. Since we are interested in the eigenvalues (frequencies) and eigenvectors (mode shapes), it is sufficient to consider only the homogeneous part of the flap equation. Another way of looking at the problem is to decompose the flap degree of freedom into two parts, which is given as

$$\beta_i = \beta_0 + \tilde{\beta}_i(t) \tag{8.8}$$

where β_0 is the equilibrium value that is a constant and $\tilde{\beta}_i(t)$ is the time-varying perturbational quantity. Substituting for β_i in Equation 8.7 and collecting the constant part and the perturbational part separately, we have

Equilibrium equation:

$$\left(\bar{\omega}_{RF}^2\right)\beta_0 = \frac{\gamma}{8}\theta_0 - \frac{\gamma}{6}\lambda_0 \tag{8.9}$$

Perturbation equation:

$$\ddot{\tilde{\beta}}_i + \frac{\gamma}{8}\dot{\tilde{\beta}}_i + \left(\bar{\omega}_{RF}^2\right)\tilde{\beta}_i = 0 \tag{8.10}$$

In this simple case, the perturbation equation is nothing but the homogeneous part of the flap equation, which describes the dynamics of the flap motion of the *i*th blade about the equilibrium flap angle β_0.

Applying the multiblade transformation (Equation 8.6) to the perturbational equation, the following four equations in multiblade coordinates are obtained. These four equations are given as

Collective flap equation:

$$\ddot{\beta}_M + \frac{\gamma}{8}\dot{\beta}_M + \left(\bar{\omega}_{RF}^2\right)\beta_M = 0 \tag{8.11}$$

1-cosine flap equation:

$$\ddot{\beta}_{1c} + 2\dot{\beta}_{1s} - \beta_{1c} + \frac{\gamma}{8}\dot{\beta}_{1c} + \frac{\gamma}{8}\beta_{1s} + \left(\bar{\omega}_{RF}^2\right)\beta_{1c} = 0 \tag{8.12}$$

1-sine flap equation:

$$\ddot{\beta}_{1s} - 2\dot{\beta}_{1c} - \beta_{1s} + \frac{\gamma}{8}\dot{\beta}_{1s} - \frac{\gamma}{8}\beta_{1c} + \left(\bar{\omega}_{RF}^2\right)\beta_{1s} = 0 \tag{8.13}$$

Differential (or alternating) flap equation:

$$\ddot{\beta}_{-M} + \frac{\gamma}{8}\dot{\beta}_{-M} + \left(\bar{\omega}_{RF}^2\right)\beta_{-M} = 0 \tag{8.14}$$

It is obvious from the above equations that the form of the collective and alternating modes remains the same as the individual blade equation of motion in the rotating frame.

The eigenvalues of the collective and differential modes can be obtained from the characteristic equation

$$S^2 + \frac{\gamma}{8} S + \left(\bar{\omega}_{RF}^2\right) = 0 \tag{8.15}$$

The roots are

$$S_{1,2} = S_{diff} = S_{coll} = -\frac{\gamma}{16} \pm i\sqrt{\bar{\omega}_{RF}^2 - \left(\frac{\gamma}{16}\right)^2}$$

The damping in collective and differential modes is $\frac{\gamma}{16}$, and the damped natural frequency is $\sqrt{\bar{\omega}_{RF}^2 - \left(\frac{\gamma}{16}\right)^2}$. These eigenvalues are the same as the eigenvalues of the flap mode in the rotating frame.

However, the cyclic mode equations (Equations 8.12 and 8.13) indicate that both the 1-cosine and the 1-sine modes are coupled. These equations can be written in the matrix form as

$$\begin{bmatrix} 1 & 0 \\ 0 & 1 \end{bmatrix} \begin{Bmatrix} \ddot{\beta}_{1c} \\ \ddot{\beta}_{1s} \end{Bmatrix} + \begin{bmatrix} \frac{\gamma}{8} & 2 \\ -2 & \frac{\gamma}{8} \end{bmatrix} \begin{Bmatrix} \dot{\beta}_{1c} \\ \dot{\beta}_{1s} \end{Bmatrix} + \begin{bmatrix} \bar{\omega}_{RF}^2 - 1 & \frac{\gamma}{8} \\ -\frac{\gamma}{8} & \bar{\omega}_{RF}^2 - 1 \end{bmatrix} \begin{Bmatrix} \beta_{1c} \\ \beta_{1s} \end{Bmatrix} = 0 \tag{8.16}$$

The eigenvalues of the coupled modes can be obtained by forming the characteristic determinant and solving for the roots.

The characteristic determinant is given as

$$\begin{vmatrix} S^2 + \frac{\gamma}{8} S + \bar{\omega}_{RF}^2 - 1 & 2S + \frac{\gamma}{8} \\ -2S - \frac{\gamma}{8} & S^2 + \frac{\gamma}{8} S + \bar{\omega}_{RF}^2 - 1 \end{vmatrix} = 0 \tag{8.17}$$

The characteristic equation is

$$\left(S^2 + \frac{\gamma}{8} S + \bar{\omega}_{RF}^2 - 1\right)^2 + \left(2S + \frac{\gamma}{8}\right)^2 = 0 \tag{8.18}$$

Equation 8.18 can be written as

$$\left(S^2+\frac{\gamma}{8}S+\bar{\omega}_{RF}^2-1\right)^2-\left(i\left\{2S+\frac{\gamma}{8}\right\}\right)^2=0 \tag{8.19}$$

Rearranging,

$$\left(S^2+\left(\frac{\gamma}{8}+2i\right)S+\bar{\omega}_{RF}^2-1+i\frac{\gamma}{8}\right)\left(S^2+\left(\frac{\gamma}{8}-2i\right)S+\bar{\omega}_{RF}^2-1-i\frac{\gamma}{8}\right)=0 \tag{8.20}$$

The two roots corresponding to the first term of Equation 8.20 can be written as

$$\begin{aligned}S_{3,4}&=\left\{-\frac{\gamma}{8}-2i\pm\sqrt{\left(\frac{\gamma}{8}+2i\right)^2-4\left(\bar{\omega}_{RF}^2-1+i\frac{\gamma}{8}\right)}\right\}\frac{1}{2}\\ S_{3,4}&=-\frac{\gamma}{16}+i\left(-1\pm\sqrt{\bar{\omega}_{RF}^2-\left(\frac{\gamma}{16}\right)^2}\right)\end{aligned} \tag{8.21}$$

The other two roots corresponding to the second term of Equation 8.20 are

$$S_{5,6}=-\frac{\gamma}{16}+i\left(1\pm\sqrt{\bar{\omega}_{RF}^2-\left(\frac{\gamma}{16}\right)^2}\right) \tag{8.22}$$

Since S_4,S_5 and S_3,S_6 are complex conjugate pairs, let us consider only S_3 and S_5:

$$S_{3,5}=-\frac{\gamma}{16}\mp i+i\sqrt{\bar{\omega}_{RF}^2-\left(\frac{\gamma}{16}\right)^2} \tag{8.23}$$

The frequencies of the 1-cosine and the 1-sine cyclic modes in the fixed frame are shifted by ±1/rev from the rotating system frequency. A comparison of frequencies and damping in the rotating and nonrotating system is given in Table 8.1.

It is important to recognize that the collective and the differential modes are uncoupled modes, whereas the cyclic modes are coupled. For example, when $\bar{\omega}_{RF}=1.1$/rev and $\gamma = 8$, the frequency in the collective and the differential modes is 0.9798/rev. The frequencies in cyclic modes are 1.9798/rev

TABLE 8.1

Comparison of Damping and Frequencies in Rotating and Nonrotating Frames

Rotating Frame		Nonrotating Frame		
All Blades Have the Same Eigenvalues		Different Modes Have Different Eigenvalues		
Damping	**Frequency**	**Mode**	**Damping**	**Frequency**
$-\frac{\gamma}{16}$	$\sqrt{\bar{\omega}_{RF}^2-(\gamma/16)^2}$	Collective	$-\frac{\gamma}{16}$	$\sqrt{\bar{\omega}_{RF}^2-(\gamma/16)^2}$
		Differential	$-\frac{\gamma}{16}$	$\sqrt{\bar{\omega}_{RF}^2-(\gamma/16)^2}$
		Cyclic[a]	$-\frac{\gamma}{16}$	$1+\sqrt{\bar{\omega}_{RF}^2-(\gamma/16)^2}$
		Cyclic[b]	$-\frac{\gamma}{16}$	$-1+\sqrt{\bar{\omega}_{RF}^2-(\gamma/16)^2}$

[a] Denoted as the high-frequency flap mode (or progressive flap).
[b] Denoted as the low-frequency flap mode (or regressive flap).

and 0.0202/rev. In each cyclic mode, the tip-path plane has both β_{1c} and β_{1s} components. These frequencies correspond to the wobbling motion of the tip–path–plane. The damping in all flap modes is very high, which is equal to $\frac{\gamma}{16}$. Hence, these are damped out quickly. In the rotating frame, all the blades have the same damping and frequency, whereas in the nonrotating frame, all rotor modes have the same damping but have different frequencies.

Similar to flap modes, lag modes also provide high-frequency and low-frequency cyclic modes, Generally, the low-frequency cyclic lag and flap modes couple with the body pitch or roll modes. When the low-frequency lag mode coincides with body frequencies, it leads to aeromechanical instability such as ground or air resonance. Sometimes, the low-frequency flap mode plays an important role in influencing the air resonance behavior of the helicopter. Since damping in lag modes is very less, the low-frequency lag mode can become unstable during operation. Hence, an external lag damper is provided to augment the lag mode damping and also to avoid lag instability.

When the multiblade coordinate transformation is applied to perturbation equations in forward flight, unlike hover, all the flap modes will be coupled. This formulation is given as an exercise to understand the formulation of the multiblade coordinate transformation applied to helicopter rotor blade dynamics. As indicated earlier, the number of rotor modes is dependent on the number of blades in the rotor system.

9

Flap Dynamics under General Motion of the Hub

In the earlier chapter, the flap dynamics of the rotor blade was developed either under hovering condition or under steady forward motion of the hub (or helicopter). For the study of the flight dynamics of the helicopter, it is essential to develop the flap dynamic equation by including a general motion of the hub and analyzing the effect of the hub motion on rotor blade flap motion. It may be noted that hub motion is created due to fuselage translational and rotational motions. While developing the equations, it is assumed that the fuselage is a rigid body and that it can undergo rigid body translation and rotation motions. The flap dynamic equation of a rotor blade is developed in a systematic manner in the following.

Coordinate Systems and Representation of Various Quantities

First, we need to define several coordinate systems for a systematic derivation of the equation of the motion of a rotor blade when the hub undergoes a general motion. Figure 9.1 shows a body-fixed ($x_{s1} - y_{s1} - z_{s1}$) coordinate system, with origin at the c.g. of the helicopter. The unit vectors are given as $\hat{e}_{xs1}, \hat{e}_{ys1}, \hat{e}_{zs1}$.

The translational velocity of the helicopter is represented as components in the body-fixed system as

$$\vec{V}_{\text{c.g.}} = u\hat{e}_{xs1} + v\hat{e}_{ys1} + w\hat{e}_{zs1} \tag{9.1}$$

The instantaneous angular velocity of the helicopter is given as

$$\vec{\omega}_{\text{heli}} = p\hat{e}_{xs1} + q\hat{e}_{ys1} + r\hat{e}_{zs1} \tag{9.2}$$

The position vector of the main rotor hub center H from the c.g. of the fuselage is given as

$$\vec{r}_{\text{H}} = h_x\hat{e}_{xs1} + h_y\hat{e}_{ys1} + h_z\hat{e}_{zs1} \tag{9.3}$$

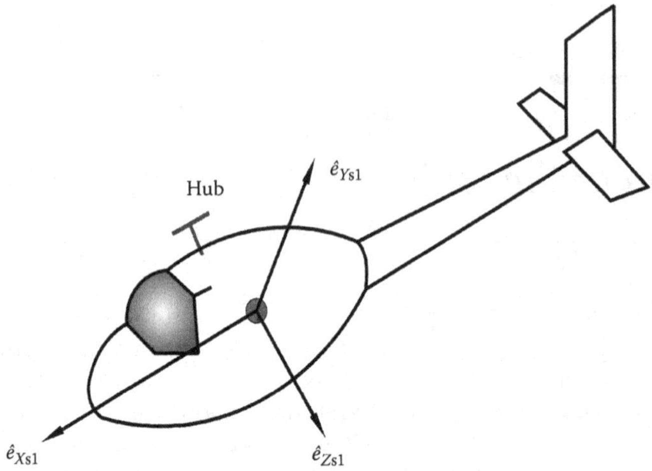

FIGURE 9.1
Body-fixed coordinate system with origin at the c.g. of the helicopter.

The velocity at the center of the hub is due to helicopter translational and rotational motions, which is defined as

$$\vec{V}_{\text{H}} = \vec{V}_{\text{c.g.}} + \vec{\omega}_{\text{heli}} X \vec{r}_{\text{H}} \tag{9.4}$$

Substituting various vector quantities from Equations 9.1 to 9.3, the velocity at the hub center is given as

$$\vec{V}_{\text{H}} = (u + qh_z - rh_y)\hat{e}_{xs1} + (v + rh_x - ph_y)\hat{e}_{ys1} + (w + ph_y - qh_x)\hat{e}_{zs1} \tag{9.5}$$

For the sake of conciseness, one can write the velocity at the hub as

$$\vec{V}_{\text{H}} = (u_{\text{H}})\hat{e}_{xs1} + (v_{\text{H}})\hat{e}_{ys1} + (w_{\text{H}})\hat{e}_{zs1} \tag{9.6}$$

Figure 9.2 shows the hub-fixed nonrotating and body-fixed coordinate systems.

The hub-fixed coordinate system ($x_{\text{H}}, y_{\text{H}}, z_{\text{H}}$) has its origin at the center of the hub, with its Z_{H} axis along the rotor shaft pointing vertically upward. The transformation relationship between the hub-fixed H system and the fuselage-fixed s1 system is given as

$$\begin{Bmatrix} \hat{e}_{x\text{H}} \\ \hat{e}_{y\text{H}} \\ \hat{e}_{z\text{H}} \end{Bmatrix} = \begin{bmatrix} -1 & 0 & 0 \\ 0 & 1 & 0 \\ 0 & 0 & -1 \end{bmatrix} \begin{Bmatrix} \hat{e}_{xs1} \\ \hat{e}_{ys1} \\ \hat{e}_{zs1} \end{Bmatrix} \tag{9.7}$$

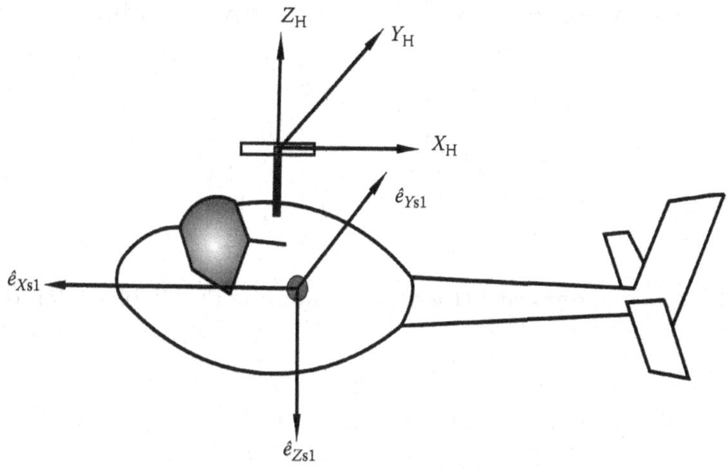

FIGURE 9.2
Hub-fixed nonrotating and body-fixed coordinate systems.

It is assumed that, in this formulation, the rotor shaft axis is parallel to the fuselage z_{s1} axis.

Figure 9.3 shows the hub-fixed nonrotating coordinate system and the hub-fixed rotating (x_{1k}, y_{1k}, z_{1k}) coordinate system rotating with the kth blade. The rotor system is rotating with an angular rate Ω about the z_1 (or z_H) axis.

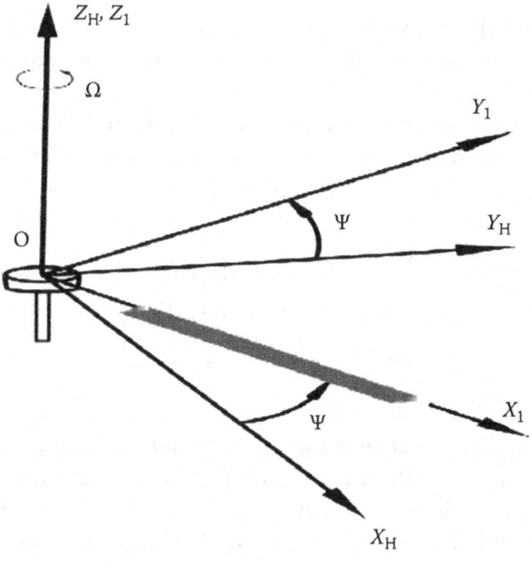

FIGURE 9.3
Hub fixed nonrotating and hub-fixed rotating coordinate systems.

The transformation relationship between the two coordinate systems is given as

$$\begin{Bmatrix} \hat{e}_{x1k} \\ \hat{e}_{y1k} \\ \hat{e}_{z1k} \end{Bmatrix} = \begin{bmatrix} \cos\psi_k & \sin\psi_k & 0 \\ -\sin\psi_k & \cos\psi_k & 0 \\ 0 & 0 & 1 \end{bmatrix} \begin{Bmatrix} \hat{e}_{xH} \\ \hat{e}_{yH} \\ \hat{e}_{zH} \end{Bmatrix} \tag{9.8}$$

where the angle ψ_k represents the azimuth angle of the kth blade. It is defined as

$$\psi_k = \psi + \frac{2\pi}{N}(k-1)$$

N is the total number of blades in the rotor system and $\psi = \Omega t$.

Combining Equations 9.7 and 9.8, one can relate the fuselage system with the hub-fixed system rotating as

$$\begin{Bmatrix} \hat{e}_{x1k} \\ \hat{e}_{y1k} \\ \hat{e}_{z1k} \end{Bmatrix} = \begin{bmatrix} -\cos\psi_k & \sin\psi_k & 0 \\ \sin\psi_k & \cos\psi_k & 0 \\ 0 & 0 & -1 \end{bmatrix} \begin{Bmatrix} \hat{e}_{xs1} \\ \hat{e}_{ys1} \\ \hat{e}_{zs1} \end{Bmatrix} \tag{9.9}$$

Next, we define the blade-fixed rotating coordinate system with its origin at the blade root, which is offset by a distance e from the hub center, as shown in Figure 9.4.

It may be noted that $1k$ and $1ek$ systems are parallel to each other, and the transformation relating is given as follows. It is assumed that the x_{1ek} axis is along the undeformed state of the kth blade.

$$\begin{Bmatrix} \hat{e}_{x1ek} \\ \hat{e}_{y1ek} \\ \hat{e}_{z1ek} \end{Bmatrix} = \begin{bmatrix} 1 & 0 & 0 \\ 0 & 1 & 0 \\ 0 & 0 & 1 \end{bmatrix} \begin{Bmatrix} \hat{e}_{x1k} \\ \hat{e}_{y1k} \\ \hat{e}_{z1k} \end{Bmatrix} \tag{9.10}$$

Figure 9.5 shows the rotating coordinate system attached to the rotor blade defining the deformed state during flap motion. It is assumed that the rotor blade is assumed to be rigid, with a root spring at the root at point A, as shown in Figure 9.4. The axis x_{2k} is along the deformed longitudinal axis of the blade.

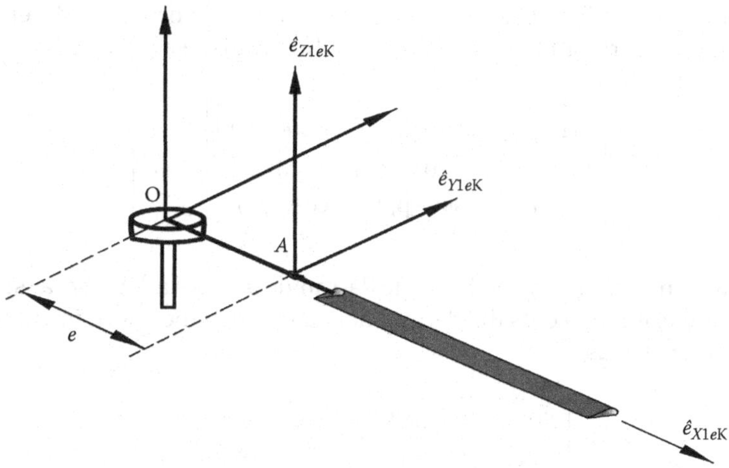

FIGURE 9.4
Blade-fixed rotating coordinate system with origin at blade root A.

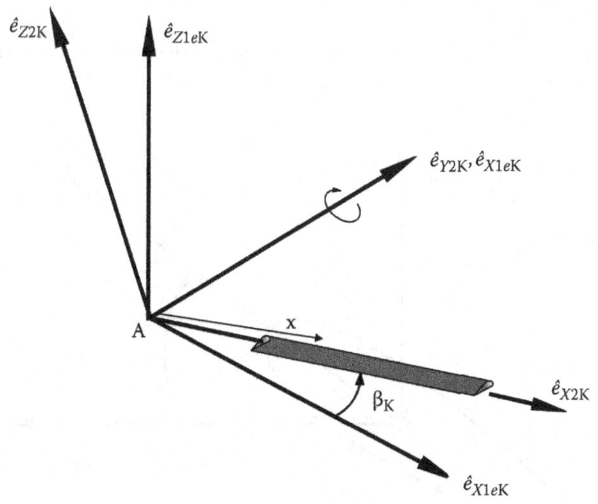

FIGURE 9.5
Rotating blade-fixed coordinate system during flap deformation.

The transformation relationship between the undeformed and deformed states of the kth rotor blade due to flap motion is given as

$$\begin{Bmatrix} \hat{e}_{x2k} \\ \hat{e}_{y2k} \\ \hat{e}_{z2k} \end{Bmatrix} = \begin{bmatrix} \cos(-\beta_k) & 0 & -\sin(-\beta_k) \\ 0 & 1 & 0 \\ +\sin(-\beta_k) & 0 & \cos(-\beta_k) \end{bmatrix} \begin{Bmatrix} \hat{e}_{x1ek} \\ \hat{e}_{y1ek} \\ \hat{e}_{z1ek} \end{Bmatrix} \tag{9.11}$$

It may be noted that, since the blade flap angle is taken as $-\beta_k$, because when β_k is positive, it represents the flap-up deformation of the blade, Equation 9.11 can be rewritten as

$$\begin{Bmatrix} \hat{e}_{x2k} \\ \hat{e}_{y2k} \\ \hat{e}_{z2k} \end{Bmatrix} = \begin{bmatrix} \cos\beta_k & 0 & \sin\beta_k \\ 0 & 1 & 0 \\ -\sin\beta_k & 0 & \cos\beta_k \end{bmatrix} \begin{Bmatrix} \hat{e}_{x1ek} \\ \hat{e}_{y1ek} \\ \hat{e}_{z1ek} \end{Bmatrix} \tag{9.12}$$

The transformation relationship between the deformed state of the rotor blade and the fuselage-fixed system can be written as

$$\begin{Bmatrix} \hat{e}_{x2k} \\ \hat{e}_{y2k} \\ \hat{e}_{z2k} \end{Bmatrix} = \begin{bmatrix} -\cos\beta_k \cos\psi_k & \cos\beta_k \sin\psi_k & -\sin\beta_k \\ \sin\psi_k & \cos\psi_k & 0 \\ \sin\beta_k \cos\psi_k & -\sin\beta_k \sin\psi_k & -\cos\beta_k \end{bmatrix} \begin{Bmatrix} x_{s1} \\ y_{s1} \\ z_{s1} \end{Bmatrix} \tag{9.13}$$

Now, let us define the blade cross-sectional coordinate system, as shown in Figure 9.6.

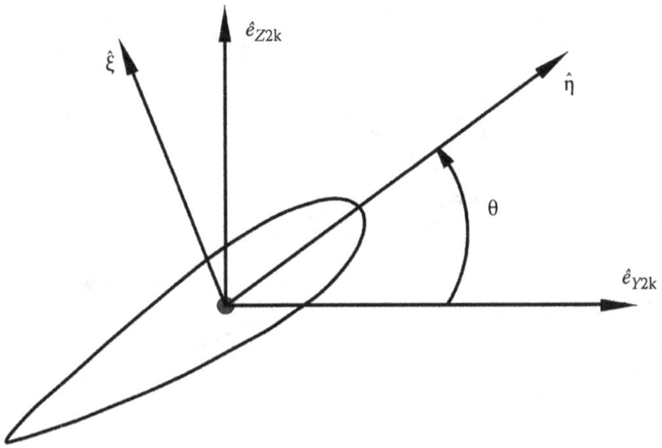

FIGURE 9.6
Blade cross-sectional coordinate system.

The transformation between the cross-sectional coordinate system at a distance r from the hinge offset location and the deformed state of the blade can be written as

$$\begin{Bmatrix} \hat{e}_{x3k} \\ \hat{\eta} \\ \hat{\xi} \end{Bmatrix} = \begin{bmatrix} 1 & 0 & 0 \\ 0 & \cos\theta & \sin\theta \\ 0 & -\sin\theta & \cos\theta \end{bmatrix} \begin{Bmatrix} \hat{e}_{x2k} \\ \hat{e}_{y2k} \\ \hat{e}_{z2k} \end{Bmatrix} \tag{9.14}$$

The angle θ represents the cross-sectional pitch angle. The reference axis of the blade is taken as the x_{2k} (or x_{3k}) axis.

It may be noted that the aerodynamic loads are evaluated based on the velocity components defined in the deformed $2k$ system.

First, let us evaluate the inertia loads and then the aerodynamic loads for formulating the equation of the motion of the kth blade under the general motion of the hub.

Kinematics

From the hub center, the position vector of any point P on the reference longitudinal axis of the blade in the deformed state can be written as (from Figures 9.4 and 9.5)

$$\vec{r}_p = R(e\hat{e}_{x1ek} + x\hat{e}_{x2k}) \tag{9.15}$$

Note that the length units are nondimensionalized with respect to the rotor radius R.

Using the coordinate transformation given in Equation 9.12, the position vector can be expressed as

$$\vec{r}_p = R(e\hat{e}_{x1ek} + x\cos\beta_k\hat{e}_{x1ek} + x\sin\beta_k\hat{e}_{z1ek}) \tag{9.16}$$

The angular velocity of the kth rotor blade is due to fuselage angular velocity (Equation 9.2) and rotor angular velocity. It is given as

$$\vec{\omega}_k = \vec{\omega}_{\text{heli}} + \Omega\hat{e}_{zH} \tag{9.17}$$

Substituting Equation 9.2 and using the coordinate transformations, the angular velocity of the kth blade can be written as

$$\vec{\omega}_k = -p\hat{e}_{xH} + q\hat{e}_{yH} - r\hat{e}_{zH} + \Omega\hat{e}_{zH} \tag{9.18}$$

Nondimensional Form

It is convenient to use nondimensional form while expressing various quantities. Time is nondimensionalized by rotor angular speed Ω, and length is nondimensionalized with rotor radius R.

We can express the angular velocity of the kth blade as

$$\vec{\omega}_K = \Omega[-p\hat{e}_{xH} + q\hat{e}_{yH} + (-r+1)\hat{e}_{zH}] \quad (9.19)$$

(Note that the quantities inside the bracket are nondimensional quantities. For the sake of convenience, the same symbol is also used in nondimensional form.) Using the transformation relationships given in Equations 9.8 to 9.10, the angular velocity vector (Equation 9.19) can be expressed in the hub-fixed rotating coordinate system as

$$\vec{\omega}_k = \Omega[(-p\cos\psi_k + q\sin\psi_k)\hat{e}_{x1ek} + (p\sin\psi_k + q\cos\psi_k)\hat{e}_{y1ek} + (-r+1)\hat{e}_{z1ek}] \quad (9.20)$$

Taking the time derivative of Equation 9.20, the angular acceleration vector can be written as

$$\dot{\vec{\omega}}_k = \Omega^2 \begin{bmatrix} (-\dot{p}\cos\psi_k + q\cos\psi_k + p\sin\psi_k + \dot{q}\sin\psi_k)\hat{e}_{x1ek} \\ +(\dot{p}\sin\psi_k - q\sin\psi_k + p\cos\psi_k + \dot{q}\cos\psi_k)\hat{e}_{y1ek} + (-\dot{r})\hat{e}_{z1ek} \end{bmatrix} \quad (9.21)$$

The velocity vector at point P on the reference longitudinal axis of the blade can be obtained from the following expression:

$$\vec{V}_P = \vec{V}_H + (\dot{\vec{r}}_P)_{rel} + \vec{\omega}_k X \vec{r}_P \quad (9.22)$$

where $\vec{V}_H$ is the velocity at the hub center O. Substituting various quantities and the taking vector cross-product, the velocity at point P can be written as

$$\begin{aligned} \vec{V}_P = {} & \Omega R\{(u + qh_z - rh_y)\hat{e}_{xs1} + (v + rh_x - ph_y)\hat{e}_{ys1} + (w + ph_y - qh_x)\hat{e}_{zs1}\} \\ & + \Omega R\{-x\sin\beta_k\dot{\beta}_k\hat{e}_{x1ek} + x\cos\beta_k\dot{\beta}_k\hat{e}_{z1ek}\} \\ & + \Omega R\{(p\sin\psi_k + q\cos\psi_k)x\sin\beta_k\}\hat{e}_{x1ek} \\ & + \Omega R\{(-r+1)(e + x\cos\beta_k) - (-p\cos\psi_k + q\sin\psi_k)x\sin\beta_k\}\hat{e}_{y1ek} \\ & + \Omega R\{-(p\sin\psi_k + q\cos\psi_k)(e + x\cos\beta_k)\}\hat{e}_{z1ek} \end{aligned} \quad (9.23)$$

It should be noted that, in Equation 9.23, the quantities inside the brackets are nondimensional quantities. Length quantities are nondimensionalized with respect to the rotor radius *R*, and time is nondimensionalized with respect to the rotor angular velocity Ω. Transforming the velocity at point P to the 2*k* system, we have

$$\begin{aligned}\vec{V}_P = &\hat{e}_{x2k}\Omega R\{(u+qh_z-rh_y)\{-\cos\beta_k\cos\psi_k\}+(v+rh_x-ph_z)\cos\beta_k\sin\psi_k \\ &+(w+ph_y-qh_x)\{-\sin\beta_k\}-(p\sin\psi_k+q\cos\psi_k)e\sin\beta_k\} \\ &+\hat{e}_{y2k}\Omega R\{(u+qh_z-rh_y)\sin\psi_k+(v+rh_x-ph_z)\cos\psi_k \\ &+(-r+1)(e+x\cos\beta_k)-(-p\cos\psi_k+q\sin\psi_k\}x\sin\beta_k\} \\ &+\hat{e}_{z2k}\Omega R\{(u+qh_z-rh_y)\sin\beta_k\cos\psi_k+(v+rh_x-ph_z)(-\sin\beta_k\sin\psi_k) \\ &+(w+ph_y-qh_x)(-\cos\beta_k)-(p\sin\psi_k+q\cos\psi_k)(x+e\cos\beta_k)+x\dot{\beta}_k\}\end{aligned} \tag{9.24}$$

The absolute acceleration of mass point P can be obtained from the expression given as

$$\vec{a}_P = \vec{a}_H + (\ddot{\vec{r}}_P)_{rel} + 2(\vec{\omega}_k X(\dot{\vec{r}}_P)_{rel}) + \dot{\vec{\omega}}_k X\vec{r}_P + \vec{\omega}_k X(\vec{\omega}_k X\ \vec{r}_P) \tag{9.25}$$

The individual components of the accelerations are given below for convenience.

Using Equations 9.2 and 9.5, the acceleration at the center of the hub can be obtained, which is given as

$$\begin{aligned}\vec{a}_H = &\Omega^2 R[(\dot{u}+\dot{q}h_z-\dot{r}h_y)\hat{e}_{xs1}+(\dot{v}+\dot{r}h_x-\dot{p}h_z)\hat{e}_{ys1}+(\dot{w}+\dot{p}h_y-\dot{q}h_x)\hat{e}_{Zs1} \\ &+\hat{e}_{xs1}\{q(w+ph_y-qh_x)-r(v+rh_x-ph_z)\} \\ &+\hat{e}_{ys1}\{r(u+qh_z-rh_y)-p(w+ph_y-qh_x)\} \\ &+\hat{e}_{zs1}\{p(v+rh_x-ph_z)-q(u+qh_z-rh_y)\}]\end{aligned} \tag{9.26}$$

The relative acceleration can be written, using Equation 9.16, as

$$(\ddot{\vec{r}}_P)_{rel} = \Omega^2 R\left[\left(-x\cos\beta_k\dot{\beta}_k^2 - x\sin\beta_k\ddot{\beta}_k\right)\hat{e}_{x1ek} + \left(-x\sin\beta_k\dot{\beta}_k^2 + x\cos\beta_k\ddot{\beta}_k\right)\hat{e}_{z1ek}\right] \tag{9.27}$$

The Coriolis acceleration term can be written as

$$\begin{aligned}2\left(\vec{\omega}_k X(\dot{\vec{r}}_P)_{rel}\right) = &2\Omega^2 R[\hat{e}_{x1ek}\{(p\sin\psi_k+q\cos\psi_k)x\cos\beta_k\dot{\beta}_k\} \\ &+\hat{e}_{y1ek}\{-(-r+1)x\sin\beta_k\dot{\beta}_k-(-p\cos\psi_k+q\sin\psi_k)x\cos\beta_k\dot{\beta}_k\} \\ &+\hat{e}_{z1ek}\{(p\sin\psi_k+q\cos\ \psi_k)x\sin\beta_k\dot{\beta}_k\}]\end{aligned} \tag{9.28}$$

The angular acceleration term is given as

$$\begin{aligned}\dot{\vec{\omega}}_k X\vec{r}_P &= \Omega^2 R[\hat{e}_{x1ek}\{\sin\psi_k(\dot{p}-q)+\cos\psi_k(p+\dot{q})\}x\sin\beta_k \\ &+\hat{e}_{y1ek}\{-\dot{r}(x\cos\beta_k+e)-\{(-\dot{p}+q)\cos\psi_k+(p+\dot{q})\sin\psi_k\}x\sin\beta_k\} \\ &+\hat{e}_{z1ek}\{-\{(\dot{p}-q)\sin\psi_k+(p+\dot{q})\cos\psi_k\}(x\cos\beta_k+e)\}]\end{aligned} \quad (9.29)$$

The centripetal acceleration term is written as

$$\begin{aligned}\vec{\omega}_k X(\vec{\omega}_k X\vec{r}_P) &= \Omega^2 R[\hat{e}_{x1ek}\{(p\sin\psi_k+q\cos\psi_k)(-(p\sin\psi_k+q\cos\psi_k)(x\cos\beta_k+e)) \\ &-(-r+1)\{(-r+1)(x\cos\beta_k+e)-(-p\cos\psi_k+q\sin\psi_k)x\sin\beta_k\}\} \\ &+\hat{e}_{y1eK}\{(-r+1)\{p\sin\psi_k+q\cos\psi_k)x\sin\beta_k\} \\ &-(-p\cos\psi_k+q\sin\psi_k)\{-(p\sin\psi_k+q\cos\psi_k)(x\cos\beta_k+e)\}\} \\ &+\hat{e}_{z1ek}\{(-p\cos\psi_k+q\sin\psi_k)\{(-r+1)(x\cos\beta_k+e) \\ &-(-p\cos\psi_k+q\sin\psi_k)x\sin\beta_k\} \\ &-(p\sin\psi_k+q\cos\psi_k)(p\sin\psi_k+q\cos\psi_k)x\sin\beta_k\}]\end{aligned} \quad (9.30)$$

Substituting all the components of acceleration given in Equations 9.26 to 9.30 in Equation 9.25, the absolute acceleration of mass point P on the reference longitudinal axis of the rotor blade can be written as

$$\begin{aligned}\vec{a}_P = \hat{e}_{x1ek}\Omega^2 R[&-(\dot{u}+\dot{q}h_z-\dot{r}h_y)\cos\psi_k+(\dot{v}+\dot{r}h_z-\dot{p}h_z)\sin\psi_k \\ &-\{q(w+ph_y-qh_x)-r(v+rh_x-ph_z)\}\cos\psi_k \\ &+\{r(u+qh_z-rh_y)-p(w+ph_y-qh_x)\}\sin\psi_k \\ &+\left\{-x\cos\beta_k\dot{\beta}_k^2-x\sin\beta_k\ddot{\beta}_k\right\} \\ &+2(p\sin\psi_k+q\cos\psi_k)x\cos\beta_k\dot{\beta}_k \\ &+\{\sin\psi_k(\dot{p}-q)+\cos\psi_k(\dot{q}+p)\}x\sin\beta_k \\ &+\{(p\sin\psi_k+q\cos\psi_k)\{-(p\sin\psi_k+q\cos\psi_k)(e+x\cos\beta_k)\} \\ &-(-r+1)\{(-r+1)(e+x\cos\beta_k)-(-p\cos\psi_k+q\sin\psi_k)x\sin\beta_k\}\}] \\ +\hat{e}_{y1ek}\Omega^2 R[&(\dot{u}+\dot{q}h_z-\dot{r}h_y)\sin\psi_k+(\dot{v}+\dot{r}h_x-\dot{p}h_z)\cos\psi_k \\ &+\{q(w+ph_y-qh_x)-r(v+rh_x-ph_z)\}\sin\psi_k\end{aligned}$$

$$+\{r(u+qh_z-rh_y)-p(w+ph_y-qh_x)\}\cos\psi_k$$
$$+2\{-(-r+1)x\sin\beta_k\dot{\beta}_k-(-p\cos\psi_k+q\sin\psi_k)x\cos\beta_k\dot{\beta}_k\}$$
$$+\{-\dot{r}(x\cos\beta_k+e)-\{(-\dot{p}+q)\cos\psi_k+(p+\dot{q})\sin\psi_k\}x\sin\beta_k\}$$
$$+\{(-r+1)(p\sin\psi_k+q\cos\psi_k)x\sin\beta_k$$
$$-(-p\cos\psi_k+q\sin\psi_k)\{-(p\sin\psi_k+q\cos\psi_k)(e+x\cos\beta_k)\}\}]$$
$$+\hat{e}_{z1ek}\Omega^2R[-(\dot{w}+\dot{p}h_y-\dot{q}h_x)$$
$$+\{p(v+rh_x-ph_z)-q(u+qh_z-rh_y)\}$$
$$+\left\{-x\sin\beta_k\dot{\beta}_k^2+x\cos\beta_k\ddot{\beta}_k\right\}$$
$$+2(p\sin\psi_k+q\cos\psi_k)x\sin\beta_k\dot{\beta}_k$$
$$+\{-\{\sin\psi_k(\dot{p}-q)+\cos\psi_k(p+\dot{q})\}(e+x\cos\beta_k)\}$$
$$+\{(-p\cos\psi_k+q\sin\psi_k)\{(-r+1)(e+x\cos\beta_k)-(-p\cos\psi_k+q\sin\psi_k)x\sin\beta_k\}$$
$$-(p\sin\psi_k+q\cos\psi_k)(p\sin\psi_k+q\cos\psi_k)\,x\sin\beta_k\}]$$
$$(9.31)$$

Note that the quantities inside the square brackets are nondimensional quantities.

Using the acceleration expression given in Equation 9.31, one can evaluate the inertia force and the inertia moment about the flap hinge by integrating the distributed inertia force acting on the rotor blade.

The inertia force and the inertia moment at the blade root are obtained from the following integrals:

$$\text{Inertia force: } \vec{P}_\text{I}=\int_0^{\frac{R-Re}{R}} m(-\vec{a}_\text{p})R\,dx \qquad (9.32)$$

$$\text{Inertia moment at blade hinge: } \vec{Q}_\text{I}=\int_0^{\frac{R-Re}{R}} Rx\hat{e}_{x2k}Xm(-\vec{a}_\text{p})R\,dx \qquad (9.33)$$

Note that the integration over the nondimensional length of the blade.

Since the acceleration expression contains terms that are either dependent or independent of the nondimensional radial location x, the following three integrals representing the mass, the first moment of mass, and the mass moment of inertia of the blade about the root hinge are defined.

$$\int_0^{\frac{R-Re}{R}} mR\,dx = M_b \quad \int_0^{\frac{R-Re}{R}} mRxR\,dx = MX_{c.g.} \quad \int_0^{\frac{R-Re}{R}} mR^2x^2R\,dx = I_b \tag{9.34}$$

where M_b is the mass of the blade, $MX_{c.g.}$ is the first moment of blade mass about the flap hinge, and I_b is the mass moment of inertia of the blade about the flap hinge.

Inertia Moment in Flap

Substituting the acceleration expression in Equations 9.32 and 9.33 and integrating, one can obtain the inertia force and moment. Since we are interested in developing only the flap equation, in the following, the component of inertia moment about the flap hinge is given. The detailed expression for other components of forces and moments can be evaluated. Using the coordinate transformation, the inertia moment expression given in Equation 9.33 can be written as

$$\vec{Q}_{I1ek} = \int_0^{\frac{R-Re}{R}} R^2\{x\cos\beta_k\hat{e}_{x1ek} + x\sin\beta_k\hat{e}_{z1ek}\}X(-m\vec{a}_p\}dx \tag{9.35}$$

The expression for flap inertia moment is given as

$$\begin{aligned}
Q_{Iy1eK} = \Omega^2 < & MX_{c.g.}R\sin\beta_K\{(\dot{u}+\dot{q}h_z-\dot{r}h_y)\cos\psi_k-(\dot{v}+\dot{r}h_x-\dot{p}h_z)\sin\psi_k \\
& +\{q(w+ph_y-qh_x)-r(v+rh_x-ph_z)\}\cos\psi_k \\
& -\{r(u+qh_z-rh_y)-p(w+ph_y-qh_x)\}\sin\psi_k\} \\
& +I_b\sin\beta_k\left\{\cos\beta_k\dot{\beta}_k^2+\sin\beta_k\ddot{\beta}_k\right\} \\
& -I_b\sin\beta_k\ 2(p\sin\psi_k+q\cos\psi_k)\cos\beta_k\dot{\beta}_k \\
& -I_b\sin\beta_k\{(\dot{p}-q)\sin\psi_k+(\dot{q}+p)\cos\psi_k\}\sin\beta_k \\
& +(p\sin\psi_k+q\cos\psi_k)^2(MX_{c.g.}Re+I_b\cos\beta_k)\sin\beta_k \\
& +(-r+1)^2(MX_{c.g.}Re+I_b\cos\beta_k)\sin\beta_k \\
& -(-r+1)(-p\cos\psi_k+q\sin\psi_k)I_b\sin^2\beta_k > \\
& +\Omega^2 < MX_{c.g.}R\cos\beta_k\{-(\dot{w}+\dot{p}h_y-\dot{q}h_x)-p(v+rh_x-ph_z)+q(u+qh_z-rh_y)\}
\end{aligned}$$

$$+ I_b \cos\beta_k \left\{-\sin\beta_k \dot{\beta}_k^2 + \cos\beta_k \ddot{\beta}_k\right\}$$
$$+ I_b \cos\beta_k \; 2(p\sin\psi_k + q\cos\psi_k)\sin\beta_k \dot{\beta}_k$$
$$- \{\sin\psi_k(\dot{p} - q) + \cos\psi_k(p + \dot{q})\}[MX_{\text{c.g.}} Re + I_b \cos\beta_k]\cos\beta_k$$
$$+ (-p\cos\psi_k + q\sin\psi_k)(-r + 1)\{MX_{\text{c.g.}} Re + I_b \cos\beta_k\}\cos\beta_k$$
$$- (-p\cos\psi_k + q\sin\psi_k)^2 I_b \sin\beta_k \cos\beta_k$$
$$- (p\sin\psi_k + q\cos\psi_k)^2 I_b \sin\beta_k \cos\beta_k > \qquad (9.36)$$

Aerodynamic Moment in Flap

Knowing the aerodynamic loads acting at every cross section of the rotor blade, the aerodynamic flap moment can be obtained. Figure 9.7 shows a typical cross section of the rotor blade and the relative air velocity components required for the evaluation of the sectional aerodynamic loads.

The velocity components of the cross section due to the motion of the blade and the hub are given in Equation 9.24. The relative air velocity component

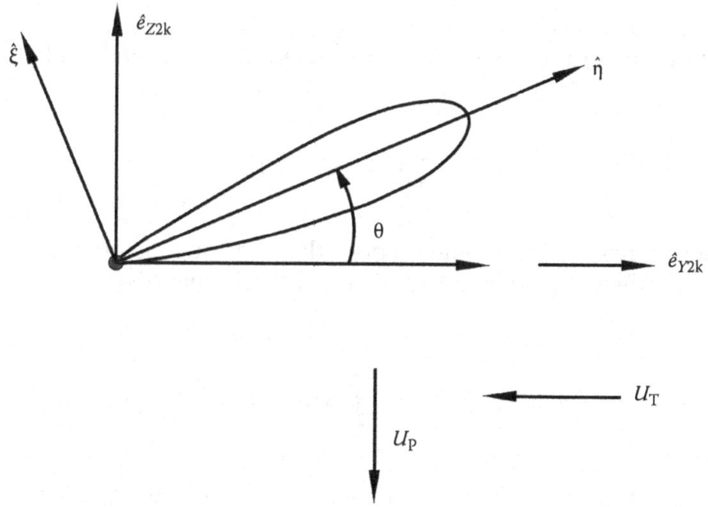

FIGURE 9.7
Cross section of the blade in the deformed state and the velocity components.

tangential to the cross section can be identified to be equal to the y component of the blade velocity, which is given as

$$U_T = +V_{Py2k} \tag{9.37}$$

Assuming the rotor inflow velocity as

$$\vec{v} = -\lambda\Omega R\,\hat{e}_{zH} \tag{9.38}$$

Using the transformation relationships, the normal component of the relative air velocity at a typical cross section of the rotor blade can be written as

$$U_P = +V_{Pz2k} + \lambda\Omega R\cos\beta_k \tag{9.39}$$

Using Equation 9.24, the expression for the tangential and normal components of the relative air velocity can be written as

$$\begin{aligned} U_T &= \Omega R\{(u+qh_z-rh_y)\sin\psi_k+(v+rh_x-ph_z)\cos\psi_k \\ &\quad +(-r+1)(e+x\cos\beta_k)-(-p\cos\psi_k+q\sin\psi_k)x\sin\beta_k\} \end{aligned} \tag{9.40}$$

and

$$\begin{aligned} U_P &= \Omega R\{(u+qh_z-rh_y)\sin\beta_k\cos\psi_k-(v+r\ h_x-ph_z)\sin\beta_k\sin\psi_k \\ &\quad -(w+ph_y-qh_x)\cos\beta_k+x\dot{\beta}_k-(p\sin\psi_k+q\cos\psi_k)(x+e\cos\beta_k)+\lambda\cos\beta_k\} \end{aligned} \tag{9.41}$$

The expression for the lift force per unit length acting on the blade can be written as

$$F_{z2} \simeq \frac{1}{2}\rho ca\left[U_T^2\theta-U_PU_T\right] \tag{9.42}$$

The aerodynamic flap moment about the flap hinge is obtained by integrating the effect of distributed lift force, and it is given as

$$Q_{Ay1ek} \cong -\int_0^{\frac{R-Re}{R}} xRF_{z2}R\,dx \tag{9.43}$$

Substituting Equation 9.42 in Equation 9.43 and using Equations 9.40 and 9.41, the aerodynamic flap moment can be obtained. While evaluating the aerodynamic flap moment, a large number of terms will have to be handled, if one retains all the terms. However, for the sake of convenience, many of the higher-order small terms can be neglected, based on a meaningful ordering scheme. In the following, various terms are assigned an order of magnitude, which will help in eliminating higher-order terms from the moment expression.

Rewriting the velocity expressions, given in Equations 9.40 and 9.41,

$$U_T = \Omega R\{A + B + C + Dx + Ex\} \tag{9.44}$$

$$U_P = \Omega R\{F + G + H + Ix + Jx + K + L\} \tag{9.45}$$

The individual terms and their leading term order can be defined, after making a small angle assumption for flap angle β_k, as

$$\begin{aligned}
&A = (u + qh_z - rh_y)\sin\psi_k && \text{Order } O(1)\\
&B = (v + rh_x - ph_z)\cos\psi_k && \text{Order } O(1)\\
&C = (-r + 1)e && \text{Order } O(1)\\
&D = (-r + 1)\cos\beta_k \simeq (-r + 1) && \text{Order } O(1)\\
&E = -(-p\cos\psi_k + q\sin\psi_k)\sin\beta_k \simeq -(-p\cos\psi_k + q\sin\psi_k)\beta_k && \text{Order } O(\epsilon^2)\\
&F = (u + qh_z - rh_y)\sin\beta_k\cos\psi_k \simeq (u + qh_z - rh_y)\beta_k\cos\psi_k && \text{Order } O(\epsilon)\\
&G = -(v + rh_x - ph_z)\sin\beta_k\sin\psi_k \simeq -(v + rh_x - ph_z)\beta_k\sin\psi_k && \text{Order } O(\epsilon)\\
&H = -(w + ph_y - qh_x)\cos\beta_k \simeq -(w + ph_y - qh_x) && \text{Order } O(\epsilon)\\
&I = \dot{\beta}_k && \text{Order } O(\epsilon)\\
&J = -(p\sin\psi_k + q\cos\psi_k) && \text{Order } O(\epsilon)\\
&K = -(p\sin\psi_k + q\cos\psi_k)e\cos\beta_k \simeq -(p\sin\psi_k + q\cos\psi_k)e && \text{Order } O(\epsilon)\\
&L = \lambda\cos\beta_k \simeq \lambda = \lambda_0 + \lambda_{1c}x\cos\psi_k + \lambda_{1s}x\sin\psi_k && \text{Order } O(\epsilon)
\end{aligned} \tag{9.46}$$

Note that the inflow velocity is assumed to be varying with azimuth and radial location.

The product of the velocity components required for evaluating the lift per unit span of the blade can be written as

$$\begin{aligned}
U_T^2 = (\Omega R)^2\{&A^2 + AB + AC + ADx + AEx\\
&+ AB + B^2 + BC + BDx + BEx\\
&+ AC + BC + C^2 + CDx + CEx\\
&+ ADx + BDx + CDx + D^2x^2 + DEx^2\\
&+ AEx + BEx + CEx + DEx^2 + E^2x^2\}
\end{aligned}$$

$$U_T^2 = (\Omega R)^2\{A^2 + 2AB + 2AC + 2ADx + 2AEx$$
$$+ B^2 + 2BC + 2BDx + 2BEx$$
$$+ C^2 + 2CDx + 2CEx$$
$$+ D^2x^2 + 2DEx^2$$
$$+ E^2x^2\}$$
$$U_P U_T = (\Omega R)^2\{AF + AG + AH + AIx + AJx + AK + AL$$
$$+ BF + BG + BH + BIx + BJx + BK + BL$$
$$+ CF + CG + CH + CIx + CJx + CK + CL$$
$$+ DFx + DGx + DHx + DIx^2 + DJx^2 + DKx + DLx$$
$$+ EFx + EGx + EHx + EIx^2 + EJx^2 + EKx + ELx\} \tag{9.47}$$

Substituting the velocity products (Equation 9.47) in Equation 9.42 and using Equation 9.43, the aerodynamic moment about the flap hinge at the blade root can be obtained. In symbolic form, it is given in the following, after expanding the expression for the blade pitch angle.

$$Q_{Ay1ek} = -\frac{1}{2}\rho ca\Omega^2 R^4 \left\{\left\{A^2\frac{l^2}{2} + 2AB\frac{l^2}{2} + 2AC\frac{l^2}{2} + 2AD\frac{l^3}{3} + 2AE\frac{l^3}{3}\right.\right.$$
$$+ B^2\frac{l^2}{2} + 2BC\frac{l^2}{2} + 2BD\frac{l^3}{3} + 2BE\frac{l^3}{3}$$
$$+ C^2\frac{l^2}{2} + 2CD\frac{l^3}{3} + 2CE\frac{l^3}{3}$$
$$+ D^2\frac{l^4}{4} + 2DE\frac{l^4}{4}$$
$$\left. + E^2\frac{l^4}{4}\right\}\{\theta_0 + \theta_{1c}\cos\psi_k + \theta_{1s}\sin\psi_k\}$$
$$-\left\{AF\frac{l^2}{2} + AG\frac{l^2}{2} + AH\frac{l^2}{2} + AI\frac{l^3}{3} + AJ\frac{l^3}{3} + AK\frac{l^2}{2} + AL\frac{l^2}{2}\right.$$
$$+ BF\frac{l^2}{2} + BG\frac{l^2}{2} + BH\frac{l^2}{2} + BI\frac{l^3}{3} + BJ\frac{l^3}{3} + BK\frac{l^2}{2} + BL\frac{l^2}{2}$$
$$+ CF\frac{l^2}{2} + CG\frac{l^2}{2} + CH\frac{l^2}{2} + CI\frac{l^3}{3} + CJ\frac{l^3}{3} + CK\frac{l^2}{2} + CL\frac{l^2}{2}$$

$$+DF\frac{l^3}{3}+DG\frac{l^3}{3}+DH\frac{l^3}{3}+DI\frac{l^4}{4}+DJ\frac{l^4}{4}+DK\frac{l^3}{3}+DL\frac{l^3}{3}$$

$$+EF\frac{l^3}{3}+EG\frac{l^3}{3}+EH\frac{l^3}{3}+EI\frac{l^4}{4}+EJ\frac{l^4}{4}+EK\frac{l^3}{3}+EL\frac{l^3}{3}\Bigg\}\Bigg\} \tag{9.48}$$

In Equation 9.48, the term l represents the nondimensional length of the blade from the hinge offset to the tip, that is, $l=\dfrac{R-Re}{R}=1-e$.

For the sake of completeness, the expanded form of the aerodynamic flap moment about the root hinge is given in the following without neglecting any term.

$$Q_{Ay1ek}=-\frac{1}{2}\rho ca^2R^4\Bigg\{\Bigg\{\frac{l^2}{2}<\sin^2\psi_k\{u^2+2u<qh_z-rh_y>+<qh_z-rh_y>^2\}>$$

$$+\frac{l^2}{2}<2\sin\psi_k\cos\psi_k\{u<v+rh_x-ph_z>+(qh_z-rh_y)(v+rh_x-ph_z)\}>$$

$$+\frac{l^2}{2}<2\sin\psi_k\{(u+qh_z-rh_y)e-er(u+qh_z-rh_y)\}>$$

$$+\frac{l^3}{3}<2\sin\psi_k\{(u+qh_z-rh_y)-r(u+qh_z-rh_y)\}>$$

$$+\frac{l^3}{3}<2\sin\psi_k(u+qh_z-rh_y)\{p\cos\psi_k-q\sin\psi_k\}\beta_k>$$

$$+\frac{l^2}{2}<\cos^2\psi_k\{v^2+2v(rh_x-ph_z)+(rh_x-ph_z)^2\}>$$

$$+\frac{l^2}{2}2\cos\psi_k\{(v+rh_x-ph_z)e-er(v+rh_x-ph_z)\}$$

$$+\frac{l^3}{3}2\cos\psi_k\{(v+rh_x-ph_z)-r(v+rh_x-ph_z)\}$$

$$+\frac{l^3}{3}2\cos\psi_k\{(v+rh_x-ph_z)\{p\cos\psi_k-q\sin\psi_k\}\beta_k\}$$

$$+\frac{l^2}{2}<e^2\{1-2r+r^2\}>$$

$$+\frac{l^3}{3}2<e\{1-2r+r^2)\}>$$

$$+\frac{l^3}{3}2<e(1-r)\{p\cos\psi_k-q\sin\psi_k\}\beta_k>$$

$$+\frac{l^4}{4}<\{1-2r+r^2\}>$$

$$+\frac{l^4}{4}2<(1-r)\{p\cos\psi_k-q\sin\psi_k\}\beta_k>$$

$$\left.+\frac{l^4}{4}<\{p\cos\psi_k-q\sin\psi_k\}^2\beta_k^2>\right\}(\theta_0+\theta_{1c}\cos\psi_k+\theta_{1s}\sin\psi_k)$$

$$-\left\{\frac{l^2}{2}<u^2+2u(qh_z-rh_y)+(qh_z-rh_y)^2>\beta_k\sin\psi_k\cos\psi_k\right.$$

$$+\frac{l^2}{2}<-\{u<v+rh_x-ph_z>+(qh_z-rh_y)(v+rh_x-ph_z)\}\beta_k\sin^2\psi_k>$$

$$+\frac{l^2}{2}<-u(w+ph_y-qh_x)-(qh_z-rh_y)(w+ph_y-qh_x)>\sin\psi_k$$

$$+\frac{l^3}{3}<(u+qh_z-rh_y)\dot{\beta}_k\sin\psi_k>$$

$$+\frac{l^3}{3}<-u(p\sin\psi_k+q\cos\psi_k)-(qh_z-rh_y)(p\sin\psi_k+q\cos\psi_k)>\sin\psi_k$$

$$+\frac{l^2}{2}e<-u(p\sin\psi_k+q\cos\psi_k)-(qh_z-rh_y)(p\sin\psi_k+q\cos\psi_k)>\sin\psi_k$$

$$+\frac{l^2}{2}<(u+qh_z-rh_y)\lambda_0\sin\psi_k>$$

$$+\frac{l^3}{3}<(u+qh_z-rh_y)\lambda_{1c}\cos\psi_k\sin\psi_k>$$

$$+\frac{l^3}{3}<(u+qh_z-rh_y)\lambda_{1s}\sin^2\psi_k>$$

$$+\frac{l^2}{2}<v(u+qh_z-rh_y)+(rh_x-ph_z)(u+qh_z-rh_y)>\beta_k\cos^2\psi_k$$

$$+\frac{l^2}{2}<-\{v^2+2\bar{v}(rh_x-ph_z)+(rh_x-ph_z)^2\}>\beta_k\cos\psi_k\sin\psi_k$$

$$+\frac{l^2}{2}<-\{v(w+ph_y-qh_x)+(rh_x-ph_z)(w+ph_y-qh_x)\}>\cos\psi_k$$

$$+\frac{l^3}{3}<(v+rh_x-ph_z)\dot{\beta}_K>\cos\psi_k$$

$$+\frac{l^3}{3}<-v(p\sin\psi_k+q\cos\psi_k)-(rh_x-ph_z)(p\sin\psi_k+q\cos\psi_k)>\cos\psi_k$$

$$+\frac{l^2}{2}e<-v(p\sin\psi_k+q\cos\psi_k)-(rh_x-ph_z)(p\sin\psi_k+q\cos\psi_k)>\cos\psi_k$$

$$+\frac{l^2}{2}<(v+rh_x-ph_z)\lambda_0\cos\psi_k>$$

$$+\frac{l^3}{3}<(v+rh_x-ph_z)\lambda_{1c}\cos^2\psi_k>$$

$$+\frac{l^3}{3}<(v+rh_x-ph_z)\lambda_{1s}\sin\psi_k\cos\psi_k>$$

$$+\frac{l^2}{2}e<(u+qh_z-rh_y)-r(u+qh_z-rh_y)>\beta_k\cos\psi_k$$

$$+\frac{l^2}{2}e<-(v+rh_x-ph_z)+r(v+rh_x-ph_z)>\beta_k\sin\psi_k$$

$$+\frac{l^2}{2}e<-(w+ph_y-qh_x)+r(w+ph_y-qh_x)>$$

$$+\frac{l^3}{3}e<(1-r)\dot{\beta}_k>$$

$$+\frac{l^3}{3}e<-(p\sin\psi_k+q\cos\psi_k)+r(p\sin\psi_k+q\cos\psi_k)>$$

$$+\frac{l^2}{2}e^2<-(p\sin\psi_k+q\cos\psi_k)+r(p\sin\psi_k+q\cos\psi_k)>$$

$$+\frac{l^2}{2}e<(1-r)\lambda_0>$$

$$+\frac{l^3}{3}e<(1-r)\lambda_{1c}\cos\psi_k>$$

$$+\frac{l^3}{3}e<(1-r)\lambda_{1s}\sin\psi_k>$$

$$+\frac{l^3}{3}<(u+qh_z-rh_y)-r(u+qh_z-rh_y)>\beta_k\cos\psi_k$$

$$+\frac{l^3}{3}<-(v+rh_x-ph_z)+r(v+rh_x-ph_z)>\beta_k\sin\psi_k$$

$$+\frac{l^3}{3}<-(w+ph_y-qh_x)+r(w+ph_y-qh_x)>$$

$$+\frac{l^4}{4}<(1-r)\dot{\beta}_k>$$

$$+\frac{l^4}{4}<-(p\sin\psi_k+q\cos\psi_k)+r(p\sin\psi_k+q\cos\psi_k)>$$

$$+\frac{l^3}{3}e<-(p\sin\psi_k+q\cos\psi_k)+r(p\sin\psi_k+q\cos\psi_k)>$$

$$+\frac{l^3}{3}<(1-r)\lambda_0>$$

$$+\frac{l^4}{4}<(1-r)\lambda_{1c}\cos\psi_k>$$

$$+\frac{l^4}{4}<(1-r)\lambda_{1s}\sin\psi_k>$$

$$+\frac{l^3}{3}<(p\cos\psi_k-q\sin\psi_k)(u+qh_z-rh_y)>\beta_k^2\cos\psi_k$$

$$+\frac{l^3}{3}<-(p\cos\psi_k-q\sin\psi_k)(v+rh_x-ph_z)>\beta_k^2\sin\psi_k$$

$$+\frac{l^3}{3}<-(p\cos\psi_k-q\sin\psi_k)(w+ph_y-qh_x)>\beta_k$$

$$+\frac{l^4}{4}<(p\cos\psi_k-q\sin\psi_k)\dot{\beta}_k>\beta_k$$

$$+\frac{l^4}{4}<-(p\cos\psi_k-q\sin\psi_k)(p\sin\psi_k+q\cos\psi_k)>\beta_k$$

$$+\frac{l^3}{3}e<-(p\cos\psi_k-q\sin\psi_k)(p\sin\psi_k+q\cos\psi_k)>\beta_k$$

$$+\frac{l^3}{3}<(p\cos\psi_k-q\sin\psi_k)\lambda_0>\beta_k$$

$$+\frac{l^4}{4}<(p\cos\psi_k-q\sin\psi_k)\lambda_{1c}\cos\psi_k>\beta_k$$

$$+\frac{l^4}{4}<(p\cos\psi_k-q\sin\psi_k)\lambda_{1s}\sin\psi_k>\beta_k\}\} \tag{9.49}$$

Combining the inertia moment (Equation 9.36) and the aerodynamic flap moment (Equation 9.49), the flap equation of the kth blade under general maneuver condition can be written as

$$Q_{Iy1ek} + Q_{Ay1ek} + K_\beta \beta_k = 0 \tag{9.50}$$

This flap equation is a nonlinear equation with time-varying coefficients. For a simplified analysis, one can make several approximations and analyze the flap dynamics of the rotor blade for perturbation in fuselage translation and rotational motions. First, let us assign an order of magnitude to various parameters and neglect the higher-order terms from the flap equation. Using the simplified flap equation, one can obtain the steady-state flap response for a given flight condition and also the transient flap response due to perturbations in the blade pitch input and in the fuselage motions. The perturbation response of the flap motion plays a significant role in determining the flight dynamic characteristics of the helicopter, that is, both the stability and the control response of the helicopter. In the following, all the parameters are represented as a combination of steady-state value and a perturbation quantity, along with an appropriate order of magnitude.

Parameter	**Order of magnitude**
$u = u_s + \tilde{u}(t)$	Order O(1)
$v = v_s + \tilde{v}(t)$	Order O(1)
$w = w_s + \tilde{w}(t)$	Order O(ϵ)
$p = p_s + \tilde{p}(t)$	Order $O\left(\epsilon^{3/2}\right)$
$q = q_s + \tilde{q}(t)$	Order $O\left(\epsilon^{3/2}\right)$
$r = r_s + \tilde{r}(t)$	Order $O\left(\epsilon^{3/2}\right)$
$\lambda_0 = \lambda_{0s} + \tilde{\lambda}_0(t)$	Order O(ϵ)
$\lambda_{1c} = \lambda_{1cs} + \tilde{\lambda}_{1c}(t)$	Order O(ϵ)
$\lambda_{1s} = \lambda_{1ss} + \tilde{\lambda}_{1s}(t)$	Order O(ϵ)
$\theta_0 = \theta_{0s} + \tilde{\theta}_0(t)$	Order O(1)
$\theta_{1c} = \theta_{1cs} + \tilde{\theta}_{1c}(t)$	Order O(1)
$\theta_{1s} = \theta_{1ss} + \tilde{\theta}_{1s}(t)$	Order O(1)
$\beta_k = \beta_{ks}(t) + \tilde{\beta}_k(t)$	Order O(ϵ)

(9.51)

In Equation 9.51, the quantities with (~) are perturbation quantities. It may be noted that, in representing the steady-state (or equilibrium) quantities, a subscript "s" is used. Substitute the terms in Equation 9.51 in the flap equation and separate the equations into two parts, after neglecting the products of the perturbation quantities. One part of the equation, containing only steady quantities, represents the steady-state nonlinear flap equation, and the other part of the equation containing perturbation quantities represents the linearized perturbation equation of the flap dynamics. As an example, let us consider the case of hover and obtain the perturbation flap equation about the hover equilibrium condition.

While obtaining the perturbation flap equation, let us assume that the hinge offset $e = 0$, and the harmonics of inflow $\lambda_{1c} = 0$ and $\lambda_{1s} = 0$.

Assuming that the helicopter is in hovering condition (i.e., u_s, v_s, $w_s = 0$, p_s, q_s, $r_s = 0$, and θ_{1cs}, $\theta_{1ss} = 0$), from Equation 9.50, the perturbation equation for flap dynamics can be written as

$$
\begin{aligned}
&\Omega^2\Big\{I_b\ddot{\tilde{\beta}}_k + I_b\tilde{\beta}_k - I_b < \left(\dot{\tilde{p}} - \tilde{q}\right)\sin\psi_k + \left(\dot{\tilde{q}} + \tilde{p}\right)\cos\psi_k > \\
&-MX_{c.g.}R\left(\dot{\tilde{w}} + \dot{\tilde{p}}h_y - \dot{\tilde{q}}h_x\right) + I_b < -\tilde{p}\cos\psi_k + \tilde{q}\sin\psi_k > \Big\} \\
&-\frac{1}{2}\rho ac\Omega^2 R^4\left\{\left\{\frac{1}{3}2\sin\psi_k(\tilde{u} + \tilde{q}h_z - \tilde{r}h_y) + \frac{1}{3}2\cos\psi_k(\tilde{v} + \tilde{r}h_x - \tilde{p}h_z) - \frac{1}{4}\,2\tilde{r}\right\}\theta_{0s}\right. \\
&+\frac{1}{4}(\tilde{\theta}_0 + \tilde{\theta}_{1c}\cos\psi_k + \tilde{\theta}_{1s}\sin\psi_k) \\
&-\left\{\frac{1}{2}(\tilde{u} + \tilde{q}h_z - \tilde{r}h_y)\lambda_{0s}\sin\psi_k + \frac{1}{2}(\tilde{v} + \tilde{r}h_x - \tilde{p}h_z)\lambda_{0s}\cos\psi_k\right. \\
&-\frac{1}{3}(\tilde{w} + \tilde{p}h_y - \tilde{q}h_x) + \frac{1}{4}\dot{\tilde{\beta}}_k - \frac{1}{4}(\tilde{p}\sin\psi_k + \tilde{q}\cos\psi_k) \\
&\left.\left.+\frac{1}{3}\tilde{\lambda}_0 - \frac{1}{3}\tilde{r}\lambda_0 + \frac{1}{4}\tilde{\lambda}_{1c}\cos\psi_k + \frac{1}{4}\tilde{\lambda}_{1s}\sin\psi_k\right\}\right\} + k_\beta\tilde{\beta}_k = 0
\end{aligned}
\tag{9.52}
$$

Dividing by $I_b\Omega^2$ and combining appropriate terms, Equation 9.52 can be written in a modified form as (it may be noted that the subscript "s" representing the steady-state quantities, in blade pitch angle and inflow, are removed for convenience, without creating any ambiguity)

$$
\begin{aligned}
&\ddot{\tilde{\beta}}_k + \frac{\gamma}{8}\dot{\tilde{\beta}}_k + \left(1 + \frac{K_\beta}{I_b\Omega^2}\right)\tilde{\beta}_k = \frac{MX_{c.g.}R}{I_b}\dot{\tilde{w}} + \left(\dot{\tilde{p}}\sin\psi_K + \frac{MX_{c.g.}R}{I_b}h_y\dot{\tilde{p}}\right) \\
&+\left(\dot{\tilde{q}}\cos\psi_k - \frac{MX_{c.g.}R}{I_b}h_x\dot{\tilde{q}}\right) + 2\tilde{p}\cos\psi_k - 2\tilde{q}\sin\psi_k
\end{aligned}
$$

$$+\frac{\gamma}{2}\left(\frac{2}{3}\theta_0-\frac{1}{2}\lambda_0\right)\tilde{u}\sin\psi_k+\frac{\gamma}{2}\left(\frac{2}{3}\theta_0-\frac{1}{2}\lambda_0\right)\tilde{v}\cos\psi_k+\frac{\gamma}{6}\tilde{w}$$

$$+\frac{\gamma}{2}\left(-\frac{2}{3}\theta_0+\frac{1}{2}\lambda_0\right)h_z\tilde{p}\cos\psi_k+\frac{\gamma}{2}\left(\frac{2}{3}\theta_0-\frac{1}{2}\lambda_0\right)h_z\tilde{q}\sin\psi_k$$

$$+\frac{\gamma}{2}\left(-\frac{2}{3}\theta_0+\frac{1}{2}\lambda_0\right)h_y\tilde{r}\sin\psi_k+\frac{\gamma}{2}\left(\frac{2}{3}\theta_0-\frac{1}{2}\lambda_0\right)h_x\tilde{r}\cos\psi_k$$

$$+\frac{\gamma}{2}\left(\frac{h_y}{3}\tilde{p}-\frac{h_x}{3}\tilde{q}\right)+\frac{\gamma}{2}\left(\frac{1}{4}\tilde{p}\sin\psi_k+\frac{1}{4}\tilde{q}\cos\psi_k\right)$$

$$+\frac{\gamma}{2}\left(-\frac{2}{4}\theta_0+\frac{1}{3}\lambda_0\right)\tilde{r}$$

$$-\frac{\gamma}{2}\left(\frac{1}{3}\tilde{\lambda}_0+\frac{1}{4}\tilde{\lambda}_{1c}\cos\psi_k+\frac{1}{4}\tilde{\lambda}_{1s}\sin\psi_k\right)$$

$$+\frac{\gamma}{2}\left(\frac{1}{4}\tilde{\theta}_0+\frac{1}{4}\tilde{\theta}_{1c}\cos\psi_k+\frac{1}{4}\tilde{\theta}_{1s}\sin\psi_k\right) \quad (9.53)$$

In Equation 9.53, γ is the Lock number, which is defined as $\gamma=\frac{\rho acR^4}{I_b}$. Equation 9.53 can be further simplified by making the assumption that the helicopter c.g. is lying on the rotor shaft axis. This assumption implies that $h_x = 0$ and $h_y = 0$. Imposing this condition, the perturbation flap equation (about the hovering condition) can be written as (noting that the nondimensional rotational natural frequency in the flap is given as $\bar{\omega}_{RF}^2=1+\frac{K_\beta}{I_b\Omega^2}$)

$$\ddot{\tilde{\beta}}_k+\frac{\gamma}{8}\dot{\tilde{\beta}}_k+\bar{\omega}_{RF}^2\tilde{\beta}_k=\frac{MX_{c.g.}R}{I_b}\dot{\tilde{w}}+\dot{\tilde{p}}\sin\psi_k+\dot{\tilde{q}}\cos\psi_k+2\tilde{p}\cos\psi_k-2\tilde{q}\sin\psi_k$$

$$+\frac{\gamma}{2}\left(\frac{2}{3}\theta_0-\frac{1}{2}\lambda_0\right)\tilde{u}\sin\psi_k+\frac{\gamma}{2}\left(\frac{2}{3}\theta_0-\frac{1}{2}\lambda_0\right)\tilde{v}\cos\psi_k+\frac{\gamma}{6}\tilde{w}$$

$$+\frac{\gamma}{2}\left(-\frac{2}{3}\theta_0+\frac{1}{2}\lambda_0\right)h_z\tilde{p}\cos\psi_k+\frac{\gamma}{2}\left(\frac{2}{3}\theta_0-\frac{1}{2}\lambda_0\right)h_z\tilde{q}\sin\psi_k$$

$$+\frac{\gamma}{2}\left(\frac{1}{4}\tilde{p}\sin\psi_k+\frac{1}{4}\tilde{q}\cos\psi_k\right)+\frac{\gamma}{2}\left(-\frac{2}{4}\theta_0+\frac{1}{3}\lambda_0\right)\tilde{r}$$

$$-\frac{\gamma}{2}\left(\frac{1}{3}\tilde{\lambda}_0+\frac{1}{4}\tilde{\lambda}_{1c}\cos\psi_k+\frac{1}{4}\tilde{\lambda}_{1s}\sin\psi_k\right)$$

$$+\frac{\gamma}{2}\left(\frac{1}{4}\tilde{\theta}_0+\frac{1}{4}\tilde{\theta}_{1c}\cos\psi_k+\frac{1}{4}\tilde{\theta}_{1s}\sin\psi_k\right) \quad (9.54)$$

Equation 9.54 clearly shows that the flap motion of the kth rotor blade is influenced by the perturbation in fuselage translation and rotational motions, the time-varying rotor inflow, and the time-varying blade pitch input given by the pilot. It is important to recognize that fuselage perturbation motion influences the dynamics of all the blades in the rotor system at the same time, but the effect depends on the azimuth location of the rotor blade. This point is observed from the presence of the azimuth location ψ_k of the kth blade.

Perturbation Flap Equations in the Multiblade Coordinate System

For the sake of convenience, let us consider a four-bladed rotor system. Applying the multiblade coordinate transformation operators given in Equation 8.6 to the perturbation flap equation in hovering condition (Equation 9.54), the following four equations are obtained. It may be noted that these equations are obtained after applying trigonometric identities.

Collective flap:

$$\ddot{\beta}_M + \frac{\gamma}{8}\dot{\beta}_M + \bar{\omega}_{RF}^2\beta_M = \dot{\tilde{w}}\frac{MX_{c.g.}R}{I_b} + \frac{\gamma}{2}\left\{\frac{\tilde{\theta}_0}{4} - \frac{\tilde{\lambda}_0}{3}\right\} + \frac{\gamma}{6}\tilde{w} - \frac{\gamma}{2}\left\{\frac{2\theta_0}{4} - \frac{\lambda_0}{3}\right\}\tilde{r} \tag{9.55}$$

Alternating flap or differential flap:

$$\ddot{\beta}_{-M} + \frac{\gamma}{8}\dot{\beta}_{-M} + \bar{\omega}_{RF}^2\beta_{-M} = 0 \tag{9.56}$$

It is obvious to note that the collective mode is influenced by perturbation in collective pitch $\tilde{\theta}_0$, vertical motion $\tilde{w}$, yaw motion $\tilde{r}$, and inflow $\tilde{\lambda}_0$. The alternating mode is uncoupled in hover.

1-cosine flap:

$$\begin{aligned}\ddot{\beta}_{1c} + 2\dot{\beta}_{1s} - \beta_{1c} + \frac{\gamma}{2}\left\{\frac{\dot{\beta}_{1c}}{4} + \frac{\beta_{1s}}{4}\right\} + \bar{\omega}_{RF}^2\beta_{1c} = 2\tilde{p} + \dot{\tilde{q}} + \frac{\gamma}{2}\left\{\frac{2}{3}\theta_0 - \frac{\lambda_0}{2}\right\}\tilde{v} \\ + \frac{\gamma}{2}\left\{-\frac{2}{3}\theta_0 + \frac{\lambda_0}{2}\right\}h_z\tilde{p} + \frac{\gamma}{2}\frac{1}{4}\tilde{q} + \frac{\gamma}{2}\frac{1}{4}\{\tilde{\theta}_{1c} - \tilde{\lambda}_{1c}\} = 0\end{aligned} \tag{9.57}$$

1-sine flap

$$\ddot{\beta}_{1s} - 2\dot{\beta}_{1c} - \beta_{1s} + \frac{\gamma}{2}\left\{\frac{\dot{\beta}_{1s}}{4} - \frac{\beta_{1c}}{4}\right\} + \bar{\omega}_{RF}^{2}\beta_{1s} = -\ 2\tilde{q} + \dot{\tilde{p}} + \frac{\gamma}{2}\left\{\frac{2}{3}\theta_0 - \frac{\lambda_0}{2}\right\}\tilde{u}$$
$$+\frac{\gamma}{2}\left\{\frac{2}{3}\theta_0 - \frac{\lambda_0}{2}\right\}h_z\tilde{q} + \frac{\gamma}{2}\frac{1}{4}\tilde{p} + \frac{\gamma}{2}\frac{1}{4}\left\{\tilde{\theta}_{1s} - \tilde{\lambda}_{1s}\right\} = 0 \tag{9.58}$$

Since 1-cosine and 1-sine equations are coupled, they can be written in matrix form as

$$\begin{bmatrix} 1 & 0 \\ 0 & 1 \end{bmatrix}\begin{Bmatrix} \ddot{\beta}_{1c} \\ \ddot{\beta}_{1s} \end{Bmatrix} + \begin{bmatrix} \frac{\gamma}{8} & 2 \\ -2 & \frac{\gamma}{8} \end{bmatrix}\begin{Bmatrix} \dot{\beta}_{1c} \\ \dot{\beta}_{1s} \end{Bmatrix} + \begin{bmatrix} \bar{\omega}_{RF}^{2} - 1 & \frac{\gamma}{8} \\ -\frac{\gamma}{8} & \bar{\omega}_{RF}^{2} - 1 \end{bmatrix}\begin{Bmatrix} \beta_{1c} \\ \beta_{1s} \end{Bmatrix}$$
$$= \frac{\gamma}{8}\begin{Bmatrix} \tilde{\theta}_{1c} \\ \tilde{\theta}_{1s} \end{Bmatrix} - \frac{\gamma}{8}\begin{Bmatrix} \tilde{\lambda}_{1c} \\ \tilde{\lambda}_{1s} \end{Bmatrix} + \frac{\gamma}{2}\left(\frac{2\theta_0}{3} - \frac{\lambda_0}{2}\right)\begin{Bmatrix} \tilde{v} - h_z\tilde{p} \\ \tilde{u} + h_z\tilde{q} \end{Bmatrix} + \begin{Bmatrix} \dot{\tilde{q}} \\ \dot{\tilde{p}} \end{Bmatrix} + \begin{bmatrix} \frac{\gamma}{8} & 2 \\ -2 & \frac{\gamma}{8} \end{bmatrix}\begin{Bmatrix} \tilde{q} \\ \tilde{p} \end{Bmatrix} \tag{9.59}$$

These equations describe the dynamics of the rotor degrees to the perturbational motion at the hub due to fuselage motion, the perturbation in inflow, and the perturbation in blade input. The cyclic mode equations are coupled; the cyclic flap motion β_{1c}, β_{1s} are influenced by perturbation in pitch ($\tilde{q}$), the roll ($\tilde{p}_y$) and translational motion ($\tilde{u}, \tilde{v}$) of the hub, inflow, and cyclic pitch input.

From Equations 9.55, 9.56, and 9.59, it is obvious that cyclic modes are uncoupled from the collective mode and the alternating mode. In addition, the influencing parameters are also different. Hence, cyclic modes in hover can be analyzed independently. (Note that, in forward flight, the collective mode will be coupled to cyclic and alternating modes.)

It was shown earlier that the rotor thrust vector is almost perpendicular to the tip-path plane. Hence, a tilt of the tip-path plane caused by hub motion or blade input essentially tilts the thrust vector. Therefore, the modified thrust vector can give rise to forces/moments about the center of the mass, leading to the perturbational motion of the vehicle. This rotor/body coupling can be understood by the following block diagram (Figure 9.8).

Since the damping in flap motion is high, the time to half amplitude is typically of the order of 0.04 to 0.05 s. This corresponds to an azimuthal motion

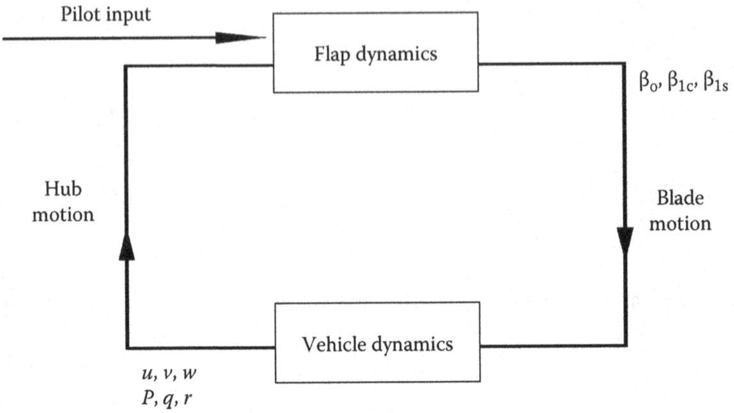

FIGURE 9.8
Schematic of the coupled rotor flap and fuselage dynamics.

of about 70° to 90°. Hence, the flap response reaches its steady-state value in a much shorter time than the perturbation input from the pilot or the perturbation in hub motion. In other words, the flap dynamics of the rotor system is much faster than the fuselage dynamics. Therefore, it is a good approximation to use the quasi-steady response of the flap motion. (This is also referred to as "low-frequency flap response.") This kind of approximation would be sufficient for the analysis of vehicle stability and control characteristics, where the frequency of motion is well below 1 Hz.

The quasi-steady approximation (or low-frequency approximation) helps in providing a good understanding of the physics of the rotor motion and also simplifies the formulation of the fuselage perturbation equations for stability and control analyses. If the quasi-steady assumption is not made, then one must include rotor dynamics in the study of vehicle dynamics. In the following, the quasi-steady response of the rotor degrees of freedom to hub motion and blade pitch input under hovering condition is presented. Neglecting the time derivative terms of the flap degrees of freedom from the left-hand side of flap equations, we have the steady-state flap response expressed as a function of hub motion.

Collective flap

$$\beta_M = \frac{1}{\bar{\omega}_{RF}^2}\left\{\frac{\gamma}{8}\left(\tilde{\theta}_0 - \frac{4}{3}\tilde{\lambda}_0\right) + \frac{\gamma}{6}\tilde{w} + \frac{MX_{c.g.}R}{I_b}\dot{\tilde{w}} - \gamma\left(\frac{\theta_0}{4} - \frac{\lambda_0}{6}\right)\tilde{r}\right\} \tag{9.60}$$

Collective flap β_M is influenced by collective pitch $\tilde{\theta}_0$, inflow $\tilde{\lambda}_0$, vertical translation $\tilde{\dot{w}}$ and $\tilde{w}$, and yaw motion $\tilde{r}$.

Cyclic flap:

$$\begin{bmatrix} \bar{\omega}_{RF}^2 - 1 & \frac{\gamma}{8} \\ -\frac{\gamma}{8} & \bar{\omega}_{RF}^2 - 1 \end{bmatrix} \begin{Bmatrix} \beta_{1c} \\ \beta_{1s} \end{Bmatrix} = \frac{\gamma}{8} \begin{Bmatrix} \tilde{\theta}_{1c} \\ \tilde{\theta}_{1s} \end{Bmatrix} - \frac{\gamma}{8} \begin{Bmatrix} \tilde{\lambda}_{1c} \\ \tilde{\lambda}_{1s} \end{Bmatrix} + \frac{\gamma}{2}\left(\frac{2\theta_0}{3} - \frac{\lambda_0}{2}\right) \begin{Bmatrix} \tilde{v} - h_z\tilde{p} \\ \tilde{u} + h_z\tilde{q} \end{Bmatrix} + \begin{Bmatrix} \dot{\tilde{q}} \\ \dot{\tilde{p}} \end{Bmatrix} + \begin{bmatrix} \frac{\gamma}{8} & 2 \\ -2 & \frac{\gamma}{8} \end{bmatrix} \begin{Bmatrix} \tilde{q} \\ \tilde{p} \end{Bmatrix} \tag{9.61}$$

Solving for β_{1c} and β_{1s} and using the notation,

$$S_c = \frac{\bar{\omega}_{RF}^2 - 1}{(\gamma/8)} \tag{9.62}$$

The parameter S_c defines the coupling between lateral and longitudinal motion due to the rotating flap frequency $\bar{\omega}_{RF}$. This parameter is also sometimes denoted as the "stiffness number."

$$\begin{Bmatrix} \beta_{1c} \\ \beta_{1s} \end{Bmatrix} = \frac{1}{\left(\frac{\gamma}{8}\right)^2 \left[\frac{\left(\bar{\omega}_{RF}^2 - 1\right)^2}{(\gamma/8)^2} + 1\right]} \left\{ \begin{bmatrix} \bar{\omega}_{RF}^2 - 1 & -\frac{\gamma}{8} \\ \frac{\gamma}{8} & \bar{\omega}_{RF}^2 - 1 \end{bmatrix} \left\{ \frac{\gamma}{8} \begin{Bmatrix} \tilde{\theta}_{1c} \\ \tilde{\theta}_{1s} \end{Bmatrix} - \frac{\gamma}{8} \begin{Bmatrix} \tilde{\lambda}_{1c} \\ \tilde{\lambda}_{1s} \end{Bmatrix} + \frac{\gamma}{2}\left(\frac{2\theta_0}{3} - \frac{\lambda_0}{2}\right) \begin{Bmatrix} \tilde{v} - h_z\tilde{p} \\ \tilde{u} + h_z\tilde{q} \end{Bmatrix} + \begin{Bmatrix} \dot{\tilde{q}} \\ \dot{\tilde{p}} \end{Bmatrix} + \begin{bmatrix} \frac{\gamma}{8} & 2 \\ -2 & \frac{\gamma}{8} \end{bmatrix} \begin{Bmatrix} \tilde{q} \\ \tilde{p} \end{Bmatrix} \right\} \right\} \tag{9.63}$$

Noting that, in hover $C_T = \frac{\sigma a}{2}\left\{\frac{\theta_0}{3} - \frac{\lambda_0}{2}\right\}$, and using Equation 9.62, the cyclic flap response to various perturbation quantities can be written as

$$\begin{Bmatrix} \beta_{1c} \\ \beta_{1s} \end{Bmatrix} = \frac{1}{S_c^2+1}\left[\begin{bmatrix} S_c & -1 \\ 1 & S_c \end{bmatrix}\begin{Bmatrix} \tilde{\theta}_{1c} \\ \tilde{\theta}_{1s} \end{Bmatrix} - \begin{bmatrix} S_c & -1 \\ 1 & S_c \end{bmatrix}\begin{Bmatrix} \tilde{\lambda}_{1c} \\ \tilde{\lambda}_{1s} \end{Bmatrix} + \begin{bmatrix} S_c & -1 \\ 1 & S_c \end{bmatrix} 8\left(\frac{2C_T}{\sigma a} + \frac{\lambda_0}{4}\right)\begin{Bmatrix} \tilde{v} - h_z\tilde{p} \\ \tilde{u} + h_z\tilde{q} \end{Bmatrix}\right.$$

$$\left. + \begin{bmatrix} S_c & -1 \\ 1 & S_c \end{bmatrix}\frac{1}{\left(\frac{\gamma}{8}\right)}\begin{Bmatrix} \dot{\tilde{q}} \\ \dot{\tilde{p}} \end{Bmatrix} + \begin{bmatrix} S_c & -1 \\ 1 & S_c \end{bmatrix}\begin{bmatrix} 1 & \frac{16}{\gamma} \\ -\frac{16}{\gamma} & 1 \end{bmatrix}\begin{Bmatrix} \tilde{q} \\ \tilde{p} \end{Bmatrix}\right] \tag{9.64}$$

Simplifying Equation 9.64,

$$\begin{Bmatrix} \beta_{1c} \\ \beta_{1s} \end{Bmatrix} = \frac{1}{S_c^2+1}\left[\begin{Bmatrix} S_c \\ 1 \end{Bmatrix}\tilde{\theta}_{1c} + \begin{Bmatrix} -1 \\ S_c \end{Bmatrix}\tilde{\theta}_{1s} - \begin{Bmatrix} S_c \\ 1 \end{Bmatrix}\tilde{\lambda}_{1c} - \begin{Bmatrix} -1 \\ S_c \end{Bmatrix}\tilde{\lambda}_{1s}\right.$$

$$+ 8\left(\frac{2C_T}{\sigma\alpha} + \frac{\lambda_0}{4}\right)\begin{Bmatrix} S_c \\ 1 \end{Bmatrix}(\tilde{v} - h_z\tilde{p}) + 8\left(\frac{2C_T}{\sigma\alpha} + \frac{\lambda_0}{4}\right)\begin{Bmatrix} -1 \\ S_c \end{Bmatrix}(\tilde{u} + h_z\tilde{q})$$

$$\left. + \frac{8}{\gamma}\begin{bmatrix} S_c \\ 1 \end{bmatrix}\dot{\tilde{q}} + \frac{8}{\gamma}\begin{Bmatrix} -1 \\ S_c \end{Bmatrix}\dot{\tilde{p}} + \begin{Bmatrix} S_c + \frac{16}{\gamma} \\ 1 - S_c\frac{16}{\gamma} \end{Bmatrix}\tilde{q} + \begin{Bmatrix} S_c\frac{16}{\gamma} - 1 \\ \frac{16}{\gamma} + S_c \end{Bmatrix}\tilde{p}\right] \tag{9.65}$$

If constant inflow is assumed, then the perturbation terms in inflow $\tilde{\lambda}_0, \tilde{\lambda}_{1c}$, and $\tilde{\lambda}_{1s}$ are all equal to 0. On the other hand, if the time variation of inflow is to be considered, then one has to formulate the appropriate equations for inflow dynamics in terms of blade and hub motion. Such a model is represented by a set of equations known as the "perturbation inflow model," which is written in the form

$$[L]^{-1}\begin{Bmatrix} \tilde{\lambda}_0 \\ \tilde{\lambda}_{1c} \\ \tilde{\lambda}_{1s} \end{Bmatrix} = \begin{Bmatrix} \text{perturbational thrust coefficient} \\ \text{perturbational roll moment} \\ \text{perturbational pitch moment} \end{Bmatrix} \tag{9.66}$$

Equation 9.66 is obtained by equating the perturbation expressions in rotor thrust, and the pitch and roll moment at the hub, obtained from the momentum theory and the blade element theory. If one takes into account the time delay in inflow dynamics, Equation 9.66 is extended by including a time derivative term, as shown:

$$[M]\begin{Bmatrix} \dot{\tilde{\lambda}}_0 \\ \dot{\tilde{\lambda}}_{1c} \\ \dot{\tilde{\lambda}}_{1s} \end{Bmatrix} + [L]^{-1}\begin{Bmatrix} \tilde{\lambda}_0 \\ \tilde{\lambda}_{1c} \\ \tilde{\lambda}_{1s} \end{Bmatrix} = \begin{Bmatrix} \text{perturbational thrust coefficient} \\ \text{perturbational roll moment} \\ \text{perturbational pitch moment} \end{Bmatrix} \tag{9.67}$$

Equation 9.67 is known as the "dynamic inflow model." For the sake of simplification, in the following, the inflow perturbations are neglected, and the inflow is taken as a constant. (Note that the details of these models can be referred to in published literature.)

Cyclic Flap Motion

Cyclic flap motion (β_{1c}, β_{1s}) is influenced by cyclic pitch input $(\tilde{\theta}_{1c}, \tilde{\theta}_{1s})$, hub translational motion $(\tilde{u}, \tilde{v})$, and rotational motion $(\tilde{p}, \tilde{q})$ and angular acceleration $(\dot{\tilde{p}}, \dot{\tilde{q}})$. The parameter $S_c = (\bar{\omega}_{RF}^2 - 1)/(\gamma/8)$ provides the coupling between lateral and longitudinal motion. The magnitude of S_c indicates the amount of cross-coupling. In the following, the relationship between flap motion derivatives and S_c is given.

Cross-coupling control derivatives:

$$\frac{\partial \beta_{1c}}{\partial \tilde{\theta}_{1c}} = \frac{\partial \beta_{1s}}{\partial \tilde{\theta}_{1s}} = \frac{S_c}{S_c^2 + 1}$$

Direct control derivatives:

$$\frac{\partial \beta_{1c}}{\partial \tilde{\theta}_{1s}} = -\frac{\partial \beta_{1s}}{\partial \tilde{\theta}_{1c}} = -\frac{1}{S_c^2 + 1}$$

Damping derivatives (cross-coupling) due to angular motion:

$$-\frac{\partial \beta_{1c}}{\partial \tilde{p}} = \frac{\partial \beta_{1s}}{\partial \tilde{q}} = \frac{1 - S_c \dfrac{16}{\gamma}}{S_c^2 + 1} + 8\left(\frac{2C_T}{\sigma\alpha} + \frac{\lambda_0}{4}\right) S_c h_z$$

Damping derivatives (direct) due to angular motion:

$$\frac{\partial\beta_{1c}}{\partial\tilde{q}} = \frac{\partial\beta_{1s}}{\partial\tilde{p}} = \frac{S_c + \dfrac{16}{\gamma}}{S_c^2+1} - 8\left(\frac{2C_T}{\sigma\alpha} + \frac{\lambda_0}{4}\right)h_z$$

Damping derivatives (cross-coupling) due to translational motion:

$$\frac{\partial\beta_{1c}}{\partial\tilde{v}} = \frac{\partial\beta_{1s}}{\partial\tilde{u}} = 8\left(\frac{2C_T}{\sigma\alpha} + \frac{\lambda_0}{4}\right)\frac{S_c}{S_c^2+1}$$

Damping derivatives (direct) due to translational motion:

$$\frac{\partial\beta_{1c}}{\partial\tilde{u}} = -\frac{\partial\beta_{1s}}{\partial\tilde{v}} = -8\left(\frac{2C_T}{\sigma\alpha} + \frac{\lambda_0}{4}\right)\frac{1}{S_c^2+1}$$

Cross-coupling derivative with angular acceleration:

$$\frac{\partial\beta_{1c}}{\partial\dot{\tilde{q}}} = \frac{\partial\beta_{1s}}{\partial\dot{\tilde{p}}} = \frac{8}{\gamma}\frac{S_c}{S_c^2+1}$$

Direct derivative with angular acceleration:

$$-\frac{\partial\beta_{1c}}{\partial\dot{\tilde{p}}} = \frac{\partial\beta_{1s}}{\partial\dot{\tilde{q}}} = \frac{8}{\gamma}\frac{1}{S_c^2+1} \tag{9.68}$$

Generally, the value of the coupling parameter S_c varies from 0 (for teetering rotor with $\bar{\omega}_{RF} = 1$) to 0.3 (for hingeless rotors). A high value of $S_c = 0.5$ is possible for heavy blades having a small value of Lock number γ. A plot of some of the derivatives is shown below in Figure 9.9.

In the range of $S_c = 0$ to 0.3 (Figure 9.9a), the direct control derivatives $(\partial\beta_{1c}/\partial\tilde{\theta}_{1s})$ and $(\partial\beta_{1s}/\partial\tilde{\theta}_{1c})$ are almost equal to unity; however, the cross-control derivatives $(\partial\beta_{1c}/\partial\tilde{\theta}_{1c})$ and $(\partial\beta_{1s}/\partial\tilde{\theta}_{1s})$ increases from 0 to 0.3. This shows that the cross-coupling becomes significant, that is, about 30% of the direct control derivative, for $S_c = 0.3$. Because of this cross-coupling, the hub moments transmitted to the fuselage will introduce a pitch–roll response coupling, which is dependent on the relative magnitude of the helicopter inertia in pitch and roll.

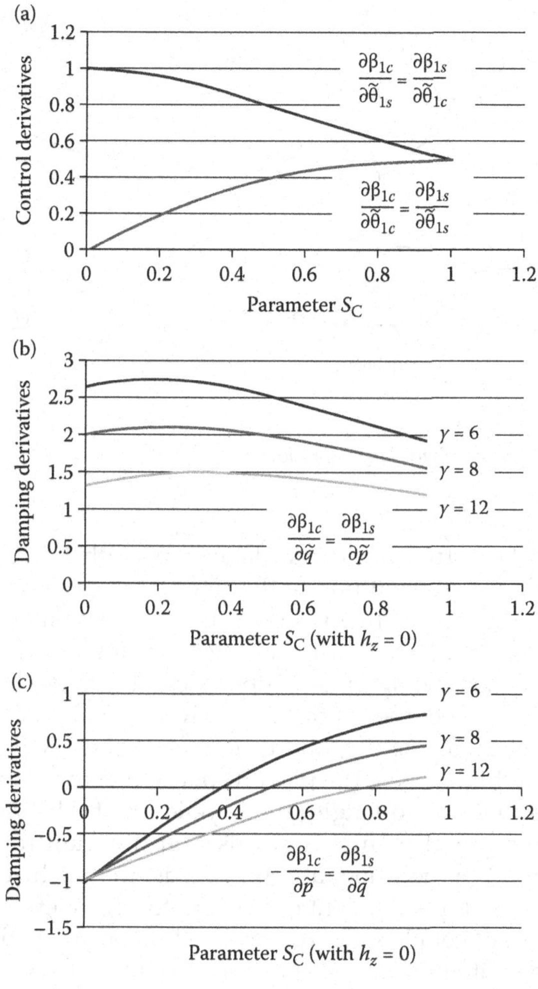

FIGURE 9.9
Important cyclic flap derivatives.

With $h_z = 0$ (Figure 9.9b), the direct damping derivative due to angular motion remains more or less a constant until $S_c = 0.5$, beyond which it decreases. The cross-damping (Figure 9.9c) varies linearly with S_c and changes its sign at high values of S_c. It is evident that the damping derivatives are highly dependent on the Lock number. For heavy blades ($\gamma = 6$), the direct damping is two-to-four times the cross-damping, whereas for light blades ($\gamma = 12$), direct damping and cross-damping are almost equal for low values of S_c. The direct damping derivative gives the angle of lag between the tip-path plane and the shaft. This lag angle provides an opposing effect due to the thrust tilt (providing an opposing moment about the fuselage c.g.) to the angular motion, as shown in Figure 9.10.

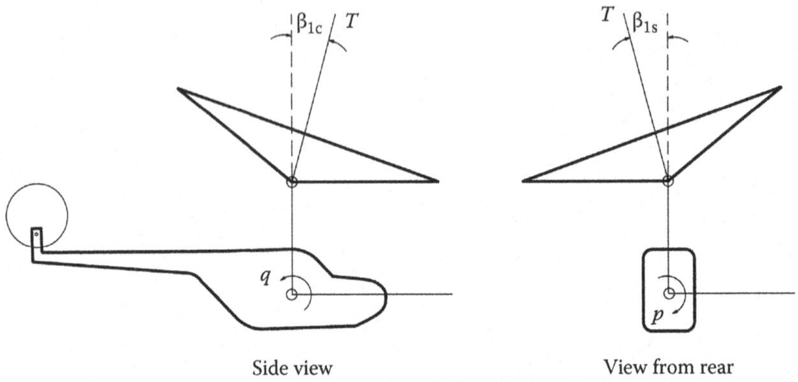

FIGURE 9.10
Tip-path plane lag due to hub (or shaft) rotation.

The damping derivatives due to translational perturbation motion depends on the operating thrust condition of the rotor. The variation with coupling parameter S_c shows a trend that is similar to the variation of control derivatives. The above formulation clearly indicates that, for hingeless rotor blades, the off-axis response due to cross-coupling can be as large as the on-axis response. At high speed, the control derivatives will also be coupled to collective input (i.e., blade coning, and differential modes will be coupled to cyclic modes). At high speed, the pitch response due to collective input can be as strong as that due to longitudinal cyclic input. The yaw response to the collective input will require compensation from tail rotor thrust variation (pedal input). In addition, at high speeds, the pitch response due to yaw motion may require different control strategies in left and right turns. These high levels of coupled motion are also influenced by the main rotor–wake interaction with the empennage control surfaces. Hence, the pilot must always stay in the loop to constantly provide corrective inputs to the vehicle for achieving the desired response. These cross-coupling effects seriously affect the performance of the helicopter. In addition, the control task of the helicopter will be further increased if the visual cues available to the pilot degrade (say, night flying, poor visibility, nap-of-the-earth flight, etc.) and also the aggressiveness of the maneuver.

It is important to recognize that, in determining the performance of the helicopter, the pilot's subjective opinion plays a significant role. The definition of Cooper and Harper (1969) on the handling quality is stated as "those qualities or characteristics of an aircraft that govern the ease and precision with which a pilot is able to perform the tasks required in support of an aircraft role." Since quantifying the ability of the pilot is more difficult, the Cooper–Harper pilot's rating scale is used (Figure 9.11) and is accepted as a measure of handling qualities.

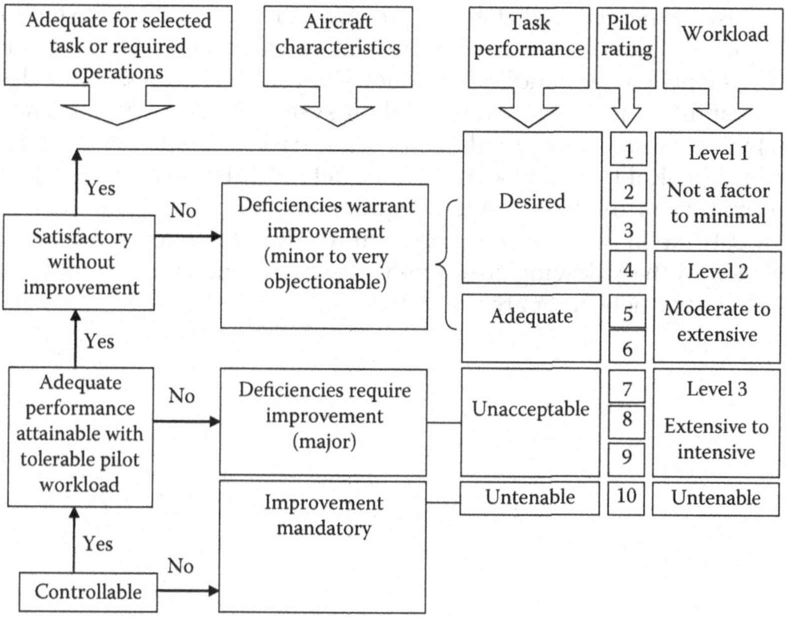

FIGURE 9.11
Cooper and Harper rating.

It is truly ideal to have a Level 1 vehicle satisfying the requirements throughout the operational flight envelope and for all mission tasks, under varying environmental conditions. However, in reality, it is almost impossible to meet the Level 1 criteria. In the absence of a Level 1 vehicle, it is acceptable to have Level 2 characteristics. It may be noted that, with handling quality rating of 6, the pilot workload is more, and he may not be able to fly for a long time due to fatigue. It may be interesting to note that a Level 2 vehicle in a good environment may become Level 3 under bad environment conditions. Although the pilot's opinion is a very important factor or may be the final decision in accepting a vehicle, quantitative criteria are necessary for setting design standards, which would be helpful for designers and certifying agencies. The most comprehensive set of requirements is provided by the U.S. Army's Aeronautical Design Standard for handling qualities (ADS-33) developed in 1982. The earlier requirements are due to the MIL-H-8501A (1961). The basic difference between the requirements of MIL-H-8501A and that of ADS-33 are related to the Mission Task Elements. The details of these can be found in the ADS-33 handling quality requirements of military rotorcraft.

Since the cross-coupling between various degrees of freedom of the helicopter deteriorates the performance of the helicopter, to improve the vehicle flying quality to a higher level and to reduce the pilot's workload, artificial stability and control augmentation is provided in the helicopter. However, before discussing artificial stability, one must analyze and understand first the flying qualities of a basic helicopter. There are several reasons to understand basic helicopter handling qualities. They are as follows: (1) the design of the stability and control augmentation system (SCAS) will become better; (2) in case of failed SCAS, the level of the basic helicopter characteristics define the criticality of the SCAS (flight safety or mission critical); (3) better flying quality of the basic vehicle requires less authority from SCAS; and (4) any saturation in SCAS authority will result in the pilot flying the basic helicopter. In the following chapter, the dynamics and stability characteristics of the basic helicopter are presented.

10

Helicopter Stability and Control

Analysis of helicopter stability and control is a very complicated dynamic problem involving the coupling of rotor motion with the fuselage degrees of freedom. In general, this problem can be simplified by assuming that blade flap dynamics is much faster than fuselage dynamics, and therefore, the steady-state flap response to fuselage perturbation motion is treated in a quasi-static (or also known as "low-frequency approximation") manner. In addition, the lead–lag and the torsion motions of the rotor blade can be neglected. The reason for this approximation can be attributed to the fact that the steady-state hub loads due to the flap motion of the blade are large compared to the hub loads due to the lag or the torsional motion. It is important to note that, for ground and air resonance problems, lag motion plays an important role, whereas in helicopter stability and control problems, lag mode is generally neglected.

In Chapter 9, we have analyzed the response of the rotor system in the flap mode (or the rotor tip-path plane) to perturbation in hub motion and also to control pitch input. Since the response of the tip-path plane to hub motion changes the orientation of the thrust vector, it gives rise not only to a force, but also to a moment about the center of mass of the fuselage. The study of helicopter dynamics is simplified by analyzing the interaction of hub loads generated due to rotor flap motion, which is due to fuselage perturbation motion. The two important aspects in the study of helicopter dynamics are related to (1) the stability of the vehicle about an operating condition, and (2) the dynamic response of the vehicle to a given control input by the pilot.

Stability

The stability of a dynamic system can be defined as the tendency of the system to return to the original state following a disturbance. Static stability is measured by the force or moment per unit disturbance, which is generated to restore the system to its original state. Dynamic stability deals with the time required to return to its original state following a unit disturbance, and it can be analyzed from the time response of the vehicle to an initial disturbance. The time histories describing various types of time response

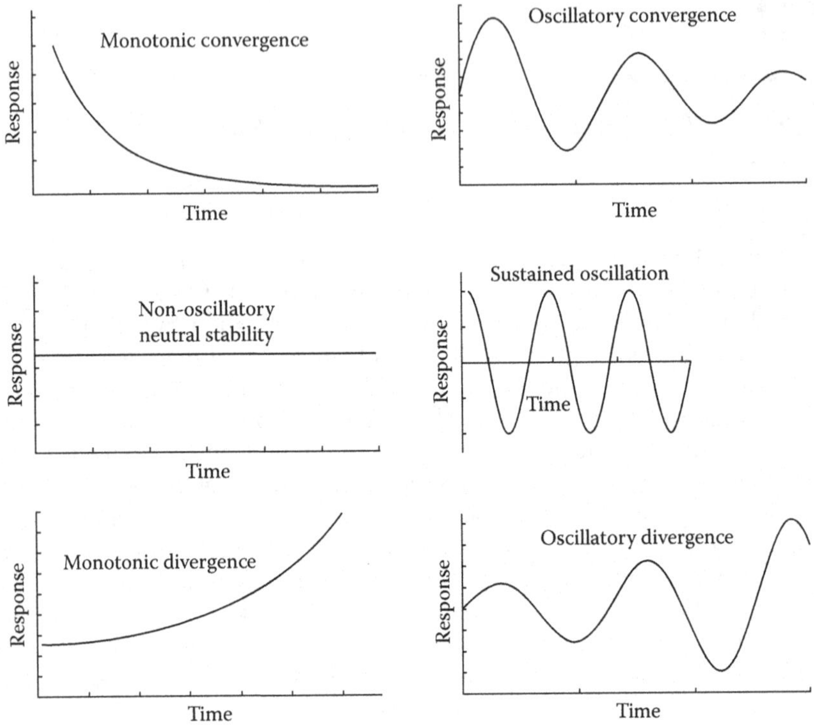

FIGURE 10.1
Types of time response describing dynamic stability.

indicating stability, namely, monotonic or oscillatory convergent motion, non-oscillatory or oscillatory neutral stable motion, and monotonic divergent or oscillatory divergent unstable motion, are shown in Figure 10.1.

Control

Control is related to the ability to generate necessary forces and moments that are required to perform a desired maneuver and/or to maintain the vehicle in the desired flight path under an external disturbance due to gust. In the following, two definitions representing the quantitative measure related to control are given.

Control power is defined as the measure of the total moment or force that can be generated in the vehicle to a given control input by the pilot to execute a maneuver.

Control sensitivity can be related to either the initial acceleration per unit stick movement (or control motion) or the steady-state velocity of the vehicle produced by unit stick motion.

The stability and control characteristics of the helicopter are analyzed using the flight dynamic equations of motion of the helicopter. While formulating the flight dynamic equations, the fuselage is treated as a rigid body undergoing a general maneuver.

Flight Dynamic Equations for a General Maneuver: Trim (Equilibrium) and Perturbation Analysis

The flight dynamic equations of motion of a helicopter under general maneuver are highly complicated. Hence, in the following, the formulations of the equations of motion and the solution procedure are described in a general symbolic manner. Subsequently, the stability characteristics of a helicopter under hovering condition will be solved using a simple model, based on the formulations presented in earlier chapters.

A very complex and general flight condition of a helicopter corresponds to a descent spin with sideslip velocity or a climb spin with sideslip velocity, as shown in Figure 10.2. The complexity arises due to two factors: one related to the kinematics of rigid body motion and the other related to describing the orientation of the vehicle and its flight path with respect to a ground-based observation coordinate system.

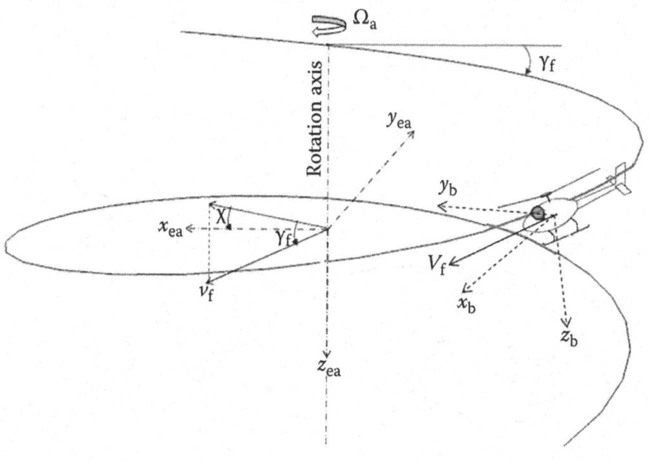

FIGURE 10.2
General maneuver of a helicopter.

The analysis of vehicle dynamics requires the development of the equations of motion of a helicopter. In this formulation, the helicopter is generally treated as a rigid body having three components of translational acceleration of the center of mass and three angular acceleration components defined along a suitable body-fixed coordinate system. The total number of degrees of freedom is 6. The corresponding six equations are three force equilibrium equations and three moment equilibrium equations. Apart from these six equilibrium equations, there are three more equations (kinematic relations) relating the instantaneous angular velocity of the helicopter to the rate of change of the orientation of the helicopter with respect to a ground-fixed coordinate system. The formulation of these equations is provided in the following.

Consider a body-fixed coordinate system ($x_b - y_b - z_b$) with origin at the center of mass of the helicopter, as shown in Figure 10.3.

Let u, v, w be the components of velocity of the center of mass of the helicopter along $x_b - y_b - z_b$, respectively, and p, q, r be the instantaneous angular velocity components along $x_b - y_b - z_b$, respectively.

The translational acceleration of the center of mass is given by

$$\vec{a} = \left(\dot{u}\hat{i}_b + \dot{v}\hat{j}_b + \dot{w}\hat{k}_b\right) + \left(u\dot{\hat{i}}_b + v\dot{\hat{j}}_b + w\dot{\hat{k}}_b\right)$$

Substituting for the time derivatives of the unit vectors,

$$\vec{a} = \left(\dot{u}\hat{i}_b + \dot{v}\hat{j}_b + \dot{w}\hat{k}_b\right) + \left(p\hat{i}_b + q\hat{j}_b + r\hat{k}_b\right) \times \left(u\hat{i}_b + v\hat{j}_b + w\hat{k}_b\right)$$

Combining the components, the acceleration at the center of mass can be written as

$$\vec{a} = \left(\dot{u} - rv + qw\right)\hat{i}_b + \left(\dot{v} - pw + ru\right)\hat{j}_b + \left(\dot{w} - qu + pv\right)\hat{k}_b \tag{10.1}$$

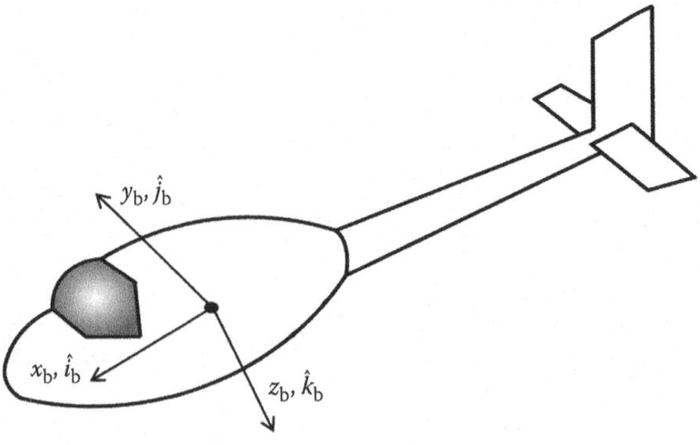

FIGURE 10.3
Body-fixed coordinate system.

The external forces acting on the helicopter fuselage are due to the following:

1. Main rotor hub loads
2. Tail rotor hub loads
3. Fuselage aerodynamic loads
4. Aerodynamic loads on horizontal tail
5. Aerodynamic loads on vertical fin
6. Gravity

Since the force equilibrium equations are written in the body-fixed ($x_b - y_b - z_b$) coordinate system, the components of gravity load along the body-fixed system have to be obtained from the coordinate transformation relationship between the earth-fixed nonrotating system and the body-fixed rotating system. This relationship essentially describes the orientation or attitude of the helicopter with respect to the ground-fixed axis system.

The attitude or orientation of the helicopter with respect to the earth-fixed system is given by three angles. Since finite rotations are not vector quantities, these angles are not unique, and the rotation sequence commonly used in flight dynamics is yaw–pitch–roll (ψ, θ, ϕ).

Let the earth-fixed coordinate system be denoted by $x_{ea} - y_{ea} - z_{ea}$. First, rotate the earth-fixed coordinate system about the z_{ea} axis through an angle ψ in a counterclockwise direction, as shown in Figure 10.4.

The transformation relationship between the earth-fixed system and the new coordinate system (x_1, y_1, z_1) can be written as

$$\begin{Bmatrix} x_1 \\ y_1 \\ z_1 \end{Bmatrix} = \begin{bmatrix} \cos\psi & \sin\psi & 0 \\ -\sin\psi & \cos\psi & 0 \\ 0 & 0 & 1 \end{bmatrix} \begin{Bmatrix} x_{ea} \\ y_{ea} \\ z_{ea} \end{Bmatrix} \tag{10.2}$$

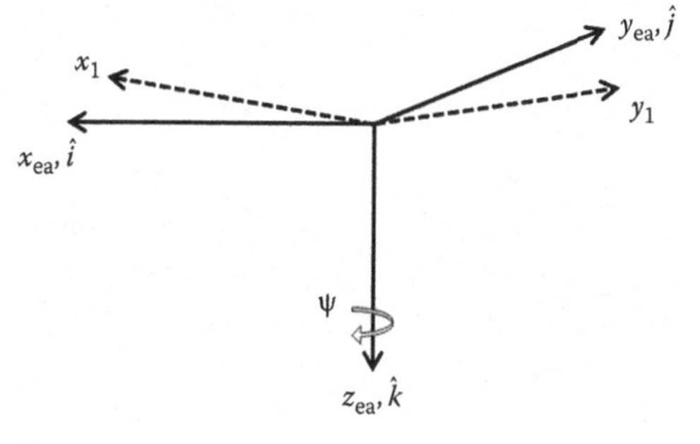

FIGURE 10.4
Rotation through a yaw angle.

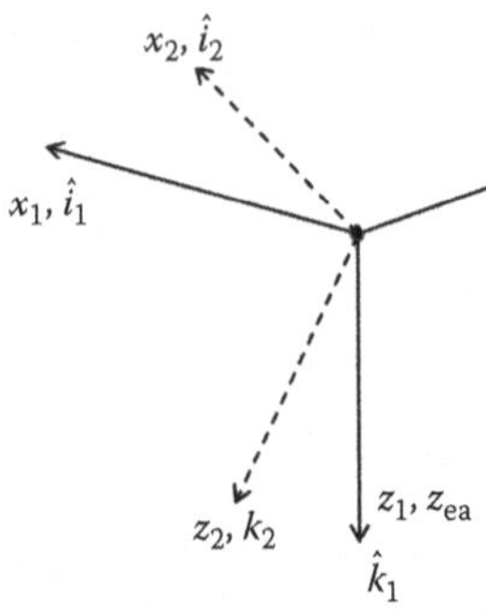

FIGURE 10.5
Rotation through a pitch angle.

Next, rotate the coordinate system about the y_1 axis through an angle θ in a counterclockwise direction, as shown in Figure 10.5.

The transformation relationship between x_1, y_1, z_1 and the new coordinate system (x_2, y_2, z_2) is given as

$$\begin{Bmatrix} x_2 \\ y_2 \\ z_2 \end{Bmatrix} = \begin{bmatrix} \cos\theta & 0 & -\sin\theta \\ 0 & 1 & 0 \\ \sin\theta & 0 & \cos\theta \end{bmatrix} \begin{Bmatrix} x_1 \\ y_1 \\ z_1 \end{Bmatrix} \tag{10.3}$$

Finally, rotate the x_2, y_2, z_2 coordinate system about the x_2 axis through an angle ϕ in a counterclockwise direction, as shown in Figure 10.6. The new coordinate system is the body-fixed system.

The transformation relationship between the two coordinate systems can be written as

$$\begin{Bmatrix} x_b \\ y_b \\ z_b \end{Bmatrix} = \begin{bmatrix} 1 & 0 & 0 \\ 0 & \cos\phi & \sin\phi \\ 0 & -\sin\phi & \cos\phi \end{bmatrix} \begin{Bmatrix} x_2 \\ y_2 \\ z_2 \end{Bmatrix} \tag{10.4}$$

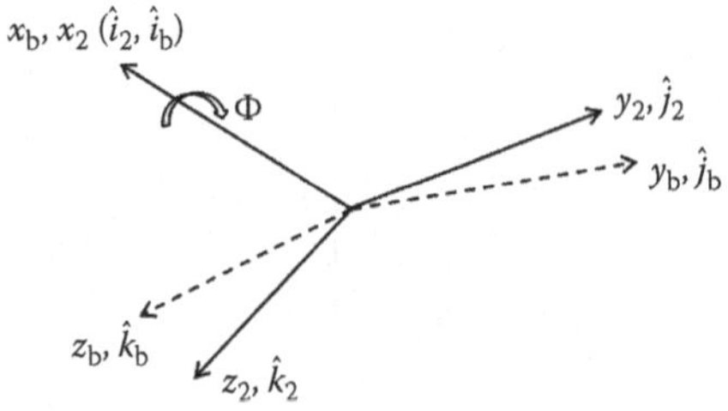

FIGURE 10.6
Rotation through a roll angle.

Combining all the transformation relationships, the final transformation relationship between the earth-fixed system and the body-fixed axis system can be written as

$$\begin{Bmatrix} x_b \\ y_b \\ z_b \end{Bmatrix} =$$

$$\begin{bmatrix} \cos\theta\cos\psi & \cos\theta\sin\psi & -\sin\theta \\ \sin\phi\sin\theta\cos\psi - \cos\phi\sin\psi & \sin\phi\sin\theta\sin\psi + \cos\phi\cos\psi & \sin\phi\cos\theta \\ \cos\phi\sin\theta\cos\psi + \sin\phi\sin\psi & \cos\phi\sin\theta\sin\psi - \sin\phi\cos\psi & \cos\phi\cos\theta \end{bmatrix} \begin{Bmatrix} x_{ea} \\ y_{ea} \\ z_{ea} \end{Bmatrix}$$

$$\begin{Bmatrix} x_b \\ y_b \\ z_b \end{Bmatrix} = [\Gamma] \begin{Bmatrix} x_{ea} \\ y_{ea} \\ z_{ea} \end{Bmatrix} \tag{10.5}$$

Since Γ is an orthogonal transformation, one has

$$\Gamma^{-1} = \Gamma^{T} \tag{10.6}$$

Let us formulate the relationship between the instantaneous angular velocity of the helicopter and the rate of change of orientation angles ψ, θ, and ϕ. The instantaneous angular velocity vector is defined in the body fixed system as

$$\vec{\omega} = p\hat{i}_b + q\hat{j}_b + r\hat{k}_b \tag{10.7}$$

Also, it may be noted that the angular velocity vector can be written in terms of the time derivatives of orientation angles as

$$\vec{\omega} = \dot{\psi}\hat{k}_{ea} + \dot{\theta}\hat{j}_1 + \dot{\phi}\hat{i}_2$$

Using the transformation relationships, given in Equations 10.3 to 10.5, one can write the instantaneous angular velocity components in terms of orientation angles and their rates as

$$p = \dot{\phi} - \dot{\psi}\sin\theta \tag{10.8}$$

$$q = \dot{\theta}\cos\phi + \dot{\psi}\sin\phi\cos\theta \tag{10.9}$$

$$r = -\dot{\theta}\sin\phi + \dot{\psi}\cos\phi\cos\theta \tag{10.10}$$

Equations 10.8 to 10.10 represent the three kinematic relationships between the instantaneous angular velocity of the helicopter and the rate of change of the helicopter orientation angles with respect to the earth-fixed system.

The components of gravity load along the $x_b - y_b - z_b$ system can be easily obtained from the transformation relationships given in Equation 10.5. Let X, Y, Z be the components of the resultant external loads, other than gravity, acting on the fuselage center of mass along the body-fixed $x_b - y_b - z_b$ coordinate system. Using Equation 10.1, the force equations can be written as

$$X = M_F(\dot{u} - rv + qw) + M_F g \sin\theta \tag{10.11}$$

$$Y = M_F(\dot{v} - pw + ru) - M_F g \sin\phi\cos\theta \tag{10.12}$$

$$Z = M_F(\dot{w} - qu + pv) - M_F g \cos\phi\cos\theta \tag{10.13}$$

where M_F is the total mass of the helicopter.

Knowing the angular velocity and the inertia tensor of the helicopter, the components of the angular momentum of the helicopter along the body-fixed axis system can be written as

$$\begin{Bmatrix} H_x \\ H_y \\ H_z \end{Bmatrix} = \begin{bmatrix} I_{xx} & -I_{xy} & -I_{xz} \\ -I_{yx} & I_{yy} & -I_{yz} \\ -I_{zx} & -I_{zy} & I_{zz} \end{bmatrix} \begin{Bmatrix} p \\ q \\ r \end{Bmatrix} \tag{10.14}$$

(Note that the inertia tensor is symmetric.)

The moment equation can be written as

$$\vec{M} = \dot{\vec{H}} + \vec{\omega} \times \vec{H} \tag{10.15}$$

where

$$\dot{\vec{H}} = \left(I_{xx}\dot{p} - I_{xy}\dot{q} - I_{xz}\dot{r}\right)\hat{i}_b + \left(-I_{yx}\dot{p} + I_{yy}\dot{q} - I_{yz}\dot{r}\right)\hat{j}_b + \left(-I_{zx}\dot{p} + I_{zy}\dot{q} + I_{zz}\dot{r}\right)\hat{k}_b \tag{10.16}$$

and

$$\begin{aligned} \vec{\omega} \times \vec{H} = &\left[q(-I_{zx}p - I_{zy}q + I_{zz}r) - r(-I_{yx}p + I_{yy}q + I_{zz}r)\right]\hat{i}_b \\ &+ \left[r(I_{xx}p - I_{xy}q - I_{xz}r) - p(-I_{zx}p - I_{zy}q + I_{zz}r)\right]\hat{j}_b \\ &+ \left[p(-I_{yx}p + I_{yy}q - I_{yz}r) - q(I_{xx}p - I_{xy}q - I_{xz}r)\right]\hat{k}_b \end{aligned} \tag{10.17}$$

Assume that the external moment acting on the helicopter center of mass can be written as

$$\vec{M} = L\hat{i}_b + M\hat{j}_b + N\hat{k}_b \tag{10.18}$$

Collecting all the terms from Equations 10.16 to 10.18 and equating the respective quantities, the three moment equations of the helicopter can be obtained, and they are given as

$$L = (I_{xx}\dot{p} - I_{xy}\dot{q} - I_{xz}\dot{r}) + q(-I_{zx}p - I_{zy}q + I_{zz}r) - r(-I_{yx}p + I_{yy}q - I_{yz}r) \tag{10.19}$$

$$M = (-I_{yx}\dot{p} + I_{yy}\dot{q} - I_{yz}\dot{r}) + r(I_{xx}p - I_{xy}q - I_{xz}r) - p(-I_{zx}p - I_{zy}q + I_{zz}r) \tag{10.20}$$

$$N = (-I_{zx}\dot{p} - I_{zy}\dot{q} + I_{zz}\dot{r}) + p(-I_{yx}p + I_{yy}q - I_{yz}r) - q(-I_{xx}p - I_{xy}q - I_{xz}r) \tag{10.21}$$

In general, the numerical value of I_{xz} is comparable to the value of I_{xx}. However, all the other cross-products of inertia can be taken as 0. Then, the simplified form of moment equations can be given as

$$L = I_{xx}\dot{p} - I_{xz}(\dot{r} + pq) - qr(I_{yy} - I_{zz}) \tag{10.22}$$

$$M = I_{yy}\dot{q} - I_{xz}(r^2 - p^2) - rp(I_{zz} - I_{xx}) \tag{10.23}$$

$$N = I_{zz}\dot{r} - I_{xz}(\dot{p} - qr) - pq(I_{xx} - I_{yy}) \tag{10.24}$$

It can be seen that, due to I_{xz}, the cyclic symmetry in the equations is lost.

From Equations 10.8 to 10.10, the kinematic relationship between the rate change of the helicopter orientation angles and the instantaneous angular velocity of the fuselage can be written as

$$\dot{\psi} = \frac{\sin\phi}{\cos\theta}q + \frac{\cos\phi}{\cos\theta}r \tag{10.25}$$

$$\dot{\theta} = \cos\phi\; q - \sin\phi\; r \tag{10.26}$$

$$\dot{\phi} = p + \tan\theta\sin\phi\; q + \tan\theta\cos\phi\; r \tag{10.27}$$

Under a steady turn maneuver, in the spin mode, the spin axis is always directed vertically. The rate of change of the orientation angles θ and ϕ are 0, and the gravitational force components along the body-fixed system are constant. This type of flight condition resembles spiral climbing or descending with the side slip.

A thorough and rigorous analysis of flight dynamics must include the blade degree of freedom in the flap, lag, and torsion modes. Inclusion of the blade equations along with the vehicle dynamic equations will present a set of highly complicated coupled equations. Generally, the lag and torsion modes of the blade are neglected since their contribution to the vehicle flying qualities is relatively less significant. The coning and the longitudinal and lateral flapping of the rotor modes have a significant influence on vehicle dynamics. Since the damping in the flap modes is large, the time constant is of the order of one-quarter to one-half of a revolution of the rotor. The dynamics of the flap mode is much faster than the fuselage dynamics, and the rotor flap reaches its steady-state value in a very short time. Hence, flap dynamics is included in vehicle dynamics in a quasi-static manner, which eliminates the blade degrees of freedom from the flight dynamic problem. It is assumed that the rotor system produces force and moments at the rotor hub instantaneously in response to vehicle motion or to pilot control inputs.

The dynamic equations of motion are nonlinear coupled differential equations. The formulation of helicopter trim equations and perturbation (stability) equations are based on the perturbation approach. The stability analysis is performed about an equilibrium (trim) flight condition, and hence, the equations of motion are linearized about the trim state of the helicopter.

The process of linearization is as follows.

Assume that each degree of freedom and control inputs can be written as two components: one representing the trim (or equilibrium) value and the other a perturbational quantity, as

$$
\begin{aligned}
\theta &= \theta_e + \tilde{\theta}(t) \qquad & u &= u_e + \tilde{u}(t) \qquad & p &= p_e + \tilde{p}(t) \\
\phi &= \phi_e + \tilde{\phi}(t) \qquad & v &= v_e + \tilde{v}(t) \qquad & q &= q_e + \tilde{q}(t) \\
\psi &= \dot{\psi}_e t \qquad & w &= w_e + \tilde{w}(t) \qquad & r &= r_e + \tilde{r}(t)
\end{aligned}
$$

$$
\theta_0 = \theta_{0e} + \tilde{\theta}_0 \qquad \theta_{1c} = \theta_{1ce} + \tilde{\theta}_{1c} \qquad \theta_{1s} = \theta_{1se} + \tilde{\theta}_{1s} \qquad \theta_{0T} = \theta_{0Te} + \tilde{\theta}_{0T} \tag{10.28}
$$

Similarly, the force and moment components acting on the helicopter are given as a sum of equilibrium and perturbation quantities. Using Taylor's theorem for analytic functions, the forces and moments can be written in the approximate form as

$$
\begin{aligned}
X = X_e &+ \left[\frac{\partial X}{\partial u} + \frac{\partial X}{\partial \beta_0}\frac{\partial \beta_0}{\partial u} + \frac{\partial X}{\partial \beta_{1c}}\frac{\partial \beta_{1c}}{\partial u} + \frac{\partial X}{\partial \beta_{1s}}\frac{\partial \beta_{1s}}{\partial u}\right]\tilde{u} + \left[\frac{\partial X}{\partial v} + \frac{\partial X}{\partial \beta_0}\frac{\partial \beta_0}{\partial v} + \frac{\partial X}{\partial \beta_{1c}}\frac{\partial \beta_{1c}}{\partial v} + \frac{\partial X}{\partial \beta_{1s}}\frac{\partial \beta_{1s}}{\partial v}\right]\tilde{v} \\
&+ \left[\frac{\partial X}{\partial w} + \frac{\partial X}{\partial \beta_0}\frac{\partial \beta_0}{\partial w} + \frac{\partial X}{\partial \beta_{1c}}\frac{\partial \beta_{1c}}{\partial w} + \frac{\partial X}{\partial \beta_{1s}}\frac{\partial \beta_{1s}}{\partial w}\right]\tilde{w} + \left[\frac{\partial X}{\partial p} + \frac{\partial X}{\partial \beta_0}\frac{\partial \beta_0}{\partial p} + \frac{\partial X}{\partial \beta_{1c}}\frac{\partial \beta_{1c}}{\partial p} + \frac{\partial X}{\partial \beta_{1s}}\frac{\partial \beta_{1s}}{\partial p}\right]\tilde{p} \\
&+ \left[\frac{\partial X}{\partial q} + \frac{\partial X}{\partial \beta_0}\frac{\partial \beta_0}{\partial q} + \frac{\partial X}{\partial \beta_{1c}}\frac{\partial \beta_{1c}}{\partial q} + \frac{\partial X}{\partial \beta_{1s}}\frac{\partial \beta_{1s}}{\partial q}\right]\tilde{q} + \left[\frac{\partial X}{\partial r} + \frac{\partial X}{\partial \beta_0}\frac{\partial \beta_0}{\partial r} + \frac{\partial X}{\partial \beta_{1c}}\frac{\partial \beta_{1c}}{\partial r} + \frac{\partial X}{\partial \beta_{1s}}\frac{\partial \beta_{1s}}{\partial r}\right]\tilde{r} \\
&+ \left[\frac{\partial X}{\partial \dot{p}} + \frac{\partial X}{\partial \beta_0}\frac{\partial \beta_0}{\partial \dot{p}} + \frac{\partial X}{\partial \beta_{1c}}\frac{\partial \beta_{1c}}{\partial \dot{p}} + \frac{\partial X}{\partial \beta_{1s}}\frac{\partial \beta_{1s}}{\partial \dot{p}}\right]\dot{\tilde{p}} + \left[\frac{\partial X}{\partial \dot{q}} + \frac{\partial X}{\partial \beta_0}\frac{\partial \beta_0}{\partial \dot{q}} + \frac{\partial X}{\partial \beta_{1c}}\frac{\partial \beta_{1c}}{\partial \dot{q}} + \frac{\partial X}{\partial \beta_{1s}}\frac{\partial \beta_{1s}}{\partial \dot{q}}\right]\dot{\tilde{q}} \\
&+ \left[\frac{\partial X}{\partial \dot{r}} + \frac{\partial X}{\partial \beta_0}\frac{\partial \beta_0}{\partial \dot{r}} + \frac{\partial X}{\partial \beta_{1c}}\frac{\partial \beta_{1c}}{\partial \dot{r}} + \frac{\partial X}{\partial \beta_{1s}}\frac{\partial \beta_{1s}}{\partial \dot{r}}\right]\dot{\tilde{r}} \\
&+ \left[\frac{\partial X}{\partial \theta} + \frac{\partial X}{\partial \beta_0}\frac{\partial \beta_0}{\partial \theta} + \frac{\partial X}{\partial \beta_{1c}}\frac{\partial \beta_{1c}}{\partial \theta} + \frac{\partial X}{\partial \beta_{1s}}\frac{\partial \beta_{1s}}{\partial \theta}\right]\tilde{\theta} + \left[\frac{\partial X}{\partial \phi} + \frac{\partial X}{\partial \beta_0}\frac{\partial \beta_0}{\partial \phi} + \frac{\partial X}{\partial \beta_{1c}}\frac{\partial \beta_{1c}}{\partial \phi} + \frac{\partial X}{\partial \beta_{1s}}\frac{\partial \beta_{1s}}{\partial \phi}\right]\tilde{\phi} \\
&+ \left[\frac{\partial X}{\partial \theta_o} + \frac{\partial X}{\partial \beta_0}\frac{\partial \beta_0}{\partial \theta_o} + \frac{\partial X}{\partial \beta_{1c}}\frac{\partial \beta_{1c}}{\partial \theta_o} + \frac{\partial X}{\partial \beta_{1s}}\frac{\partial \beta_{1s}}{\partial \theta_o}\right]\tilde{\theta}_0 + \left[\frac{\partial X}{\partial \theta_{1c}} + \frac{\partial X}{\partial \beta_0}\frac{\partial \beta_0}{\partial \theta_{1c}} + \frac{\partial X}{\partial \beta_{1c}}\frac{\partial \beta_{1c}}{\partial \theta_{1c}} + \frac{\partial X}{\partial \beta_{1s}}\frac{\partial \beta_{1s}}{\partial \theta_{1c}}\right]\tilde{\theta}_{1c} \\
&+ \left[\frac{\partial X}{\partial \theta_{1s}} + \frac{\partial X}{\partial \beta_0}\frac{\partial \beta_0}{\partial \theta_{1s}} + \frac{\partial X}{\partial \beta_{1c}}\frac{\partial \beta_{1c}}{\partial \theta_{1s}} + \frac{\partial X}{\partial \beta_{1s}}\frac{\partial \beta_{1s}}{\partial \theta_{1s}}\right]\tilde{\theta}_{1s} + \left[\frac{\partial X}{\partial \theta_{0T}} + \frac{\partial X}{\partial \beta_0}\frac{\partial \beta_0}{\partial \theta_{0T}} + \frac{\partial X}{\partial \beta_{1c}}\frac{\partial \beta_{1c}}{\partial \theta_{0T}} + \frac{\partial X}{\partial \beta_{1s}}\frac{\partial \beta_{1s}}{\partial \theta_{0T}}\right]\tilde{\theta}_{0T}
\end{aligned}
\tag{10.29}
$$

In writing the above expansion, it is assumed that only the flap mode of the rotor is included in a quasi-static form in the evaluation of rotor loads.

For convenience, the above expansion can be written in a compact form as

$$
\begin{aligned}
X = X_e &+ X_u\tilde{u} + X_v\tilde{v} + X_w\tilde{w} + X_p\tilde{p} + X_q\tilde{q} + X_r\tilde{r} + X_{\dot{p}}\dot{\tilde{p}} + X_{\dot{q}}\dot{\tilde{q}} + X_{\dot{r}}\dot{\tilde{r}} + X_\theta\tilde{\theta} + X_\phi\tilde{\phi} \\
&+ X_{\theta_0}\tilde{\theta}_0 + X_{\theta_{1c}}\tilde{\theta}_{1c} + X_{\theta_{1s}}\tilde{\theta}_{1s} + X_{\theta_{0T}}\tilde{\theta}_{0T}
\end{aligned}
\tag{10.30}
$$

In general, the derivative terms associated with $\dot{\tilde{p}}$, $\dot{\tilde{q}}$, and $\dot{\tilde{r}}$ are generally very small compared to the inertia of the helicopter, and hence, they can be neglected. Also, the terms associated with $\tilde{\theta}$ and $\tilde{\phi}$ are generally 0; hence, these terms are not usually included in the Taylor series expansion.

Similar to Equation 10.30, other force and moment components are expanded.

In Equation 10.30, the quantity with subscript "e" corresponds to the equilibrium quantity in steady-state maneuver, and it is a constant (i.e., it is time invariant). Substituting these types of expressions and Equation 10.28 in the equations of motion (Equations 10.11 to 10.13, 10.22 to 10.24, 10.8 to 10.10) and collecting terms corresponding to the equilibrium state and those corresponding to the perturbation quantities (after neglecting the product of perturbation quantities), and equating them separately to 0, two sets of equations are formed. The time invariant set corresponds to the trim state, and the time variant set represents the linearized perturbation equations.

The time variant equations correspond to the stability equations. The trim equations are nonlinear algebraic equations. They are given as follows:

Force equations:

$$m(-r_e v_e + q_e w_e) + mg \sin \theta_e = X_e \tag{10.31}$$

$$m(-p_e w_e + r_e u_e) - mg \sin \phi_e \cos \theta_e = Y_e \tag{10.32}$$

$$m(-q_e u_e + p_e v_e) - mg \cos \phi_e \cos \theta_e = Z_e \tag{10.33}$$

Moment equations:

$$-I_{xz}(p_e q_e) - (I_{yy} - I_{zz}) q_e r_e = L_e \quad \text{(roll)} \tag{10.34}$$

$$-I_{xz}\left(r_e^2 - p_e^2\right) - (I_{zz} - I_{xx}) r_e p_e = M_e \quad \text{(pitch)} \tag{10.35}$$

$$-I_{xz}(-q_e r_e) - (I_{xx} - I_{yy}) p_e q_e = N_e \quad \text{(yaw)} \tag{10.36}$$

where X_e, Y_e, Z_e and L_e, M_e, N_e are the steady-state forces and moments acting on the helicopter at the center of mass location. The steady-state roll, pitch, and yaw rates (p_e, q_e, r_e) are related to the steady-state spin rate $\dot{\psi}_e$ as (from Equations 10.8–10.10)

$$p_e = -\dot{\psi}_e \sin \theta_e \tag{10.37}$$

$$q_e = \dot{\psi}_e \sin \phi_e \cos \theta_e \tag{10.38}$$

$$r_e = \dot{\psi}_e \cos \phi_e \cos \theta_e \tag{10.39}$$

In level flight, $p_e = q_e = r_e = 0$. Since there is no angular motion of the fuselage, the trim equations reduce to three force and three moment equilibrium equations (Equations 10.31 to 10.36). On the other hand, if the helicopter is performing a steady turn, p_e, q_e, and r_e will have finite values. The trim equations are nonlinear algebraic equations and can be solved by the iterative numerical scheme.

Equations 10.31 to 10.39 comprise a set of nine equations. Under equilibrium condition, the total number of unknown quantities is 13. They are given as follows:

Translation velocity components: u_e, v_e, w_e

$$\text{Angular velocity components: } p_e, q_e, r_e$$

$$\text{Steady spin rate: } \dot{\psi}_e$$

$$\text{Orientation angles: } \theta_e, \phi_e$$

$$\text{Pilot input angles: } \theta_0, \theta_{1c}, \theta_{1s}, \theta_{0TR} \tag{10.40}$$

Since there are only nine equations, for a unique mathematical solution, four quantities must be viewed as independent, and they must be prescribed. It may be noted that any four quantities can be prescribed, and the remaining nine quantities can be evaluated by solving the nine equations. For a general flight condition, the following four quantities are usually prescribed. They are

$$V_{fe}\text{: flight speed}$$

$$\gamma_{fe}\text{: flight path angle}$$

$$\dot{\psi}_e = \Omega_e \text{ : turn rate or spin rate}$$

$$\beta_e\text{: side-slip angle} \tag{10.41}$$

It is interesting to note that the variables of the problem are different from the given four quantities. One has to formulate a proper relationship between them. In the following, the relationships between trim parameters V_{fe}, β_e, and γ_{fe}, and the velocity components u_e, v_e, and w_e are obtained. The incidence and side-slip angles are defined as

$$\alpha_e = \tan^{-1}\left(\frac{w_e}{u_e}\right) \text{ : incidence angle} \tag{10.42}$$

$$\beta_e = \sin^{-1}\left(\frac{v_e}{V_{fe}}\right) \text{ : side slip angle} \tag{10.43}$$

Note that the components of V_{fe} are u_e, v_e, w_e along the body-fixed coordinate system. This velocity vector of the helicopter is turning at a steady spin rate Ω_e, as shown in Figure 10.7.

Let us assume that the velocity vector $\vec{V}_{fe}$ makes an angle γ_{fe} (flight path angle) with respect to the horizontal plane ($x_{ea} - y_{ea}$ plane) of the earth-fixed system, as shown in Figure 10.8.

From Figure 10.8, the velocity components of V_{fe} along the earth-fixed system can be written as

$$U_{ea} = V_{fe} \cos \gamma_{fe} \cos \chi \tag{10.44}$$

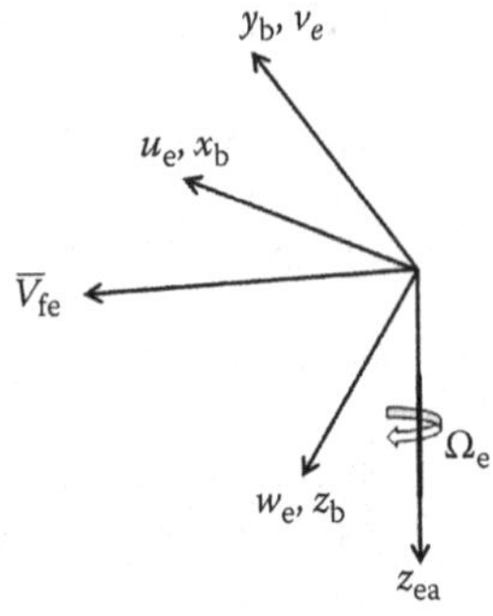

FIGURE 10.7
Spin rate and the components of the velocity vector.

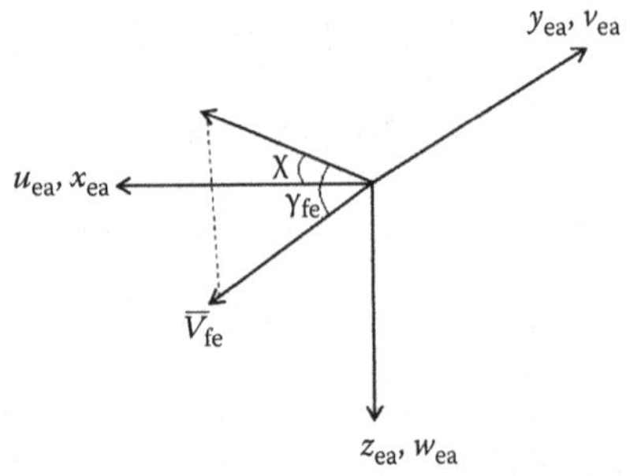

FIGURE 10.8
Flight path angle defined with respect to the earth-fixed horizontal plane.

$$V_{ea} = V_{fe} \cos \gamma_{fe} \sin \chi \tag{10.45}$$

$$W_{ea} = V_{fe} \sin \gamma_{fe} \tag{10.46}$$

The angle χ represents the angle between the earth-fixed x axis (x_{ea}) and the projected velocity vector in the horizontal plane.

Using the transformation relationship between the earth-fixed system and the body fixed system (Equation 10.5), the velocity components in the body-fixed system can be obtained as

$$u_e = V_{fe}[\cos \gamma_{fe} \cos \theta_e\{\cos \chi \cos \psi_e + \sin \chi \sin \psi_e\} - \sin \gamma_{fe} \sin \theta_e] \tag{10.47}$$

$$\begin{aligned} v_e = V_{fe}\big[&\cos\gamma_{fe} \cos\chi\{\sin\phi_e \sin\theta_e \cos\psi_e - \cos\phi_e \sin\psi_e\} \\ &+ \cos\gamma_{fe} \sin\chi\{\sin\phi_e \sin\theta_e \sin\psi_e + \cos\phi_e \cos\psi_e\} + \sin\gamma_{fe} \sin\phi_e \cos\theta_e\big] \end{aligned} \tag{10.48}$$

$$w_e = V_{fe}[\cos\gamma_{fe}\cos\chi\{\cos\phi_e\sin\theta_e\cos\psi_e + \sin\phi_e\sin\psi_e\}$$
$$+\cos\gamma_{fe}\sin\chi\{\cos\phi_e\sin\theta_e\sin\psi_e - \sin\phi_e\cos\psi_e\} + \sin\gamma_{fe}\cos\phi_e\cos\theta_e] \tag{10.49}$$

Combining the two angles χ and ψ_e, and representing $\chi_e = \chi - \psi_e$, the velocity components can be simplified as

$$u_e = V_{fe}[\cos\gamma_{fe}\cos\theta_e\cos\chi_e - \sin\gamma_{fe}\sin\theta_e] \tag{10.50}$$

$$v_e = V_{fe}[\cos\gamma_{fe}\sin\phi_e\sin\theta_e\cos\chi_e + \cos\gamma_{fe}\cos\phi_e\sin\chi_e + \sin\gamma_{fe}\sin\phi_e\cos\theta_e] \tag{10.51}$$

$$w_e = V_{fe}[\cos\gamma_{fe}\cos\phi_e\sin\theta_e\cos\chi_e - \cos\gamma_{fe}\sin\phi_e\sin\chi_e + \sin\gamma_{fe}\cos\phi_e\cos\theta_e] \tag{10.52}$$

where $\chi_e = \chi - \psi_e$ is denoted as the track angle. Note that χ and ψ_e $(= \Omega_e t)$ are time-dependent quantities.

During steady maneuver, u_e, v_e, w_e, θ_e, ϕ_e, γ_{fe} are constants. Hence, χ_e must also be a constant, even though χ and ψ_e are time varying, but their difference will be a constant under steady maneuver. From a geometric point of view, one can approximately say that the track angle χ_e represents the angle between the velocity vector and the body-fixed x axis (x_b axis) projected on the horizontal earth plane. In steady maneuver, this track angle remains a constant, and it is related to the side-slip angle, which is described in the following.

Since the side-slip angle is defined as $\sin\beta_e = \dfrac{v_e}{V_{fe}}$, there is a relationship between the track angle and the side-slip angle. From Equation 10.49, one can write the expression for side-slip angle in terms of the equilibrium angle in the pitch and roll orientation angles of the helicopter, the flight path angle, and the track angle as

$$\sin\beta_e = \cos\gamma_{fe}\sin\phi_e\sin\theta_e\cos\chi_e + \cos\gamma_{fe}\cos\phi_e\sin\chi_e + \sin\gamma_{fe}\sin\phi_e\cos\theta_e \tag{10.53}$$

Denoting

$$\begin{aligned} K_1 &= \cos\gamma_{fe}\sin\phi_e\sin\theta_e \\ K_2 &= \cos\gamma_{fe}\cos\phi_e \\ K_3 &= \sin\beta_e - \sin\gamma_{fe}\sin\phi_e\cos\theta_e \end{aligned} \tag{10.54}$$

Substituting the new variables in Equation 10.53, one has

$$K_1\cos\chi_e + K_2\sin\chi_e - K_3 = 0 \tag{10.55}$$

Rearranging Equation 10.55 and squaring both sides yields

$$K_1^2 \cos^2 \chi_e = K_3^2 - 2K_2K_3 \sin \chi_e + K_2^2 \sin^2 \chi_e \tag{10.56}$$

Rewriting Equation 10.56 as

$$\left(K_2^2 + K_1^2\right)\sin^2 \chi_e - 2K_2K_3 \sin \chi_e + K_3^2 - K_1^2 = 0 \tag{10.57}$$

Solving for the track angle,

$$\sin \chi_e = \frac{2K_2K_3 \pm \sqrt{4K_2^2K_3^2 - 4\left(K_3^2 - K_1^2\right)\left(K_2^2 + K_1^2\right)}}{2\left(K_2^2 + K_1^2\right)} \tag{10.58}$$

Simplifying the expression,

$$\sin \chi_e = \frac{K_2K_3 \pm \sqrt{K_1^4 + K_1^2K_2^2 - K_3^2K_1^2}}{\left(K_1^2 + K_2^2\right)} \tag{10.59}$$

Of the two solutions, only one of the angles will be physically meaningful, and it should satisfy Equation 10.51. This track angle χ_e is defined in terms of β_e, θ_e, ϕ_e, and γ_{fe}.

The iterative procedure for trim analysis during steady maneuvering flight is as follows:

1. Given all the parameters of the vehicle, blade data, flight speed, flight path angle, side-slip angle, and turn rate.
2. Assume θ_e, ϕ_e, θ_{0e}, θ_{1ce}, θ_{1se}, θ_{0Te}.
3. Compute track angle χ_e.
4. Obtain u_e, v_e, w_e, p_e, q_e, r_e.
5. Solve for the rotor inflow.
6. Solve the blade equations for the response and hub loads.
7. Obtain fuselage and other surface aerodynamic loads.
8. Balance the fuselage force and moment equations.
9. Obtain new estimates θ_{0e}, θ_{1ce}, θ_{1se}, θ_{0Te}, ϕ_e, θ_e.
10. Go to step 2. Iterate until convergence is reached. The converged set corresponds to the trim condition in steady maneuver of the helicopter.

Linearized Perturbation Equations

Along with the trim equations, after neglecting the product of the perturbation quantities, linearized stability equations are obtained. For example, let us consider the linearized force equation in the X direction (using Equations 10.11, 10.28, and 10.30).

$$M_F\dot{\tilde{u}} - M_F(r_e\tilde{v} + v_e\tilde{r}) + M_F(q_e\tilde{w} + w_e\tilde{q}) + M_F g\cos\theta_e\tilde{\theta} =$$

$$X_u\tilde{u} + X_v\tilde{v} + X_w\tilde{w} + X_p\tilde{p} + X_q\tilde{q} + X_r\tilde{r} + X_{\dot{p}}\dot{\tilde{p}} + X_{\dot{q}}\dot{\tilde{q}} + X_{\dot{r}}\dot{\tilde{r}} + X_\theta\tilde{\theta} + X_\phi\tilde{\phi} \tag{10.60}$$

$$+X_{\theta_0}\tilde{\theta}_0 + X_{\theta_{1c}}\tilde{\theta}_{1c} + X_{\theta_{1s}}\tilde{\theta}_{1s} + X_{\theta_{0T}}\tilde{\theta}_{0T}$$

Similarly, all the force, moment, and kinematic equations are written in this linearized form. The quantities X_u, X_v, X_w...... etc. are known as stability derivatives, and X_{θ_0}, $X_{\theta_{1c}}$..... etc. are known as control derivatives. Before writing the linearized equations in matrix form, one must establish a relationship between the orientation rates of the angles in θ (pitch), ϕ (roll), and ψ (yaw), and the instantaneous angular rates of the helicopter. For finite angular motion, one must formulate this relationship using the Euler angle transformation. If the pitch and roll angles are small, one can resort to the approximation that $\dot{\tilde{\phi}} = \tilde{p}$ (roll), $\dot{\tilde{\theta}} = \tilde{q}$ (pitch), and $\dot{\tilde{\psi}} = \tilde{r}$ (yaw). In the following, only this approximate relationship is used, which is sufficient for the purpose of explaining the fundamentals of helicopter dynamics.

It is important to note that the perturbation equations have to be nondimensionalized before proceeding to perform stability or control response calculations. Dividing the force equations by $M_b\Omega^2 R$ and moment equations by $M_b\Omega^2 R^2$, and performing some simple mathematical manipulations, the perturbation equations of motion can be written in matrix form as (Note: 6 dynamical equations + 2 kinematic relations. The yaw rate relationship is generally not included since the yaw angle can be directly obtained from the yaw rate.)

$$\dot{\boldsymbol{X}} = \boldsymbol{A}\boldsymbol{X} + \boldsymbol{B}\boldsymbol{u} \tag{10.61}$$

The state vector: $\{\boldsymbol{X}\} = \left[\tilde{u}, \tilde{v}, \tilde{w}, \tilde{p}, \tilde{q}, \tilde{r}, \tilde{\theta}, \tilde{\phi}\right]^T$

$$\{\boldsymbol{u}\} = \left[\tilde{\theta}_0, \tilde{\theta}_{1s}, \tilde{\theta}_{1c}, \tilde{\theta}_{0T}\right]^T$$

The vector $\boldsymbol{X}$ is the nondimensional state vector, $\boldsymbol{u}$ is the control vector, $\boldsymbol{A}$ is the system matrix, and $\boldsymbol{B}$ is the control matrix. The equations can be arranged in a manner such that longitudinal dynamics (involving only $\tilde{u}$, $\tilde{q}$, and $\tilde{\theta}$) can be separated from lateral dynamics (involving $\tilde{v}$, $\tilde{p}$, $\tilde{r}$, and $\tilde{\phi}$) and heave dynamics (involving $\tilde{w}$). Even though there is cross-coupling between all the motions, for the purpose of analysis and understanding, and also for the preliminary design of the control laws of automatic flight control system, longitudinal dynamics, lateral dynamics, and heave dynamics are analyzed separately.

For steady turn case, the equilibrium quantities p_e, q_e, and r_e will be present in the linearized perturbation equations, as shown in Equation 10.60. For level flight, these quantities are 0. The quantities u_e, v_e, and w_e represent the steady velocity components along the three body-fixed axes, with origin at the center of mass of the helicopter. The component of velocity w_e represents descent or climb, and v_e corresponds to side slip.

The dynamic characteristics of the helicopter are analyzed by identifying the eigenvalues of the system matrix $\boldsymbol{A}$. Solving the full system may be computationally easy, but analyzing the results becomes a little challenging because of coupling effects. Therefore, to gain a good understanding, different motions are decoupled and solved separately.

Stability Characteristics

The linearized perturbation equation (Equation 10.61) is used to obtain the stability behavior of the base helicopter by solving an eigenvalue problem of the homogeneous part of the equation.

$$\dot{\boldsymbol{X}} - \boldsymbol{A}\boldsymbol{X} = 0 \tag{10.62}$$

Assuming a solution as $\boldsymbol{X} = \bar{\boldsymbol{X}} e^{st}$, the eigenvalue problem is obtained.

$$[s\boldsymbol{I} - \boldsymbol{A}]\{\boldsymbol{X}\} = 0 \tag{10.63}$$

The characteristic determinant is

$$|s\boldsymbol{I} - \boldsymbol{A}| = 0 \tag{10.64}$$

Since system matrix $\boldsymbol{A}$ is of size 8 × 8, there are eight eigenvalues. If all the roots are complex, then there will be four pairs of complex conjugate roots. In general, for a helicopter, the root s_i (i = 1 to 8) will contain both complex and real roots. One can represent the complex roots by $s_i = \sigma_i \pm i\omega_i$, where σ_i represents the modal damping in the ith mode and ω_i represents the frequency of the ith mode. When σ_i is negative, the mode is stable, and when σ_i is positive, the mode is unstable. When ω_i is equal to 0, the root is a real root, indicating that the mode is either a pure divergent mode (if σ_i

is positive) or a pure convergent mode (if σ_i is negative). The eigen analysis of an 8 × 8 matrix is mathematically a very simple task to perform in a computer. However, an analytical solution will be a formidable task. The reason for a simplified analytical approach for stability analysis is that it will provide a substantial understanding of the phenomenon and information about which stability derivatives influence a mode and how the stability of the system can be improved by design modifications. The eigenvalues of the eighth-order (coupled) system for a typical helicopter is shown in Figure 10.9 as a root locus plot.

The eight roots contain two complex conjugates (four roots) plus four real roots. Of these, the root represented as "phugoid" is initially unstable at hover and becomes stable at midrange of forward speeds and again becomes unstable at high forward speeds. The eigenvector of the phugoid root indicates predominant participation from pitch, forward velocity, and vertical velocity states. The other complex root is known as the "Dutch roll," which has predominant participation from roll, side-slip velocity, and yaw. This root, although oscillatory, is generally stable. The other four roots are real roots, and they represent a stable convergent motion over the entire speed range. They are pitch subsidence, heave subsidence, roll subsidence, and spiral subsidence. As the name suggests, these modes have predominant contribution from the respective motion. The spiral mode is predominantly yaw motion in hover, with roll and side-slip contribution increasing with forward flight. A thorough understanding of the stability of the helicopter can be obtained by analyzing the system in subsets. By separating the dynamics into subsets, longitudinal dynamics, lateral dynamics, and heave dynamics are mathematically decoupled, even though, physically, there is coupling. In the following, a highly simplified analysis of helicopter stability in longitudinal, lateral, and heave dynamics is presented, using the various expressions derived in earlier chapters. Through these simplified formulations, the essential features of helicopter stability in hover are captured.

Simplified Treatment of Helicopter Dynamics in Hover

Using the mathematical formulation developed in earlier chapters for hub loads and rotor flap response to hub perturbation motions, a highly simplified analysis of helicopter stability can be developed. The results of this simplified analysis provide a fundamental understanding on the stability of the helicopter.

A simplified model of a helicopter is shown in Figure 10.10. The body axis system ($X_b - Y_b - Z_b$) has its origin at the center of mass of the helicopter. The main rotor is assumed to be rotating counterclockwise when viewed from

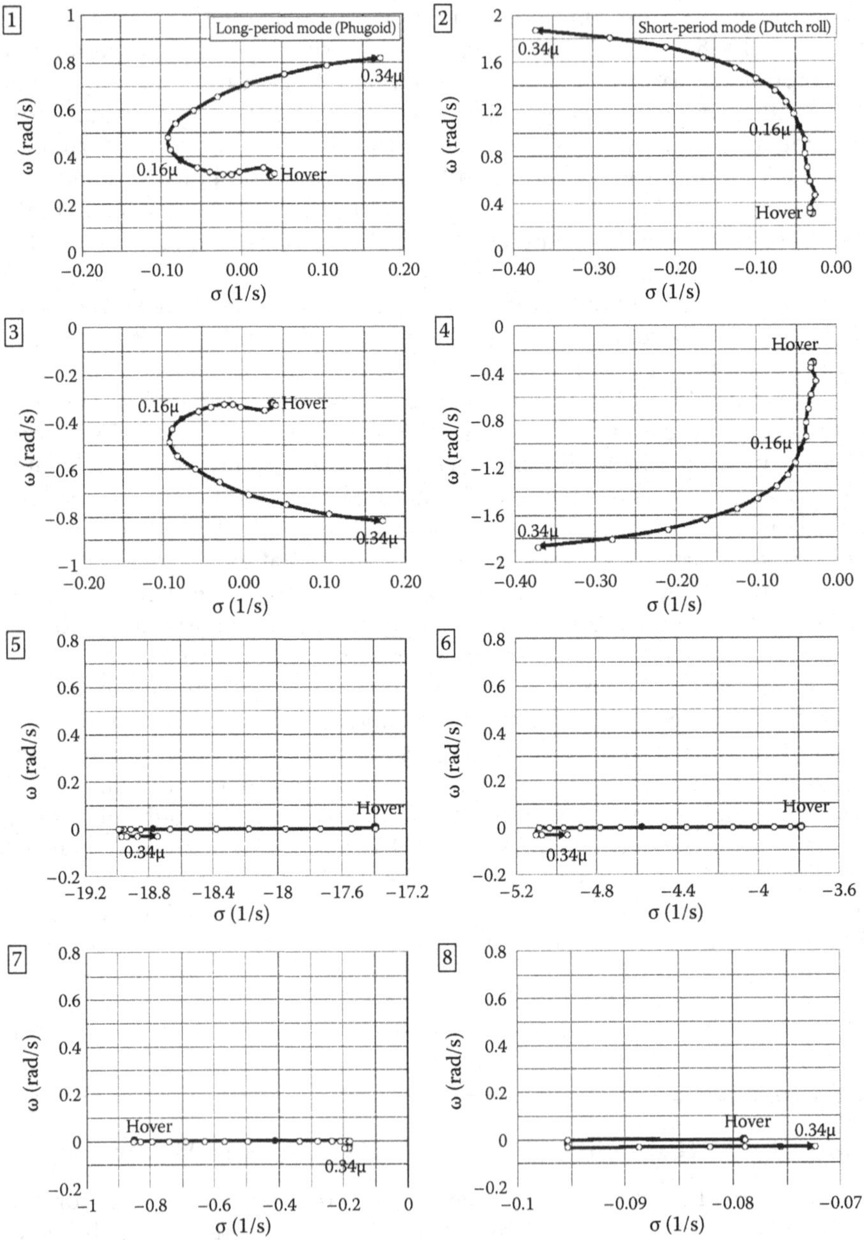

FIGURE 10.9
Loci of eigenvalues as a function of forward speed.

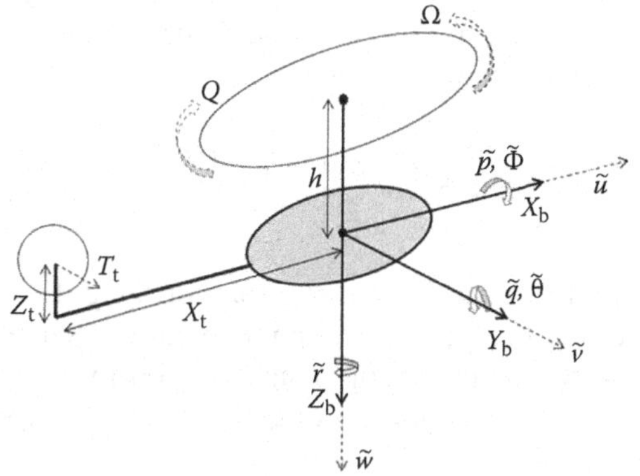

FIGURE 10.10
Simple model of helicopter with reference axis system.

top. The perturbation motion of the helicopter along the body axes is shown in Figure 10.10.

Longitudinal Dynamics in Hover

For the purpose of discussion, let us consider the longitudinal dynamics of a hovering helicopter, as shown in Figure 10.10. In Chapter 4, the expressions for hub forces and moments in terms blade flapping are derived under hovering condition (Equations 4.73 to 4.82). Using these expressions, explicit formulation of perturbation in hub loads will be derived (assuming zero twist for the blade) and used in the formulation of helicopter dynamics in hover. It may be noted that the proper change in sign will be incorporated by taking into consideration the direction of axes system indicated in Figures 4.5 and 10.10.

As the first example, let us write the relevant equations representing the longitudinal dynamics of the helicopter, consisting of the nondimensional variables: pitch angle ($\tilde{\theta}$), pitch rate ($\tilde{q}$), and longitudinal velocity ($\tilde{u}$).

From Chapter 4, for a rotor under hovering condition, the longitudinal hub force due to the rotor flap is derived and is given in Equation 4.78. (Note that the negative sign is changed because of the change in the direction of x axis.) The longitudinal force in dimensional form can be written as

$$X_F = \rho\pi R^2(\Omega R)^2 \frac{\sigma a}{2}\left[\frac{\theta_0}{3} - \frac{\lambda}{2}\right]\beta_{1c} \tag{10.65}$$

Similarly, the pitching moment at the center of mass of the helicopter due to the main rotor hub force in longitudinal direction and the pitch moment at the hub is given (from Equation 4.82) as

$$M_{yF} = \rho\pi R^2(\Omega R)^2 R\,\frac{\sigma a}{2}\left[\frac{\varpi_{RF}^2 - 1}{r} - \frac{C_T}{\sigma a}\frac{h}{R}\right]\{-\beta_{1c}\} \tag{10.66}$$

(Note that the quantity h denotes the rotor hub height from the center of mass of the helicopter. h should be taken as negative when the rotor hub is above the center of mass due to proper sign convention.)

The perturbation force equation along the body X_b axis can be written (from Equations 10.11, 10.28, and 10.60) as

$$\Omega^2 R M_F \dot{\tilde{u}} + M_F g\tilde{\theta} = \Delta X_F \tag{10.67}$$

(Note that the perturbation acceleration is nondimensionalised with respect to $\Omega^2 R$. The equilibrium angle in pitch θ_e is assumed as 0 during hover.)

In Equation 10.67, the first term on the left-hand side represents the inertia force of the fuselage, the second term represents the component of gravity force along the longitudinal direction, and the term on the right-hand side represents the perturbation in longitudinal force due to main rotor aerodynamic loads, which is to be obtained from Equation 10.65, and it is expressed in terms of the perturbation motion of the helicopter and the rotor blade cyclic input as

$$\Delta X_F = \frac{\partial X_F}{\partial \beta_{1c}}\left[\frac{\partial \beta_{1c}}{\partial \tilde{u}}\tilde{u} + \frac{\partial \beta_{1c}}{\partial \tilde{q}}\tilde{q} + \frac{\partial \beta_{1c}}{\partial \dot{\tilde{q}}}\dot{\tilde{q}} + \frac{\partial \beta_{1c}}{\partial \tilde{\theta}_{1c}}\tilde{\theta}_{1c} + \frac{\partial \beta_{1c}}{\partial \tilde{\theta}_{1s}}\tilde{\theta}_{1s}\right] \tag{10.68}$$

The perturbation moment equation for pitching motion can be written as

$$\Omega^2 I_{YY}\dot{\tilde{q}} = \Delta M_{yF} \tag{10.69}$$

where the right hand–side term represents the perturbation in fuselage pitching moment due to main rotor aerodynamic loads, which is to be obtained from Equation 10.66, and it is expressed as

$$\Delta M_{yF} = \frac{\partial M_{yF}}{\partial \beta_{1c}}\left[\frac{\partial \beta_{1c}}{\partial \tilde{u}}\tilde{u} + \frac{\partial \beta_{1c}}{\partial \tilde{q}}\tilde{q} + \frac{\partial \beta_{1c}}{\partial \dot{\tilde{q}}}\dot{\tilde{q}} + \frac{\partial \beta_{1c}}{\partial \tilde{\theta}_{1c}}\tilde{\theta}_{1c} + \frac{\partial \beta_{1c}}{\partial \tilde{\theta}_{1s}}\tilde{\theta}_{1s}\right] \tag{10.70}$$

The kinematic relationship between the rate of change of the pitch angle and the pitch rate is given as

$$\dot{\tilde{\theta}} = \tilde{q} \tag{10.71}$$

The set of Equations 10.65 to 10.71 represents the longitudinal dynamics of the helicopter in hover. One can simplify these equations further by substituting for various derivative terms and rearranging the equations in state space form.

For the sake of simplicity, let us express Equations 10.65 and 10.66 as

$$X_F = \alpha_1 \beta_{1c} \tag{10.72}$$

$$M_{yF} = -\alpha_2 \beta_{1c} \tag{10.73}$$

where α_1 and α_2 are given as

$$\alpha_1 = \rho \pi R^2 (\Omega R)^2 \frac{\sigma a}{2}\left[\frac{\theta_0}{3} - \frac{\lambda}{2}\right] \tag{10.74}$$

and

$$\alpha_2 = \rho \pi R^2 (\Omega R)^2 R \frac{\sigma a}{2}\left[\frac{\varpi_{RF}^2 - 1}{r} - \frac{C_T}{\sigma a}\frac{h}{R}\right] \tag{10.75}$$

Substituting Equation 10.68 in Equation 10.67, and Equation 10.70 in Equation 10.69, and using Equations 10.74 and 10.75, and nondimensionalizing the force equation with $M_b \Omega^2 R$ and the moment equation by $M_b \Omega^2 R^2$, the perturbation equations in longitudinal translation and pitch can be written, respectively, as

Perturbation force equation in the X direction:

$$\frac{M_F}{M_b}\dot{\tilde{u}} + \frac{M_F}{M_b}\frac{g}{\Omega^2 R}\tilde{\theta} - \bar{\alpha}_1 \frac{\partial \beta_{1c}}{\partial \dot{\tilde{q}}}\dot{\tilde{q}} = \bar{\alpha}_1 \frac{\partial \beta_{1c}}{\partial \tilde{u}}\tilde{u} + \bar{\alpha}_1 \frac{\partial \beta_{1c}}{\partial \tilde{q}}\tilde{q} + \bar{\alpha}_1 \frac{\partial \beta_{1c}}{\partial \tilde{\theta}_{1c}}\tilde{\theta}_{1c} + \bar{\alpha}_1 \frac{\partial \beta_{1c}}{\partial \tilde{\theta}_{1s}}\tilde{\theta}_{1s} \tag{10.76}$$

where

$$\bar{\alpha}_1 = \frac{\rho \pi R^3}{M_b}\frac{\sigma a}{2}\left[\frac{\theta_0}{3} - \frac{\lambda}{2}\right] \tag{10.77}$$

Perturbation pitching moment equation:

$$\frac{I_{YY}}{M_b R^2}\dot{\tilde{q}} + \bar{\alpha}_2 \frac{\partial \beta_{1c}}{\partial \dot{\tilde{q}}}\dot{\tilde{q}} = -\bar{\alpha}_2 \frac{\partial \beta_{1c}}{\partial \tilde{u}}\tilde{u} - \bar{\alpha}_2 \frac{\partial \beta_{1c}}{\partial \tilde{q}}\tilde{q} - \bar{\alpha}_2 \frac{\partial \beta_{1c}}{\partial \tilde{\theta}_{1c}}\tilde{\theta}_{1c} - \bar{\alpha}_2 \frac{\partial \beta_{1c}}{\partial \tilde{\theta}_{1s}}\tilde{\theta}_{1s} \tag{10.78}$$

where

$$\bar{\alpha}_2 = \frac{\rho \pi R^3}{M_b}\frac{\sigma a}{2}\left[\frac{\varpi_{RF}^2 - 1}{r} - \frac{C_T}{\sigma a}\frac{h}{R}\right] \tag{10.79}$$

Combining Equations 10.71, 10.76, and 10.78, the longitudinal dynamic equations of the helicopter in hover can be written in matrix form as

$$\begin{bmatrix} \frac{M_F}{M_b} & 0 & -\bar{\alpha}_1\frac{\partial\beta_{1c}}{\partial\dot{\tilde{q}}} \\ 0 & 1 & 0 \\ 0 & 0 & \frac{I_{YY}}{M_bR^2}+\bar{\alpha}_2\frac{\partial\beta_{1c}}{\partial\dot{\tilde{q}}} \end{bmatrix}\begin{Bmatrix} \dot{\tilde{u}} \\ \dot{\tilde{\theta}} \\ \dot{\tilde{q}} \end{Bmatrix} = \begin{bmatrix} \bar{\alpha}_1\frac{\partial\beta_{1c}}{\partial\tilde{u}} & -\frac{M_F}{M_b}\frac{g}{\Omega^2R} & \bar{\alpha}_1\frac{\partial\beta_{1c}}{\partial\tilde{q}} \\ 0 & 0 & 1 \\ -\bar{\alpha}_2\frac{\partial\beta_{1c}}{\partial\tilde{u}} & 0 & -\bar{\alpha}_2\frac{\partial\beta_{1c}}{\partial\tilde{q}} \end{bmatrix}\begin{Bmatrix} \tilde{u} \\ \tilde{\theta} \\ \tilde{q} \end{Bmatrix}$$

$$+\begin{bmatrix} \bar{\alpha}_1\frac{\partial\beta_{1c}}{\partial\tilde{\theta}_{1c}} & \bar{\alpha}_1\frac{\partial\beta_{1c}}{\partial\tilde{\theta}_{1s}} \\ 0 & 0 \\ -\bar{\alpha}_2\frac{\partial\beta_{1c}}{\partial\tilde{\theta}_{1c}} & -\bar{\alpha}_2\frac{\partial\beta_{1c}}{\partial\tilde{\theta}_{1s}} \end{bmatrix}\begin{Bmatrix} \tilde{\theta}_{1c} \\ \tilde{\theta}_{1s} \end{Bmatrix} \tag{10.80}$$

The various derivative terms in the matrices, involving longitudinal flap and the perturbation motion of the fuselage, can be obtained in terms of coupling parameter S_c, derived from Chapter 9 (derivatives given in Equation 9.68). The longitudinal stability of the system can be analyzed by neglecting the excitation term from the right-hand side of Equation 10.80 and solving the eigenvalues of the homogeneous part of the equation. One can notice that the magnitude of the derivative terms (inertia terms) on the left-hand side of Equation 10.80 will usually be very small, and hence, these terms are often neglected from the formulation. (This aspect is shown in the following by solving an example problem.)

Rearranging the homogenous part of Equation 10.80 (in symbolic form), after neglecting the derivative terms in the inertia matrix, as

$$\frac{d}{dt}\begin{Bmatrix} \tilde{u} \\ \tilde{\theta} \\ \tilde{q} \end{Bmatrix} - \begin{bmatrix} X_u & -\frac{g}{\Omega^2R} & X_q \\ 0 & 0 & 1 \\ M_u & 0 & M_q \end{bmatrix}\begin{Bmatrix} \tilde{u} \\ \tilde{\theta} \\ \tilde{q} \end{Bmatrix} = 0 \tag{10.81}$$

The eigenvalue problem becomes

$$\begin{vmatrix} s-X_u & \frac{g}{\Omega^2R} & -X_q \\ 0 & s & -1 \\ -M_u & 0 & s-M_q \end{vmatrix} = 0 \tag{10.82}$$

The characteristic equation is given by

$$s^3 - s^2(X_u + M_q) + s(X_u M_q - M_u X_q) + \frac{g}{\Omega^2 R} M_u = 0 \tag{10.83}$$

One can solve for the three roots and establish the stability behavior of the helicopter in longitudinal dynamics. Using the above derivation, the stability of the helicopter in longitudinal dynamics is analyzed for the sample data given in Table 10.1. The stability of the helicopter was analyzed for different values of the coupling parameter S_c, which predominantly influences the stability of the helicopter.

For various values of the coupling parameter, the eigenvalue problem for longitudinal dynamics is solved, and the results are given in the following. (Note: This problem can be treated as an exercise for a clear understanding.)

$S_c = 0.0$:

$$\begin{bmatrix} 62.003 & 0 & 0 \\ 0 & 1 & 0 \\ 0 & 0 & 6.2057 \end{bmatrix} \begin{Bmatrix} \dot{\tilde{u}} \\ \dot{\tilde{\theta}} \\ \dot{\tilde{q}} \end{Bmatrix} = \begin{bmatrix} -0.0313 & -0.09 & 0.3059 \\ 0 & 0 & 1 \\ 0.0038 & 0 & -0.0371 \end{bmatrix} \begin{Bmatrix} \tilde{u} \\ \tilde{\theta} \\ \tilde{q} \end{Bmatrix}$$

TABLE 10.1

Given and Calculated Data of the Various Parameters of a Helicopter

Parameter	Value
Weight of the helicopter, W	45,000 N
g	9.81 m/s²
Mass moment of inertia of the helicopter in pitch, I_{yy}	20,000 kgm²
Main rotor radius, R	6.6 m
Main rotor blade chord, c	0.5 m
Number of blades in the main rotor system, N	4
Rotor angular velocity, Ω	32 rad/s
Density of air, ρ	0.954 kg/m³
Height of the rotor hub above center of gravity (c.g.), h	−1.6 m
Mass per unit length of the blade, ρ_b	11.21 kg/m
Lift curve slope, a	5.73/rad
Mass of the fuselage (W/g), M_F	4587.2 kg
Thrust coefficient, C_T	0.007727
Mass of the rotor blade, M_b	73.99 kg
Mass moment of inertia of the blade about its root, I_b	1074.3 kgm²
Inflow ratio, λ_o	0.06216
Rotor solidity, σ	0.09646
Lock number, γ	4.8276

$$\begin{Bmatrix} \dot{\tilde{u}} \\ \dot{\tilde{\theta}} \\ \dot{\tilde{q}} \end{Bmatrix} = \begin{bmatrix} -0.0005 & -0.0015 & 0.0049 \\ 0 & 0 & 1 \\ 0.006 & 0 & -0.006 \end{bmatrix} \begin{Bmatrix} \tilde{u} \\ \tilde{\theta} \\ \tilde{q} \end{Bmatrix}$$

Eigenvalues:

$$\begin{Bmatrix} -0.0123 \\ 0.0029 + 0.0080i \\ 0.0029 - 0.0080i \end{Bmatrix}$$

Eigenvectors:

$$\begin{bmatrix} 0.1269 & -0.0605 + 0.1519i & -0.0605 - 0.1519i \\ 0.9918 & 0.9865 & 0.9865 \\ -0.0122 & 0.0029 + 0.0079i & 0.0029 - 0.0079i \end{bmatrix}$$

$S_c = 0.1$:

$$\begin{bmatrix} 62.003 & 0 & -0.0148 \\ 0 & 1 & 0 \\ 0 & 0 & 6.2141 \end{bmatrix} \begin{Bmatrix} \dot{\tilde{u}} \\ \dot{\tilde{\theta}} \\ \dot{\tilde{q}} \end{Bmatrix} = \begin{bmatrix} -0.0310 & -0.09 & 0.3118 \\ 0 & 0 & 1 \\ 0.0176 & 0 & -0.1772 \end{bmatrix} \begin{Bmatrix} \tilde{u} \\ \tilde{\theta} \\ \tilde{q} \end{Bmatrix}$$

$$\begin{Bmatrix} \dot{\tilde{u}} \\ \dot{\tilde{\theta}} \\ \dot{\tilde{q}} \end{Bmatrix} = \begin{bmatrix} -0.0005 & -0.0015 & 0.0050 \\ 0 & 0 & 1 \\ 0.0028 & 0 & -0.0285 \end{bmatrix} \begin{Bmatrix} \tilde{u} \\ \tilde{\theta} \\ \tilde{q} \end{Bmatrix}$$

Eigenvalues:

$$\begin{Bmatrix} -0.0328 \\ 0.0019 + 0.0110i \\ 0.0019 - 0.0110i \end{Bmatrix}$$

Eigenvectors:

$$\begin{bmatrix} -0.0499 & -0.0223 + 0.1248i & -0.0223 - 0.1248i \\ -0.9982 & 0.9919 & 0.9919 \\ 0.0328 & 0.0019 + 0.0109i & 0.0019 - 0.0109i \end{bmatrix}$$

$S_c = 0.2$:

$$\begin{bmatrix} 62.003 & 0 & -0.0287 \\ 0 & 1 & 0 \\ 0 & 0 & 6.2348 \end{bmatrix} \begin{Bmatrix} \dot{\tilde{u}} \\ \dot{\tilde{\theta}} \\ \dot{\tilde{q}} \end{Bmatrix} = \begin{bmatrix} -0.0301 & -0.09 & 0.3117 \\ 0 & 0 & 1 \\ 0.0306 & 0 & -0.3165 \end{bmatrix} \begin{Bmatrix} \tilde{u} \\ \tilde{\theta} \\ \tilde{q} \end{Bmatrix}$$

$$\begin{Bmatrix} \dot{\tilde{u}} \\ \dot{\tilde{\theta}} \\ \dot{\tilde{q}} \end{Bmatrix} = \begin{bmatrix} -0.0005 & -0.0015 & 0.0050 \\ 0 & 0 & 1 \\ 0.0049 & 0 & -0.0508 \end{bmatrix} \begin{Bmatrix} \tilde{u} \\ \tilde{\theta} \\ \tilde{q} \end{Bmatrix}$$

Eigenvalues:

$$\begin{Bmatrix} -0.0537 \\ 0.0012 + 0.0114i \\ 0.0012 - 0.0114i \end{Bmatrix}$$

Eigenvectors:

$$\begin{bmatrix} 0.0323 & -0.0135 + 0.1233i & -0.0135 - 0.1233i \\ 0.9980 & 0.9922 & 0.9922 \\ -0.0536 & 0.0012 + 0.0114i & 0.0012 - 0.0114i \end{bmatrix}$$

$S_c = 0.3$:

$$\begin{bmatrix} 62.003 & 0 & -0.0410 \\ 0 & 1 & 0 \\ 0 & 0 & 6.2657 \end{bmatrix} \begin{Bmatrix} \dot{\tilde{u}} \\ \dot{\tilde{\theta}} \\ \dot{\tilde{q}} \end{Bmatrix} = \begin{bmatrix} -0.0287 & -0.09 & 0.3060 \\ 0 & 0 & 1 \\ 0.0420 & 0 & -0.4475 \end{bmatrix} \begin{Bmatrix} \tilde{u} \\ \tilde{\theta} \\ \tilde{q} \end{Bmatrix}$$

$$\begin{Bmatrix} \dot{\tilde{u}} \\ \dot{\tilde{\theta}} \\ \dot{\tilde{q}} \end{Bmatrix} = \begin{bmatrix} -0.0005 & -0.0015 & 0.0049 \\ 0 & 0 & 1 \\ 0.0067 & 0 & -0.0714 \end{bmatrix} \begin{Bmatrix} \tilde{u} \\ \tilde{\theta} \\ \tilde{q} \end{Bmatrix}$$

Eigenvalues:

$$\begin{Bmatrix} -0.0737 \\ 0.0009 + 0.0115i \\ 0.0009 - 0.0115i \end{Bmatrix}$$

Eigenvectors:

$$\begin{bmatrix} 0.0247 & -0.0098+0.1241i & -0.0098-0.1241i \\ 0.9970 & 0.9922 & 0.9922 \\ -0.0734 & 0.0009+0.0114i & 0.0009-0.0114i \end{bmatrix}$$

It can be seen from the eigenvalues that one root is a stable real root and the other two roots are complex roots with a positive real part indicating instability. Since the eigenvalues are given in nondimensional form, they can be dimensionalised by multiplying with rotor angular velocity. It is well known that the helicopter is unstable in the phugoid mode in hover. The root locus plot is shown in Figure 10.11.

The above treatment of solving the longitudinal dynamics of the helicopter does not provide an understanding of the influence of various derivative terms in Equation 10.81 (or Equation 10.80) of the problem. Hence, the problem can be further simplified by decoupling the pitch rate motion from the phugoid (longitudinal and pitch) by using the following technique.

Let us write the longitudinal motion as

$$\begin{Bmatrix} \dot{X}_1 \\ \dot{X}_2 \end{Bmatrix} = \begin{bmatrix} A_{11} & A_{12} \\ A_{21} & A_{22} \end{bmatrix} \begin{Bmatrix} X_1 \\ X_2 \end{Bmatrix} \tag{10.84}$$

From the above equation, the two partition equations from the eigenvalue form can be written as

$$(sI - A_{11})X_1 - A_{12}X_2 = 0 \tag{10.85}$$

$$[sI - A_{22}]X_2 - A_{21}X_1 = 0 \tag{10.86}$$

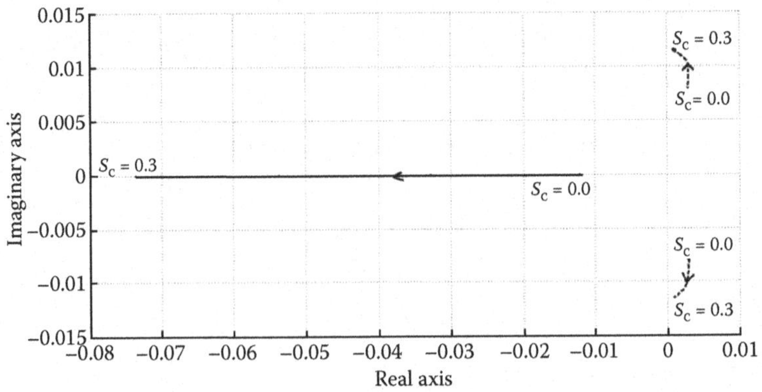

FIGURE 10.11
Variation in eigenvalues with change in coupling parameter S_c.

Substituting for X_1 from Equation 10.85 in Equation 10.86, the characteristic equation becomes

$$f_2(s) = |sI - A_{22} - A_{21}[sI - A_{11}]^{-1}A_{12}| = 0 \tag{10.87}$$

On the other hand, if we substitute for X_2 from Equation 10.86 in Equation 10.85, one obtains

$$f_1(s) = |sI - A_{11} - A_{12}[sI - A_{22}]^{-1}A_{21}| = 0 \tag{10.88}$$

The two equations, Equations 10.87 and 10.88, can be solved to obtain the eigenvalues of the system. In a weakly coupled system, the eigenvalues corresponding to the degrees of freedom (X_1) can be obtained from the equation $f_1(s) = 0$. The eigenvalues corresponding to X_2 can be obtained from the equation $f_2(s) = 0$. However, in solving for the eigenvalues, one may have to approximate the inverse matrix in the characteristic equation, which is given in a general form as

$$[sI - A]^{-1} = -A^{-1}[I + sA^{-1} + s^2A^{-2} + \ldots] \tag{10.89}$$

When the coupling is weak, this expansion can be approximated to two terms.

Applying the above approach to longitudinal dynamics, one has (from Equation 10.81)

$$\begin{aligned} A_{11} &= \begin{bmatrix} X_u & -\dfrac{g}{\Omega^2 R} \\ 0 & 0 \end{bmatrix} \quad A_{21} = [\, M_u \;\; 0 \,] \\ A_{12} &= \begin{bmatrix} X_q \\ 1 \end{bmatrix} \qquad\quad A_{22} = M_q \end{aligned} \tag{10.90}$$

Substituting the respective terms in Equation 10.88, the eigenvalues of vector $X_1\ (\tilde{u}, \tilde{\theta})^T$ can be obtained from the simplified characteristic equation given as

$$f_1(s) = \left| \begin{bmatrix} s - X_u & \dfrac{g}{\Omega^2 R} \\ 0 & s \end{bmatrix} - \begin{bmatrix} X_q \\ 1 \end{bmatrix} \left(-\frac{1}{M_q}\left(1 + \frac{s}{M_q}\right)\right) [\, M_u \;\; 0 \,] \right| = 0 \tag{10.91}$$

Simplifying Equation 10.91 as

$$f_1(s) = \left| \begin{bmatrix} s - X_u & \dfrac{g}{\Omega^2 R} \\ 0 & s \end{bmatrix} - \begin{bmatrix} X_q M_u & 0 \\ M_u & 0 \end{bmatrix} \left(-\frac{1}{M_q}\left(1 + \frac{s}{M_q}\right)\right) \right| = 0 \tag{10.92}$$

Rearranging the terms, the characteristic equation can be written as

$$\begin{vmatrix} s - X_u + \dfrac{X_q M_u}{M_q}\left(1 + \dfrac{s}{M_q}\right) & \dfrac{g}{\Omega^2 R} \\ \dfrac{M_u}{M_q}\left(1 + \dfrac{s}{M_q}\right) & s \end{vmatrix} = 0 \tag{10.93}$$

Expanding,

$$s^2\left(1 + \frac{X_q M_u}{M_q^2}\right) - s\left(X_u - \frac{X_q M_u}{M_q} + \frac{M_u}{M_q^2}\frac{g}{\Omega^2 R}\right) - \frac{M_u}{M_q}\frac{g}{\Omega^2 R} = 0 \tag{10.94}$$

Neglecting the terms with X_q (based on approximate order) in the first term, the natural frequency in the phugoid mode can be given as

$$\omega_p^2 = -\frac{M_u}{M_q}\frac{g}{\Omega^2 R} \tag{10.95}$$

The damping in the phugoid mode is given by

$$2\zeta_p \omega_p = -\left(X_u - \frac{X_q M_u}{M_q} + \frac{M_u}{M_q^2}\frac{g}{\Omega^2 R}\right) \tag{10.96}$$

These expressions clearly indicate the influence of the ratio of the pitching moment due to speed (speed stability) and the pitching moment due to pitch rate (direct damping in pitch) on both the frequency and the damping in the phugoid mode. As described earlier, the speed stability M_u depends on the amount of flapping due to perturbational motion in $\tilde{u}$. This derivative is the major source of instability and also plays a dominant role in damping. Hence, augmenting or improving speed stability will have a significant effect in improving the damping in the phugoid mode. This is the reason for having a tail plate, which can give rise to a pitch moment about the c.g. However, the tail plate becomes effective in forward flight only. In addition, the c.g. location also plays a significant role in phugoid mode stability.

General Notes

The longitudinal stability of a hingeless helicopter is generally of inferior quality compared to that of an articulated rotor helicopter. The distinction becomes more pronounced at high forward speeds. Aft. c.g. location deteriorates the longitudinal stability. Because of the unstable phugoid mode and also due to its severity in hingeless helicopters, stability augmentation systems are provided to improve the quality of flight. The positive aspect of this poor stability of hingeless rotors is that they provide higher control moments for improved maneuverability.

The other mode, namely pitch, is a damped mode. The contribution to pitch damping is due to M_q.

It is important to recognize that the above reduced order modeling holds good for hovering condition. However, in forward flight, due to the coupling between vertical velocity ($\dot{z}$), pitch, and forward velocity ($\dot{x}$), the reduced order modeling cannot be directly applicable.

Vertical Dynamics (Heave Dynamics)

The uncoupled vertical dynamics equation can be derived from the fundamental theory of a rotor having a small (or perturbational) axial velocity.

With reference to Figure 10.10, assuming a small vertical velocity of the helicopter $\tilde{w}$, the expression for rotor thrust from the blade element theory can be written as (assuming that the vertical velocity changes the rotor inflow uniformly)

$$T = \rho\pi R^2(\Omega R)^2 \frac{\sigma a}{2}\left[\frac{\theta_0}{3} - \frac{\lambda}{2} + \frac{\tilde{w}}{2}\right] \tag{10.97}$$

Similarly, the expression for thrust from the momentum theory can be expressed as

$$T = \rho\pi R^2(\Omega R)^2(\lambda - \tilde{w})2\lambda \tag{10.98}$$

It is important to note that perturbation in vertical velocity changes the thrust, which, in turn, can influence the rotor inflow. Therefore, it is essential to evaluate the change in inflow due to a small perturbation in vertical velocity of the helicopter first, before we formulate the equation of the helicopter in heave dynamics.

From Equation 10.97, change in thrust due to change in vertical velocity and the collective pitch can be expressed as

$$\Delta T = \frac{\partial T}{\partial \lambda}\frac{\partial \lambda}{\partial \tilde{w}}\tilde{w} + \frac{\partial T}{\partial \tilde{w}}\tilde{w} + \frac{\partial T}{\partial \tilde{\theta}}\tilde{\theta}_0 \tag{10.99}$$

From Equation 10.97, it can be noted that

$$\frac{\partial T}{\partial \lambda} = \rho \pi R^2 (\Omega R)^2 \frac{\sigma a}{2}\left[-\frac{1}{2}\right] \tag{10.100}$$

and

$$\frac{\partial T}{\partial \tilde{w}} = \rho \pi R^2 (\Omega R)^2 \frac{\sigma a}{2}\left[\frac{1}{2}\right] \tag{10.101}$$

Total change in thrust due to perturbation velocity $\tilde{w}$ can be obtained from Equation 10.97 as

$$\frac{dT}{d\tilde{w}} = \rho \pi R^2 (\Omega R)^2 \frac{\sigma a}{2}\left[-\frac{1}{2}\frac{\partial \lambda}{\partial \tilde{w}} + \frac{1}{2}\right] \tag{10.102}$$

Similarly, the total change in thrust can also be obtained from Equation 10.98 as

$$\frac{dT}{d\tilde{w}} = \rho \pi R^2 (\Omega R)^2 \left[4\lambda \frac{\partial \lambda}{\partial \tilde{w}} - 2\lambda - 2\tilde{w}\frac{\partial \lambda}{\partial \tilde{w}}\right] \tag{10.103}$$

Equating the expressions in Equations 10.102 and 10.103, one obtains

$$-\frac{\sigma a}{4}\frac{\partial \lambda}{\partial \tilde{w}} + \frac{\sigma a}{4} = 4\lambda \frac{\partial \lambda}{\partial \tilde{w}} - 2\lambda - 2\tilde{w}\frac{\partial \lambda}{\partial \tilde{w}} \tag{10.104}$$

Rearranging the terms, one obtains

$$\left(4\lambda + \frac{\sigma a}{4} - 2\tilde{w}\right)\frac{\partial \lambda}{\partial \tilde{w}} = \frac{\sigma a}{4} + 2\lambda \tag{10.105}$$

From Equation 10.105, one can obtain an expression for the variation in inflow velocity due to perturbation in vertical velocity of the helicopter at hovering condition (i.e., at $\tilde{w} = 0$) as

$$\frac{\partial \lambda}{\partial \tilde{w}} = \frac{2\lambda + \frac{\sigma a}{4}}{4\lambda + \frac{\sigma a}{4}} \tag{10.106}$$

Substituting Equations 10.100, 10.101, and 10.106 in Equation 10.99, the change in thrust at hover due to perturbation in vertical motion and collective input can be written as

$$\Delta T = \rho\pi R^2(\Omega R)^2 \frac{\sigma a}{2}\left[-\frac{1}{2}\frac{2\lambda + \frac{\sigma a}{4}}{4\lambda + \frac{\sigma a}{4}} + \frac{1}{2}\right]\tilde{w} + \rho\pi R^2(\Omega R)^2 \frac{\sigma a}{6}\tilde{\theta}_0 \qquad (10.107)$$

On simplification, Equation 10.107 becomes

$$\Delta T = \rho\pi R^2(\Omega R)^2 \frac{\sigma a}{4}\left[\frac{-2\lambda - \frac{\sigma a}{4} + 4\lambda + \frac{\sigma a}{4}}{4\lambda + \frac{\sigma a}{4}}\right]\tilde{w} + \rho\pi R^2(\Omega R)^2 \frac{\sigma a}{6}\tilde{\theta}_0 \qquad (10.108)$$

The variation thrust can be expressed as

$$\Delta T = \rho\pi R^2(\Omega R)^2 \sigma a \cdot \frac{2\lambda}{16\lambda + \sigma a}\tilde{w} + \rho\pi R^2(\Omega R)^2 \frac{\sigma a}{6}\tilde{\theta}_0 \qquad (10.109)$$

The force balance equation representing the vertical dynamics of the helicopter can be written as (with reference to Figure 10.10)

$$M_F g - (T + \Delta T) = M_F \Omega^2 R \dot{\tilde{w}} \qquad (10.110)$$

Substituting for ΔT from Equation 10.109 and cancelling the steady-state value of thrust equal to the weight of the helicopter, the vertical dynamics of the helicopter can be written as

$$-\frac{\rho\pi R^2 R}{M_F}\frac{\sigma a}{6}\tilde{\theta}_0 - \frac{\rho\pi R^2 \cdot R}{M_F}\frac{\sigma a 2\lambda}{16\lambda + \sigma a}\tilde{w} = \dot{\tilde{w}} \qquad (10.111)$$

Equation 10.111 can be written in symbolic form as

$$\dot{\tilde{w}} = Z_w \tilde{w} + Z_{\theta_0}\tilde{\theta}_0 \qquad (10.112)$$

The derivative Z_w and Z_{θ_0} can be obtained from the expressions given in Equation 10.111. Knowing the thrust coefficient C_T, the inflow λ in hover can be obtained. It is important to recognize that Z_w is always negative, and hence, the heave mode is always stable in hover.

Lateral–Directional Dynamics

The lateral–directional dynamics of the helicopter involves motion in roll, yaw, and side slip. According to the eighth-order model, the degrees of freedom contributing to lateral dynamics are side-slip velocity $\tilde{v}$, roll rate $\tilde{p}$, yaw rate $\tilde{r}$, and roll angle $\tilde{\phi}$. The reduced fourth-order system will have four eigenvalues. One of them will be a complex eigenvalue representing Dutch roll oscillation, and two real roots will represent periodic motion in the roll and spiral modes. In general, the roll mode is heavily damped and the spiral mode is lightly damped, but both are real stable roots. The roll subsidence is due to roll damping L_p, and the yaw subsidence is due to the yaw derivative N_r. The oscillatory Dutch roll mode is similar to the oscillatory phugoid mode, but the major difference is that, in general, the Dutch roll mode is stable, while the phugoid mode is unstable.

In lateral dynamics, the coupling between yaw–roll–side slip is relatively strong; hence, reduced order modeling will not be accurate enough to predict the eigenvalues. However, there are certain coordinate transformations that can be used to obtain a weakly coupled system. Then, the order of the system can be reduced. However, the validity of such a procedure depends on the magnitude of coupling between yaw and roll. When the coupling is high, the validity of order reduction breaks down. The key derivatives of yaw–roll coupling are N_p and L_r.

In the following, the dynamics of the helicopter in hover under lateral and directional motions is derived using the mathematical formulation presented in earlier chapters. This derivation is very similar to the one presented for longitudinal dynamics. Keeping Figure 10.10 as a reference, the perturbation equations for lateral–directional dynamics can be written for each degree of freedom.

The perturbation equation in lateral motion can be written as

$$M_F\Omega^2 R\dot{\tilde{v}} = \Delta T_T + \Delta Y_F + M_F g\tilde{\phi} \tag{10.113}$$

where ΔT_T represents the tail rotor thrust perturbation and ΔY_F is the perturbation in side force due to the main rotor.

The perturbation equation in roll motion can be written as

$$I_{XX}\Omega^2\dot{\tilde{p}} = \Delta M_{xF} - \Delta T_T Z_T \tag{10.114}$$

where ΔM_{xF} represents the roll moment at the center of mass due to the main rotor hub loads. Note that the quantity Z_T is negative as per the coordinate system shown in Figure 10.10.

The perturbation in yaw dynamics can be written as

$$I_{ZZ}\Omega^2\dot{\tilde{r}} = \Delta T_T X_T \tag{10.115}$$

Note that the quantity X_T is negative as per the coordinate system shown in Figure 10.10.

The kinematic relationship between the time derivative of the roll angle and roll rate can be written as

$$\dot{\tilde{\phi}} = \tilde{p} \tag{10.116}$$

The set of equations, given in Equations 10.113 to 10.116, represents the coupled lateral–directional dynamics of the helicopter in hover. Let us now substitute the various perturbation quantities in these equations to obtain the final set of complete equations.

Following the argument given in deriving Equation 10.109, the perturbation in tail rotor thrust can be expressed (in terms of quantities related to tail rotor) as

$$\Delta T_T = -\rho\pi R_T^2(\Omega_T R_T)^2\sigma_T a \cdot \frac{2\lambda_T}{16\lambda_T + \sigma_T a}(\tilde{v} + X_T\tilde{r} - Z_T\tilde{p}) + \rho\pi R_T^2(\Omega_T R_T)^2\frac{\sigma_T a}{6}\tilde{\theta}_{0T} \tag{10.117}$$

In Equation 10.117, the inflow λ_T corresponds to the tail rotor inflow. It can be evaluated from the main rotor torque coefficient (same as the power coefficient) as given in the following.

The main rotor power (or torque) coefficient in hover can be expressed as (from Equations 2.42 to 2.45)

$$C_P = C_Q = \kappa\frac{C_T^{3/2}}{\sqrt{2}} + \frac{\sigma C_{d0}}{8} \tag{10.118}$$

Using Equation 10.118, the main rotor torque can be written as

$$Q = \rho\pi R^2(\Omega R)^2 R\left[\kappa\frac{C_T^{3/2}}{\sqrt{2}} + \frac{\sigma C_{d0}}{8}\right] \tag{10.119}$$

Knowing the main rotor torque, the tail rotor thrust can be obtained from torque balance as

$$T_T = -\rho\pi R^2(\Omega R)^2 R\left[\kappa\frac{C_T^{3/2}}{\sqrt{2}} + \frac{\sigma C_{d0}}{8}\right]\frac{1}{X_T} \tag{10.120}$$

Note that the quantity X_T is a negative as per the coordinate system shown in Figure 10.10.

Knowing the tail rotor thrust, the tail rotor inflow during hovering condition can be obtained as

$$\nu_T = \sqrt{\frac{T_T}{2\rho\pi R_T^2}} \quad \lambda_T = \frac{\nu_T}{\Omega_T R_T} \tag{10.121}$$

Next, consider the side force. Using the expression for main rotor side force in terms of the lateral flapping given in Equation 4.79,

$$Y_F = -\rho\pi R^2(\Omega R)^2 \frac{\sigma a}{2}\left[\frac{\theta_0}{3} - \frac{\lambda}{2}\right]\beta_{1s} \tag{10.122}$$

This equation can be written as

$$Y_F = -\alpha_1\beta_{1s} \tag{10.123}$$

where $\alpha_1 = \rho\pi R^2(\Omega R)^2 \frac{\sigma a}{2}\left[\frac{\theta_0}{3} - \frac{\lambda}{2}\right]$.

From Equation 10.123, the perturbation in the main rotor side force can be expressed as

$$\Delta Y_F = \frac{\partial Y_F}{\partial \beta_{1s}}\left[\frac{\partial \beta_{1s}}{\partial \tilde{v}}\tilde{v} + \frac{\partial \beta_{1s}}{\partial \tilde{p}}\tilde{p} + \frac{\partial \beta_{1s}}{\partial \dot{\tilde{p}}}\dot{\tilde{p}} + \frac{\partial \beta_{1s}}{\partial \tilde{\theta}_{1c}}\tilde{\theta}_{1c} + \frac{\partial \beta_{1s}}{\partial \tilde{\theta}_{1s}}\tilde{\theta}_{1s}\right] \tag{10.124}$$

The roll moment at the center of mass due to the main rotor load is given as (Equation 4.82)

$$M_{XF} = \rho\pi R^2(\Omega R)^2 R \frac{\sigma a}{2}\left[\frac{\varpi_{RF}^2 - 1}{r} - \frac{C_T}{\sigma a}\frac{h}{R}\right]\{-\beta_{1s}\} = -\alpha_2\beta_{1s} \tag{10.125}$$

where $\alpha_2 = \rho\pi R^2(\Omega R)^2 R \frac{\sigma a}{2}\left[\frac{\varpi_{RF}^2 - 1}{r} - \frac{C_T}{\sigma a}\frac{h}{R}\right]$.

From Equation 10.125, the perturbation in roll moment can be written as

$$\Delta M_{XF} = \frac{\partial M_{XF}}{\partial \beta_{1s}}\left[\frac{\partial \beta_{1s}}{\partial \tilde{v}}\tilde{v} + \frac{\partial \beta_{1s}}{\partial \tilde{p}}\tilde{p} + \frac{\partial \beta_{1s}}{\partial \dot{\tilde{p}}}\dot{\tilde{p}} + \frac{\partial \beta_{1s}}{\partial \tilde{\theta}_{1c}}\tilde{\theta}_{1c} + \frac{\partial \beta_{1s}}{\partial \tilde{\theta}_{1s}}\tilde{\theta}_{1s}\right] \tag{10.126}$$

Substituting various expressions for the perturbation loads, the perturbation equation in side slip (Equation 10.113) can be written as

$$\Omega^2 R M_F \dot{\tilde{v}} =$$

$$-\rho\pi R_T^2 (\Omega_T R_T)^2 \sigma_T a \, . \, \frac{2\lambda_T}{16\lambda_T + \sigma_T a}(\tilde{v} + X_T \tilde{r} - Z_T \tilde{p}) + \rho\pi R_T^2 (\Omega_T R_T)^2 \, \frac{\sigma_T a}{6} \tilde{\theta}_{0T}$$

$$-\frac{\partial Y_F}{\partial \beta_{1s}}\left[\frac{\partial \beta_{1s}}{\partial \tilde{v}}\tilde{v} + \frac{\partial \beta_{1s}}{\partial \tilde{p}}\tilde{p} + \frac{\partial \beta_{1s}}{\partial \dot{\tilde{p}}}\dot{\tilde{p}} + \frac{\partial \beta_{1s}}{\partial \tilde{\theta}_{1c}}\tilde{\theta}_{1c} + \frac{\partial \beta_{1s}}{\partial \tilde{\theta}_{1s}}\tilde{\theta}_{1s}\right] + M_F g \tilde{\phi} \tag{10.127}$$

Note that the velocity and rate quantities are nondimensionalized with respect to the main rotor angular velocity Ω.

Dividing Equation 10.127 by $M_b \Omega^2 R$ and rearranging the terms, the side-slip equation can be written as

$$\frac{M_F \Omega^2 R}{M_b \Omega^2 R}\dot{\tilde{v}} = -\bar{\alpha}_3 \frac{\sigma_T a 2\lambda_T}{16\lambda_T + \sigma_T a}\left[\tilde{v} + \frac{X_T}{R}\tilde{r} - \frac{Z_T}{R}\tilde{p}\right]\frac{\Omega R}{\Omega_T R_T} - \bar{\alpha}_3 \frac{\sigma_T a}{6}\tilde{\theta}_{0T}$$

$$-\bar{\alpha}_1\left[\frac{\partial \beta_{1s}}{\partial \tilde{v}}\tilde{v} + \frac{\partial \beta_{1s}}{\partial \tilde{p}}\tilde{p} + \frac{\partial \beta_{1s}}{\partial \dot{\tilde{p}}}\dot{\tilde{p}} + \frac{\partial \beta_{1s}}{\partial \tilde{\theta}_{1c}}\tilde{\theta}_{1c} + \frac{\partial \beta_{1s}}{\partial \tilde{\theta}_{1s}}\tilde{\theta}_{1s}\right] + \frac{M_F g}{M_b \Omega^2 R}\tilde{\phi} \tag{10.128}$$

where $\bar{\alpha}_1$ and $\bar{\alpha}_3$ are given, respectively, as

$$\bar{\alpha}_1 = \frac{\rho\pi R^3}{M_b}\frac{\sigma a}{2}\left[\frac{\theta_0}{3} - \frac{\lambda}{2}\right] \quad \text{and} \quad \bar{\alpha}_3 = \frac{\rho\pi R_T^2 (\Omega_T R_T)^2}{M_b \Omega^2 R} \tag{10.129}$$

The perturbation equation for rolling rate (from Equation 10.114) can be written as

$$I_{xx}\Omega^2 \dot{\tilde{p}} = \frac{\partial M_{XF}}{\partial \beta_{1s}}\left[\frac{\partial \beta_{1s}}{\partial \tilde{v}}\tilde{v} + \frac{\partial \beta_{1s}}{\partial \tilde{p}}\tilde{p} + \frac{\partial \beta_{1s}}{\partial \dot{\tilde{p}}}\dot{\tilde{p}} + \frac{\partial \beta_{1s}}{\partial \tilde{\theta}_{1c}}\tilde{\theta}_{1c} + \frac{\partial \beta_{1s}}{\partial \tilde{\theta}_{1s}}\tilde{\theta}_{1s}\right]$$

$$+\rho\pi R_T^2 (\Omega_T R_T)^2 \sigma_T a \, . \, \frac{2\lambda_T}{16\lambda_T + \sigma_T a}\left(\tilde{v} + \frac{X_T}{R}\tilde{r} - \frac{Z_T}{R}\tilde{p}\right)\frac{\Omega R}{\Omega_T R_T} Z_T$$

$$-\rho\pi R_T^2 (\Omega_T R_T)^2 \, \frac{\sigma_T a}{6}\frac{\Omega R}{\Omega_T R_T} Z_T \tilde{\theta}_{0T} \tag{10.130}$$

Nondimensionalizing the equation by dividing with $M_b\Omega^2R^2$, one obtains

$$\frac{I_{XX}\Omega^2}{M_b\Omega^2R^2}\dot{\tilde{p}}=-\bar{\alpha}_2\left[\frac{\partial\beta_{1s}}{\partial\tilde{v}}\tilde{v}+\frac{\partial\beta_{1s}}{\partial\tilde{p}}\tilde{p}+\frac{\partial\beta_{1s}}{\partial\dot{\tilde{p}}}\dot{\tilde{p}}+\frac{\partial\beta_{1s}}{\partial\tilde{\theta}_{1c}}\tilde{\theta}_{1c}+\frac{\partial\beta_{1s}}{\partial\tilde{\theta}_{1s}}\tilde{\theta}_{1s}\right]$$
$$+\bar{\alpha}_3\frac{Z_T}{R}\frac{\sigma_Ta2\lambda_T}{16\lambda_T+\sigma_Ta}\left[\tilde{v}+\frac{X_T}{R}\tilde{r}-\frac{Z_T}{R}\tilde{p}\right]\frac{\Omega R}{\Omega_TR_T}-\bar{\alpha}_3\frac{Z_T}{R}\frac{\sigma_Ta}{6}\tilde{\theta}_{0T} \tag{10.131}$$

The yaw dynamic equation (Equation 10.115) can be written as

$$I_{zz}\Omega^2\dot{\tilde{r}}=-\rho\pi R_T^2(\Omega_TR_T)^2\sigma_Ta\cdot\frac{2\lambda_T}{16\lambda_T+\sigma_Ta}\left(\tilde{v}+\frac{X_T}{R}\tilde{r}-\frac{Z_T}{R}\tilde{p}\right)\frac{\Omega R}{\Omega_TR_T}X_T$$
$$+\rho\pi R_T^2(\Omega_TR_T)^2\frac{\sigma_Ta}{6}\frac{\Omega R}{\Omega_TR_T}X_T\tilde{\theta}_{0T} \tag{10.132}$$

Dividing by $M_b\Omega^2R^2$ and nondimensionalizing, the yaw equation becomes

$$\frac{I_{zz}\Omega^2\dot{\tilde{r}}}{M_b\Omega^2R^2}=-\frac{\rho\pi R_T^2(\Omega_TR_T)^2X_T}{M_b\Omega^2R^2}\frac{\sigma_Ta2\lambda_T}{16\lambda_T+\sigma_Ta}\left(\tilde{v}+\frac{X_T}{R}\tilde{r}-\frac{Z_T}{R}\tilde{p}\right)\frac{\Omega R}{\Omega_TR_T}$$
$$+\frac{\rho\pi R_T^2(\Omega_TR_T)^2X_T}{M_b\Omega^2R^2}\frac{\sigma_Ta}{6}\tilde{\theta}_{0T} \tag{10.133}$$

Rewriting the above equation in a compact form as

$$\frac{I_{zz}\Omega^2\dot{\tilde{r}}}{M_b\Omega^2R^2}=-\bar{\alpha}_3\frac{X_T}{R}\frac{\sigma_Ta2\lambda_T}{16\lambda_T+\sigma_Ta}\left(\tilde{v}+\frac{X_T}{R}\tilde{r}-\frac{Z_T}{R}\tilde{p}\right)\frac{\Omega R}{\Omega_TR_T}+\bar{\alpha}_3\frac{X_T}{R}\frac{\sigma_Ta}{6}\tilde{\theta}_{0T} \tag{10.134}$$

Equations 10.116, 10.128, 10.131, and 10.134 represent the coupled lateral–yaw dynamic equations. These equations can be written in state space form as, after defining (for compactness)

$$\bar{\alpha}_4=\frac{\sigma_Ta2\lambda_T}{16\lambda_T+\sigma_Ta}\frac{\Omega R}{\Omega_TR_T} \tag{10.135}$$

$$\begin{bmatrix} \frac{M_F}{M_b} & 0 & \bar{\alpha}_1 \frac{\partial \beta_{1s}}{\partial \dot{\tilde{p}}} & 0 \\ 0 & 1 & 0 & 0 \\ 0 & 0 & \frac{I_{xx}}{M_b R^2} + \bar{\alpha}_2 \frac{\partial \beta_{1s}}{\partial \dot{\tilde{p}}} & 0 \\ 0 & 0 & 0 & \frac{I_{zz}}{M_b R^2} \end{bmatrix} \begin{Bmatrix} \dot{\tilde{v}} \\ \dot{\tilde{\phi}} \\ \dot{\tilde{p}} \\ \dot{\tilde{r}} \end{Bmatrix} =$$

$$\begin{bmatrix} -\bar{\alpha}_3\bar{\alpha}_4 - \bar{\alpha}_1 \frac{\partial \beta_{1s}}{\partial \tilde{v}} & \frac{M_F g}{M_b \Omega^2 R} & \bar{\alpha}_3\bar{\alpha}_4 \frac{Z_T}{R} - \bar{\alpha}_1 \frac{\partial \beta_{1s}}{\partial \tilde{p}} & -\bar{\alpha}_3\bar{\alpha}_4 \frac{X_T}{R} \\ 0 & 0 & 1 & 0 \\ -\bar{\alpha}_2 \frac{\partial \beta_{1s}}{\partial \tilde{v}} + \bar{\alpha}_3\bar{\alpha}_4 \frac{Z_T}{R} & 0 & -\bar{\alpha}_2 \frac{\partial \beta_{1s}}{\partial \tilde{p}} - \bar{\alpha}_3\bar{\alpha}_4 \left(\frac{Z_T}{R}\right)^2 & \bar{\alpha}_3\bar{\alpha}_4 \frac{Z_T}{R}\frac{X_T}{R} \\ -\bar{\alpha}_3\bar{\alpha}_4 \frac{X_T}{R} & 0 & \bar{\alpha}_3\bar{\alpha}_4 \frac{Z_T}{R}\frac{X_T}{R} & -\bar{\alpha}_3\bar{\alpha}_4 \left(\frac{X_T}{R}\right) \end{bmatrix} \begin{Bmatrix} \tilde{v} \\ \tilde{\phi} \\ \tilde{p} \\ \tilde{r} \end{Bmatrix} +$$

$$\begin{bmatrix} -\bar{\alpha}_1 \frac{\partial \beta_{1s}}{\partial \tilde{\theta}_{1c}} & -\bar{\alpha}_1 \frac{\partial \beta_{1s}}{\partial \tilde{\theta}_{1s}} & \bar{\alpha}_3 \frac{\sigma_T a}{6} \\ 0 & 0 & 0 \\ -\bar{\alpha}_2 \frac{\partial \beta_{1s}}{\partial \tilde{\theta}_{1c}} & -\bar{\alpha}_2 \frac{\partial \beta_{1s}}{\partial \tilde{\theta}_{1s}} & -\bar{\alpha}_3 \frac{Z_T}{R}\frac{\sigma_T a}{6} \\ 0 & 0 & \bar{\alpha}_3 \frac{X_T}{R}\frac{\sigma_T a}{6} \end{bmatrix} \begin{Bmatrix} \tilde{\theta}_{1c} \\ \tilde{\theta}_{1s} \\ \tilde{\theta}_{0T} \end{Bmatrix} \tag{10.136}$$

Noting that the flap derivative terms in the inertia matrix are very small in magnitude, they can be neglected. Inverting the inertia matrix, Equation 10.136 can be written in the first-order state space form, which is given as

$$\begin{Bmatrix} \dot{\tilde{v}} \\ \dot{\tilde{\phi}} \\ \dot{\tilde{p}} \\ \dot{\tilde{r}} \end{Bmatrix} = \begin{bmatrix} Y_v & \frac{g}{\Omega^2 R} & Y_p & Y_r \\ 0 & 0 & 1 & 0 \\ L_v & 0 & L_p & L_r \\ N_v & 0 & N_p & N_r \end{bmatrix} \begin{Bmatrix} \tilde{v} \\ \tilde{\phi} \\ \tilde{p} \\ \tilde{r} \end{Bmatrix} + \begin{bmatrix} Y_{\theta 1c} & Y_{\theta 1s} & Y_{\theta 0T} \\ 0 & 0 & 0 \\ L_{\theta 1c} & L_{\theta 1s} & L_{\theta 0T} \\ N_{\theta 1c} & N_{\theta 1s} & N_{\theta 0T} \end{bmatrix} \begin{Bmatrix} \tilde{\theta}_{1c} \\ \tilde{\theta}_{1s} \\ \tilde{\theta}_{0T} \end{Bmatrix} \tag{10.137}$$

One can solve the stability of the helicopter in lateral–yaw dynamics by solving the eigenvalue problem of the homogeneous part of Equation 10.137.

One can solve for the four roots and establish the stability behavior of the helicopter. Using the above formulation, the stability of the helicopter in the coupled lateral–yaw dynamics is analyzed for the sample data given in Table 10.2. The stability of the helicopter was analyzed for different values of the coupling parameter S_c.

The results of the stability analysis are provided in the following for various cases of coupling parameter S_c.

TABLE 10.2

Given and Calculated Data of Various Parameters

Parameter	Value
Weight of the helicopter, W	45,000 N
g	9.81 m/s^2
Roll inertia of helicopter, I_{xx}	5000 kgm^2
Main rotor radius, R	6.6 m
Tail rotor radius, R_T	1.275 m
Main rotor blade chord, c	0.5 m
Tail rotor blade chord, c_T	0.19 m
Number of blades in the main rotor, N	4
Number of blades in the tail rotor, N_T	4
Main rotor angular velocity, Ω	32 rad/s
Tail rotor angular velocity, Ω_T	160 rad/s
Density of air, ρ	0.954 kg/m^3
Height of the rotor hub above the center of mass, h	−1.6 m
Height of the tail rotor hub center above the c.g., Z_T	−2 m
Tail rotor hub location, X_T	−7.9 m
Mass per unit length of the main rotor blade, M_b	11.21 kg/m
Lift curve slope, a	5.73/rad
Drag coefficient, C_{do}	0.01
Empirical loss factor, κ	1.15
Helicopter mass, M_F	4587.2 kg
Main rotor thrust coefficient, C_T	0.007727
Main rotor torque coefficient, C_Q	0.000658
Main rotor torque, Q	25311 Nm
Tail rotor thrust, T_T	3204 N
Main rotor blade mass, M_b	73.99 kg
Mass moment of inertia of the main rotor blade about the flap hinge, I_b	1074.3 kg m^2
Main rotor inflow in hover, λ_o	0.06216
Tail rotor inflow in hover, λ_T	0.08890
Main rotor solidity ratio, σ	0.09646
Tail rotor solidity ratio, σ_T	0.1897
Main rotor Lock number, γ	4.8276

$S_c = 0.0$:

$$\begin{bmatrix} 62.003 & 0 & 0 & 0 \\ 0 & 1 & 0 & 0 \\ 0 & 0 & 1.5514 & 0 \\ 0 & 0 & 0 & 4.9646 \end{bmatrix} \begin{Bmatrix} \dot{\tilde{v}} \\ \dot{\tilde{\varphi}} \\ \dot{\tilde{p}} \\ \dot{\tilde{r}} \end{Bmatrix} = \begin{bmatrix} -0.0637 & -0.09 & 0.3157 & 0.0387 \\ 0 & 0 & 1 & 0 \\ -0.0136 & 0 & -0.04 & 0.0117 \\ 0.0387 & 0 & 0.0117 & -0.0463 \end{bmatrix} \begin{Bmatrix} \tilde{v} \\ \tilde{\varphi} \\ \tilde{p} \\ \tilde{r} \end{Bmatrix}$$

$$\begin{Bmatrix} \dot{\tilde{v}} \\ \dot{\tilde{\varphi}} \\ \dot{\tilde{p}} \\ \dot{\tilde{r}} \end{Bmatrix} = \begin{bmatrix} -0.0010 & 0.0015 & -0.0051 & 0.0006 \\ 0 & 0 & 1 & 0 \\ -0.0088 & 0 & -0.0258 & 0.0076 \\ 0.0078 & 0 & 0.0024 & -0.0093 \end{bmatrix} \begin{Bmatrix} \tilde{v} \\ \tilde{\varphi} \\ \tilde{p} \\ \tilde{r} \end{Bmatrix}$$

Eigenvalues:

$$\begin{Bmatrix} -0.0387 \\ 0.0026 + 0.0177i \\ 0.0026 - 0.0177i \\ -0.0027 \end{Bmatrix}$$

Eigenvectors:

$$\begin{bmatrix} -0.0439 & 0.0102 + 0.0781i & 0.0102 - 0.0781i & 0.4468 \\ 0.9982 & 0.9964 & 0.9964 & -0.7253 \\ -0.0386 & 0.0026 + 0.0177i & 0.0026 - 0.0177i & 0.0019 \\ 0.0148 & -0.0198 - 0.0182i & -0.0198 + 0.0182i & 0.5238 \end{bmatrix}$$

$S_c = 0.1$:

$$\begin{bmatrix} 62.003 & 0 & 0.0148 & 0 \\ 0 & 1 & 0 & 0 \\ 0 & 0 & 1.5598 & 0 \\ 0 & 0 & 0 & 4.9646 \end{bmatrix} \begin{Bmatrix} \dot{\tilde{v}} \\ \dot{\tilde{\varphi}} \\ \dot{\tilde{p}} \\ \dot{\tilde{r}} \end{Bmatrix} = \begin{bmatrix} -0.0633 & -0.09 & 0.3216 & 0.0387 \\ 0 & 0 & 1 & 0 \\ -0.0274 & 0 & -0.1802 & 0.0117 \\ 0.0387 & 0 & 0.0117 & -0.0463 \end{bmatrix} \begin{Bmatrix} \tilde{v} \\ \tilde{\varphi} \\ \tilde{p} \\ \tilde{r} \end{Bmatrix}$$

$$\begin{Bmatrix} \dot{\tilde{v}} \\ \dot{\tilde{\varphi}} \\ \dot{\tilde{p}} \\ \dot{\tilde{r}} \end{Bmatrix} = \begin{bmatrix} -0.0010 & 0.0015 & -0.0052 & 0.0006 \\ 0 & 0 & 1 & 0 \\ -0.0176 & 0 & -0.1155 & 0.0075 \\ 0.0078 & 0 & 0.0024 & -0.0093 \end{bmatrix} \begin{Bmatrix} \tilde{v} \\ \tilde{\varphi} \\ \tilde{p} \\ \tilde{r} \end{Bmatrix}$$

Eigenvalues:

$$\begin{Bmatrix} -0.1184 \\ -0.0005+0.0141i \\ -0.0005-0.0141i \\ -0.0027 \end{Bmatrix}$$

Eigenvectors:

$$\begin{bmatrix} 0.0175 & -0.0023-0.0101i & -0.0023+0.0101i & 0.1751 \\ -0.9929 & 0.9937 & 0.9937 & -0.8539 \\ 0.1175 & -0.0005+0.0140i & -0.0005-0.0140i & 0.0056 \\ -0.0038 & -0.0391-0.0232i & -0.0391+0.0232i & 0.49 \end{bmatrix}$$

$S_c = 0.2$:

$$\begin{bmatrix} 62.003 & 0 & 0.0287 & 0 \\ 0 & 1 & 0 & 0 \\ 0 & 0 & 1.5805 & 0 \\ 0 & 0 & 0 & 4.9646 \end{bmatrix} \begin{Bmatrix} \dot{\tilde{v}} \\ \dot{\tilde{\varphi}} \\ \dot{\tilde{p}} \\ \dot{\tilde{r}} \end{Bmatrix} = \begin{bmatrix} -0.0624 & -0.09 & 0.3215 & 0.0387 \\ 0 & 0 & 1 & 0 \\ -0.0404 & 0 & -0.3194 & 0.0117 \\ 0.0387 & 0 & 0.0117 & -0.0463 \end{bmatrix} \begin{Bmatrix} \tilde{v} \\ \tilde{\varphi} \\ \tilde{p} \\ \tilde{r} \end{Bmatrix}$$

$$\begin{Bmatrix} \dot{\tilde{v}} \\ \dot{\tilde{\varphi}} \\ \dot{\tilde{p}} \\ \dot{\tilde{r}} \end{Bmatrix} = \begin{bmatrix} -0.0010 & 0.0015 & -0.0051 & 0.0006 \\ 0 & 0 & 1 & 0 \\ -0.0255 & 0 & -0.2021 & 0.0074 \\ 0.0078 & 0 & 0.0024 & -0.0093 \end{bmatrix} \begin{Bmatrix} \tilde{v} \\ \tilde{\varphi} \\ \tilde{p} \\ \tilde{r} \end{Bmatrix}$$

Eigenvalues:

$$\begin{Bmatrix} -0.2037 \\ -0.0005+0.0129i \\ -0.0005-0.0129i \\ -0.0077 \end{Bmatrix}$$

Eigenvectors:

$$\begin{bmatrix} -0.0120 & -0.0023-0.01094i & -0.0023+0.01094i & 0.1241 \\ 0.9798 & 0.9925 & 0.9925 & -0.7995 \\ -0.1996 & -0.0005+0.0128i & -0.0005-0.0128i & 0.0061 \\ 0.0029 & -0.0441-0.0286i & -0.0441+0.0286i & 0.5876 \end{bmatrix}$$

$S_c = 0.3$:

$$\begin{bmatrix} 62.003 & 0 & 0.0410 & 0 \\ 0 & 1 & 0 & 0 \\ 0 & 0 & 1.6115 & 0 \\ 0 & 0 & 0 & 4.9646 \end{bmatrix} \begin{Bmatrix} \dot{\tilde{v}} \\ \dot{\tilde{\varphi}} \\ \dot{\tilde{p}} \\ \dot{\tilde{r}} \end{Bmatrix} = \begin{bmatrix} -0.0611 & -0.09 & 0.3158 & 0.0387 \\ 0 & 0 & 1 & 0 \\ -0.0518 & 0 & -0.4504 & 0.0117 \\ 0.0387 & 0 & 0.0117 & -0.0463 \end{bmatrix} \begin{Bmatrix} \tilde{v} \\ \tilde{\varphi} \\ \tilde{p} \\ \tilde{r} \end{Bmatrix}$$

$$\begin{Bmatrix} \dot{\tilde{v}} \\ \dot{\tilde{\varphi}} \\ \dot{\tilde{p}} \\ \dot{\tilde{r}} \end{Bmatrix} = \begin{bmatrix} -0.0010 & 0.0015 & -0.0049 & 0.0006 \\ 0 & 0 & 1 & 0 \\ -0.0322 & 0 & -0.2795 & 0.0073 \\ 0.0078 & 0 & 0.0024 & -0.0093 \end{bmatrix} \begin{Bmatrix} \tilde{v} \\ \tilde{\varphi} \\ \tilde{p} \\ \tilde{r} \end{Bmatrix}$$

Eigenvalues:

$$\begin{Bmatrix} -0.2807 \\ -0.0005 + 0.0124i \\ -0.0005 - 0.0124i \\ -0.0082 \end{Bmatrix}$$

Eigenvectors:

$$\begin{bmatrix} -0.0097 & -0.0018 - 0.1139i & -0.0018 + 0.1139i & 0.0979 \\ 0.9627 & 0.9918 & 0.9918 & -0.7456 \\ -0.2703 & -0.0005 + 0.0123i & -0.0005 - 0.0123i & 0.0061 \\ 0.0026 & -0.0464 - 0.0319i & -0.0464 + 0.0319i & 0.6592 \end{bmatrix}$$

The eigenvalues are shown in the root locus plot in Figure 10.12. From the sample results, it can be seen that the coupled lateral–yaw dynamics in hover is stable for S_c not equal to 0. Only when S_c is 0, the Dutch roll mode is unstable. The non-oscillatory stable modes correspond to the pure roll mode and yaw mode. The more stable mode corresponds to the roll mode, and the less stable mode corresponds to the yaw mode. This observation is made by analyzing the eigenvectors of the modes.

Control Characteristics/Control Response

The control characteristics of helicopters deal with the problem of evaluating the response of the system to a given input, which is the blade pitch

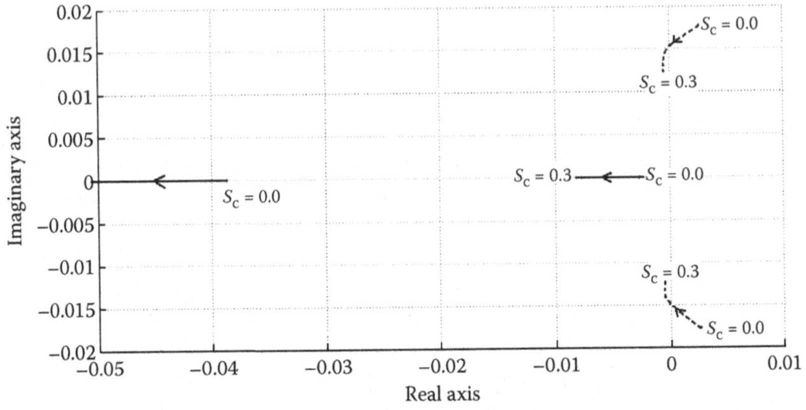

FIGURE 10.12
Variation in eigenvalues with change in coupling parameter S_c.

input. Since there are four inputs, generally, the study involves analyzing the helicopter behavior to each one of them independently. The control characteristics or handling quality requirements of a helicopter are specified in MIL-H-8501A or by ADS-33. In the following, a simplified treatment of the control response is provided.

Pitch, Roll, and Yaw Response to Control Inputs

In analyzing the control response of the helicopter, one can address the problem of (1) initial response to control input and/or (2) steady-state response. Generally, for accurate maneuvering, it is the initial response that plays a major role. The response behavior of the vehicle is always analyzed using the linearized perturbation equations developed earlier.

$$\dot{X} = AX + Bu \tag{10.138}$$

In general, the pilot's opinion plays a major role in deciding whether the helicopter has a good/adequate/poor control response behavior. If the response of the vehicle to unit pilot input is very small, then the pilot will find the vehicle too sluggish, and if the response is too large, then the vehicle is oversensitive. Hence, there is a trade-off between sluggishness and oversensitivity. It must be noted that the opinion of the pilot will also change with experience.

In the handling quality requirements, two parameters of the helicopter are used. One is control power and the other is control damping. These will be explained in the following by considering the uncoupled motion in any one axis (say roll).

The uncoupled roll equation can be written as

$$\dot{\tilde{p}} = \left\{ \frac{\partial L}{\partial p} \tilde{p} + \frac{\partial L}{\partial \theta_{1c}} \tilde{\theta}_{1c} \right\} \frac{1}{I_{XX}} \quad (10.139)$$

Rewriting Equation 10.139 as

$$\dot{\tilde{p}} - \frac{1}{I_{XX}} \frac{\partial L}{\partial p} \tilde{p} = \frac{1}{I_{XX}} \frac{\partial L}{\partial \theta_{1c}} \tilde{\theta}_{1c} \quad (10.140)$$

Note: $L_p = \dfrac{\partial L}{\partial p} \dfrac{1}{I_{XX}}; L_{\theta 1c} = \dfrac{\partial L}{\partial \theta_{1c}} \dfrac{1}{I_{XX}} \rightarrow L_p = \bar{L}_p / I_{XX}; L_{\theta 1c} = \bar{L}_{\theta 1c} / I_{XX}$

Equation 10.140 is a first-order differential equation, and its solution can be written as

$$\tilde{p} = -\frac{\left(\dfrac{\bar{L}_{\theta 1c}}{I_{XX}} \right)}{\left(\dfrac{\bar{L}_p}{I_{XX}} \right)} \left[1 - e^{\left(\frac{\bar{L}_p}{I_{XX}} \right) \tau} \right] \tilde{\theta}_{1c} \quad (10.141)$$

Note that $\tau = \Omega t$ is the nondimensional time. The solution can also be expressed as

$$\tilde{p} = \frac{\left(\dfrac{\text{control moment/unit control input}}{\text{roll inertia}} \right)}{\left(\dfrac{\text{roll damping}}{\text{roll inertia}} \right)} \left[1 - e^{-\left(\frac{\text{roll damping}}{\text{roll inertia}} \right) \tau} \right] \quad (10.142)$$

The handling quality requirements are based on the two parameters representing the initial angular acceleration of roll motion for unit control input and damping in roll, that is,

$$\frac{\text{control moment/unit control input}}{\text{roll inertia}} \quad \text{vs.} \quad \frac{\text{roll damping}}{\text{roll inertia}}$$

$$\text{or} \quad \frac{\bar{L}_{\theta 1c}}{I_{XX}} \quad \text{vs.} \quad \frac{\bar{L}_p}{I_{XX}}$$

or

$$L_{\theta 1c} \text{ vs. } L_p$$

Note that, in addition to the rotor aerodynamic characteristics in generating control moment, the inertia of the helicopter also plays a major role in determining the control characteristics of the helicopter. Similar parameters are used for pitch and yaw with the corresponding control moment, damping, and inertia.

For pitch, the two parameters are $M_{\theta_{1s}}$ vs. M_q.

For yaw, the two parameters are $N_{\theta_{0T}}$ vs. N_T.

An alternate form of representing the handling qualities is by using the parameters: steady-state rate per unit control moment and the time constant, that is,

For roll, $\dfrac{L_{\theta 1c}}{L_p}$ vs. $\dfrac{1}{L_p}$.

For pitch, $\dfrac{M_{\theta 1s}}{M_q}$ vs. $\dfrac{1}{M_q}$.

For yaw, $\dfrac{N_{\theta 0T}}{N_T}$ vs. $\dfrac{1}{N_r}$.

Gust Response

Assume a vertical upward gust w_g, which can be added to the vertical motion of the helicopter, and write the uncoupled vertical dynamic equation as

$$\dot{\tilde{w}} = Z_w \tilde{w} + Z_w w_g \tag{10.143}$$

The heave damping derivative defines the transient response as well as the gain for the gust input. The expression for Z_w in hover is given as (Equation 10.111)

$$Z_w = -\frac{2\sigma a \rho \pi R^2 R\lambda}{(16\lambda + \sigma a)(M_F)}$$

Note that $\sigma \pi R^2 = NCR =$ blade area (A_b).

An important parameter in the damping derivative is blade loading ($M_F/\pi R^2$). Since the blade loading is much higher than the wing loading of a fixed-wing aircraft, helicopters are less sensitive to gust than a fixed-wing aircraft of the same weight.

Helicopter gust response problem is quite complex, which involves the characterization and modeling of atmospheric disturbances, the analysis of helicopter response, and the formulation of the ride qualities. These aspects need to be addressed while considering operating environments, such as nap-of-the-earth flight, flying close to and landing in a high-rise building, etc.

11

Ground Resonance–Aeromechanical Instability: A Simple Model

During the operation of the helicopter, the low-frequency lag mode of the rotor system can couple with body roll or pitch to cause instability. When the instability happens on ground, it is called "ground resonance," and when it happens in air, it is called "air resonance." When the helicopter is on ground, it has a well-defined pitch and roll frequencies determined by the landing gear stiffness and the inertia properties of the helicopter. The rotor blade is flexible in flap, lag, and torsion modes, and the natural frequencies of these modes vary with the rotor speed of the rotation. Under a particular condition, the rotor low-frequency cyclic lag mode couples with the body mode, leading to a resonance condition. When the helicopter is in flight, the pitch and roll modes of the fuselage are coupled to both the low-frequency flap and the low-frequency lag modes. Sometimes, the nature of coupling will be quite different in air than in ground. Since ground resonance is potentially more dangerous than air resonance, avoiding these instabilities is an important design consideration. Generally, an external damper is provided in the lead–lag mode to improve damping in the lag mode and avoid the instability. The resonance condition above 120% of the nominal rotor revolutions per minute (rpm) and below 30% of the rotor rpm is acceptable because the rotor rarely operates above 100% rpm. At a low rotor speed, the rotor has less energy, so one can pass through this resonance region without creating any large amplitude motion. Therefore, for a fairly large range of rotor operating condition, resonance has to be avoided. Analytical treatment of coupled rotor–fuselage problem for ground and air resonance is quite a complex problem because the model must properly take into account blade dynamics, aerodynamic effects, and the surface condition of the ground. However, for the purpose of understanding, one can use a very simple model of a rotor supported on a platform having lateral and longitudinal translational motions, as shown in Figure 11.1.

Rotor aerodynamics can be neglected for ground resonance simulation because it is assumed that the blade does not generate any lift on ground. Damping in rotor and support structure are usually of mechanical type, as shown in Figures 11.1 and 11.2. Such a simplified model captures the essential features of ground resonance phenomenon. The basic analysis of a simplified model was due to Coleman and Feingold (1958).

Consider a four-bladed rotor system supported on a platform (Figure 11.1), which can have longitudinal and lateral translational perturbation motions.

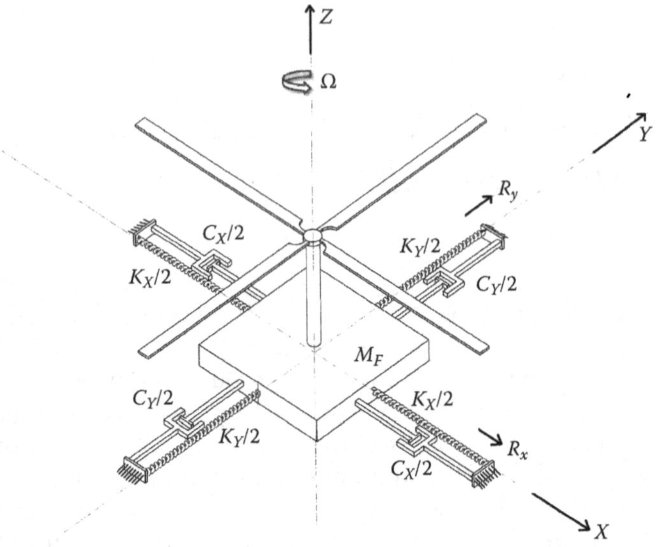

FIGURE 11.1
Simple model of rotor–fuselage system for ground resonance.

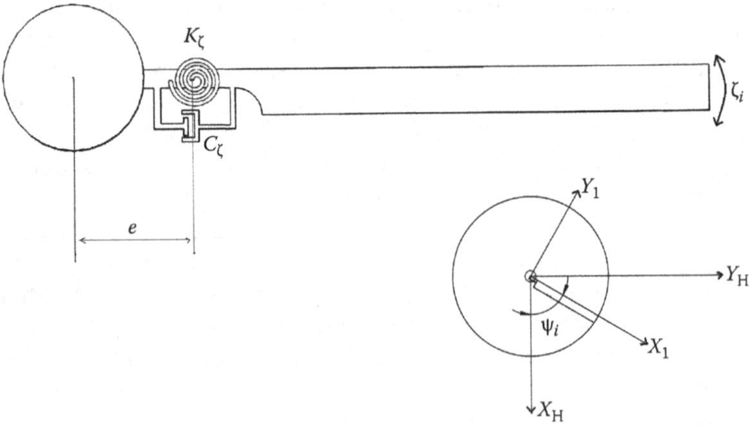

FIGURE 11.2
Idealized model of the rotor blade in the lag mode.

The blades are assumed to have only lag motion. The mass of the platform is M_F; the stiffness along the x-direction is K_x and along the y-direction is K_y; and the damping along x,y directions, assumed to be of viscous type, are C_x and C_y, respectively. The rotor blade can be idealized as a uniform rigid blade having a root offset with a root spring K_ς and a root damper C_ς, as shown in Figure 11.2. Following the mathematical procedure given in section on

Isolated Lag Dynamics in Chapter 6 and using hub translational motions in longitudinal and lateral directions, the equations of motion of the coupled rotor lag and fuselage dynamics can be derived.

The equation of the motion of the ith blade in the lag mode can be written as

$$\ddot{\zeta}_i + \frac{C_\zeta}{I_b\Omega}\dot{\zeta}_i + \left(\frac{K_\zeta}{I_b\Omega^2} + \frac{MX_{c.g}e}{I_b}\right)\zeta_i - \frac{MX_{c.g}R}{I_b}\left[\ddot{\bar{R}}_x \sin\psi_i - \ddot{\bar{R}}_y \cos\psi_i\right] = 0 \quad (11.1)$$

where

$$M = \text{mass of blade} = \int_e^R m\,dr$$

$$MX_{c.g} = \int_e^R r\ m\ dr$$

$$I_b = \int_e^R r^2\ m\ dr$$

$$\bar{R}_x = \frac{R_x}{R}, \bar{R}_y = \frac{R_y}{R}$$

$$\psi_i = \psi + \frac{2\pi}{N}(i-1) \text{ is azimuth location of the } i\text{th blade}$$

The equations of the motion of the platform are obtained by summing the root shears due to all the blades and applying to the platform. The force along the x-direction is given by

$$P_{Hx} = \sum_{i=1}^{N} \cos\psi_i P_{xi} - \sin\psi_i P_{yi} \quad (11.2)$$

where P_{xi} and P_{yi} are the root shear loads at the root of the ith blade. Substituting the root shear loads, Equation 11.2 can be written as

$$P_{Hx} = \sum_{i=1}^{N} \cos\psi_i \left\{M\Omega^2\left\{X_{c.g} + e + 2X_{c.g}\dot{\zeta}_i - \ddot{R}_x \cos\psi_i - \ddot{R}_y \sin\psi_i\right\}\right\}$$
$$- \sin\psi_i \left\{M\Omega^2\left\{X_{c.g}\zeta_i - X_{c.g}\ddot{\zeta}_i + \ddot{R}_x \sin\psi_i - \ddot{R}_y \cos\psi_i\right\}\right\} \quad (11.3)$$

The force along the y-direction is given as

$$P_{Hy} = \sum_{i=1}^{N} \sin\psi_i P_{xi} + \cos\psi_i P_{yi} \tag{11.4}$$

Substituting the root inertia loads of the blade, Equation 11.4 can be written as

$$\begin{aligned} P_{Hy} = \sum_{i=1}^{N} & \sin\psi_i \left\{ M\Omega^2 \left\{ X_{c.g} + e + 2X_{c.g}\dot{\zeta}_i - \ddot{R}_x \cos\psi_i - \ddot{R}_y \sin\psi_i \right\}\right\} \\ & + \cos\psi_i \left\{ M\Omega^2 \left\{ X_{c.g}\zeta_i - X_{c.g}\ddot{\zeta}_i + \ddot{R}_x \sin\psi_i - \ddot{R}_y \cos\psi_i \right\}\right\} \end{aligned} \tag{11.5}$$

Using multiblade coordinate transformation, the hub loads can be simplified as

$$P_{Hx} = M\Omega^2 \left\{ X_{c.g} \frac{N}{2} \ddot{\zeta}_{1s} - R\ddot{\bar{R}}_x N \right\} \tag{11.6}$$

$$P_{Hy} = M\Omega^2 \left\{ -X_{c.g} \frac{N}{2} \ddot{\zeta}_{1c} - R\ddot{\bar{R}}_y N \right\} \tag{11.7}$$

Assuming that the number of the blades $N = 4$, applying multiblade coordinate transformation to the blade lag equation (Equation 11.1), and writing only the cyclic mode equations (since collective and differential modes do not couple with body motion), we have four equations. They are two cyclic lag mode equations and two translational motion equations of the platform.

1 cosine cyclic lag mode:

$$\ddot{\zeta}_{1c} + \frac{C_\zeta}{I_b\Omega}\dot{\zeta}_{1c} + 2\dot{\zeta}_{1s} + \left(\bar{\omega}_{RL}^2 - 1\right)\zeta_{1c} + \frac{C_\zeta}{I_b\Omega}\zeta_{1s} + \frac{MX_{c.g}R}{I_b}\ddot{\bar{R}}_y = 0 \tag{11.8}$$

1 sine cyclic lag mode:

$$\ddot{\zeta}_{1s} + \frac{C_\zeta}{I_b\Omega}\dot{\zeta}_{1s} + 2\dot{\zeta}_{1c} + \left(\bar{\omega}_{RL}^2 - 1\right)\zeta_{1s} + \frac{C_\zeta}{I_b\Omega}\zeta_{1c} - \frac{MX_{c.g}R}{I_b}\ddot{\bar{R}}_x = 0 \tag{11.9}$$

where

$$\bar{\omega}_{RL} = \left\{ \frac{K_\zeta}{I_b\Omega^2} + \frac{MX_{c.g}e}{I_b} \right\}^{1/2}$$

is the rotating natural frequency of the blade in the lag mode.

Translational equations of the platform:

x-direction:

$$\Omega^2 M_F R\ddot{\bar{R}}_x + \Omega C_x R\dot{\bar{R}}_x + K_x R\bar{R}_x = M\Omega^2 \left\{ X_{c.g} \frac{N}{2} \ddot{\zeta}_{1s} - R\ddot{\bar{R}}_x N \right\} \tag{11.10}$$

y-direction:

$$\Omega^2 M_F R\ddot{\bar{R}}_y + \Omega C_y R\dot{\bar{R}}_y + K_y R\bar{R}_y = M\Omega^2 \left\{ -X_{c.g} \frac{N}{2} \ddot{\zeta}_{1c} - R\ddot{\bar{R}}_y N \right\} \tag{11.11}$$

Writing the Equations 11.8 to 11.11 in matrix form,

$$[M]\begin{Bmatrix} \ddot{\zeta}_{1c} \\ \ddot{\zeta}_{1s} \\ \ddot{\bar{R}}_x \\ \ddot{\bar{R}}_y \end{Bmatrix} + [C]\begin{Bmatrix} \dot{\zeta}_{1c} \\ \dot{\zeta}_{1s} \\ \dot{\bar{R}}_x \\ \dot{\bar{R}}_y \end{Bmatrix} + [K]\begin{Bmatrix} \zeta_{1c} \\ \zeta_{1s} \\ \bar{R}_x \\ \bar{R}_y \end{Bmatrix} = 0 \tag{11.12}$$

where the mass, damping, and stiffness matrices are given by

$$[M] = \begin{bmatrix} 1 & 0 & 0 & \dfrac{MX_{c.g}R}{I_b} \\ 0 & 1 & -\dfrac{MX_{c.g}R}{I_b} & 0 \\ 0 & -\dfrac{N}{2}\dfrac{X_{c.g}}{R} & \dfrac{M_F}{M}+N & 0 \\ \dfrac{X_{c.g}}{R}\dfrac{N}{2} & 0 & 0 & \dfrac{M_F}{M}+N \end{bmatrix}$$

$$[C] = \begin{bmatrix} \dfrac{C_\zeta}{I_b\Omega} & 2 & 0 & 0 \\ -2 & \dfrac{C_\zeta}{I_b\Omega} & 0 & 0 \\ 0 & 0 & \dfrac{C_x}{I_b\Omega} & 0 \\ 0 & 0 & 0 & \dfrac{C_y}{I_b\Omega} \end{bmatrix}$$

$$[K]=\begin{bmatrix} \bar{\omega}_{RL}^2-1 & \frac{C_\zeta}{I_b\Omega} & 0 & 0 \\ -\frac{C_\zeta}{I_b\Omega} & \bar{\omega}_{RL}^2-1 & 0 & 0 \\ 0 & 0 & \frac{K_x}{M\Omega^2} & 0 \\ 0 & 0 & 0 & \frac{K_y}{M\Omega^2} \end{bmatrix}$$

These are a set of homogeneous coupled differential equations. The eigenvalue of the system provides information about the stability of the system. Writing Equation 11.12 in a state-space form,

$$\{\dot{q}\}=[A]\{q\} \tag{11.13}$$

where the state vector and the system matrix are written as

$$\{q\}=\begin{Bmatrix} \zeta_{1c} \\ \zeta_{1s} \\ \bar{R}_x \\ \bar{R}_y \\ \dot{\zeta}_{1c} \\ \dot{\zeta}_{1s} \\ \dot{\bar{R}}_x \\ \dot{\bar{R}}_y \end{Bmatrix}$$

$$A=\begin{bmatrix} 0 & I \\ -M^{-1}K & -M^{-1}C \end{bmatrix}$$

Solving for the eigenvalues of matrix $[A]$ from the characteristic determinant,

$$|A-sI|=0$$

one can obtain information about the stability of the system. The eigenvalues s_i appear as complex conjugates $s_i=\sigma_i \pm i\,\omega_i$, where σ_i represents the damping and ω_i represents the frequency of the *i*th mode. When σ_i is negative, the mode is stable, and when σ_i is positive, the mode is unstable.

Ground resonance studies indicate that instability occurs when the rotor low-frequency lag mode couples with the body pitch mode or the roll mode. In addition, instability occurs only for soft in-plane or articulated rotor system and not for stiff in-plane rotors. (However, it may be noted that a stiff in-plane blade in one range of rotor speed will become a soft in-plane blade at a high range of rotor speed.) Using the data given in Table 11.1, ground resonance study can be performed. In the following, a frequency diagram depicting ground resonance phenomenon is shown in Figure 11.3. These results have been generated by assuming zero damping in the blade and the fuselage. Therefore, when the system is stable, the eigenvalues will be complex pairs with a zero real part ($\sigma_i = 0$) for all modes. When the system becomes unstable, some eigenvalues will be complex with a positive real part (σ_i), indicating instability. When the low-frequency lag mode coincides with body frequency, there is instability. The range of instability will shift to a high range of rotor speed of rotation as the lag frequency in the nonrotating state is increased, that is, $\left(\omega_{\mathrm{NRL}} = \sqrt{\dfrac{K_\zeta}{I_\mathrm{b}}}\right)$.

This simple example problem depicts the nature of ground resonance in helicopters. A detailed analysis must include the aerodynamic loads and the flap modes, in addition to body modes and lag modes.

TABLE 11.1

Data Used in the Example Problem

Parameter	Value	Units
M_F: fuselage mass	3500	kg
M: blade mass	60	kg
N: number of rotor blades	4	–
R: radius of the rotor disk	6	m
e: hinge offset	0	m
$Xc.g$: center of gravity	3	m
I_b: blade moment of inertia	720	kg–m^2
C_x: damping coefficient in the longitudinal mode	0	N/(m/s)
C_y: damping coefficient in the lateral mode	0	N/(m/s)
C_z: lag damping coefficient (in rotating frame)	0	N.m/(rad/s)
Nonrotating lag frequency	3	Hz
Rotor angular speed	50–2000	rpm
Kf_x: stiffness factor in the longitudinal mode	(ω_x**2*Mf)	kg–(rad/sec)^2
Kf_y: stiffness factor in the lateral mode	(ω_y**2*Mf)	kg–(rad/sec)^2
ω_x: natural frequency in the longitudinal mode	2	Hz
ω_y: natural frequency in the lateral mode	4	Hz

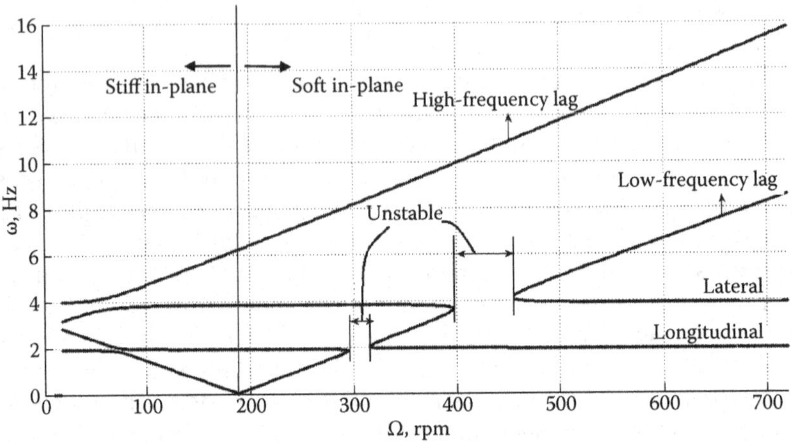

FIGURE 11.3
Variation of an imaginary part of eigenvalues as a function of rotor rpm.

Several studies on ground and air resonance problems address the effect of aeroelastic couplings on lag mode damping and how aeroelastic couplings can be tailored to improve the inherent damping in the lag mode. There are also several other problems associated with the practical aspect of operating the helicopter under different ground conditions, such as concrete, grass, snow, etc.

Bibliography*

Bauchau, O.A. 1985. A beam theory for anisotropic materials. *Journal of Applied Mechanics* 52:416–422.

Bauchau, O.A. and C.H. Hong. 1987a. Finite element approach to rotor blade modeling. *Journal of American Helicopter Society* 32:60–67.

Bauchau, O.A. and C.H. Hong. 1987b. Large displacement analysis of naturally curved and twisted composite beams. *AIAA Journal* 25:1469–1475.

Bauchau, O.A. and C.H. Hong. 1988. Non-linear composite beam theory. *Journal of Applied Mechanics* 55:156–163.

Bielawa, R.L. 1992. Rotary wing structural dynamics and aeroelasticity. *AIAA Series.*

Bousman, W.G. 1981. An experimental investigation of the effects of aeroelastic couplings on aeromechanical stability of a hingeless rotor helicopter. *Journal of the American Helicopter Society* 26:46–54.

Bramwell, A.R.S. 1976. *Helicopter Dynamics.* London, United Kingdom: Edward Arnold Publications.

Celi, R. 1991. Helicopter rotor dynamics in coordinated turns. *Journal of the American Helicopter Society* 36(4).

Cheeseman, I.C. and W.E. Bennett. 1955. The effect of the ground on a helicopter rotor in forward flight. *ARC R&M* 3021.

Chen, R.T.N. 1984. Flight dynamics of rotorcraft in steep high-g turns. *Journal of Aircraft* 21(1).

Chen, R.T.N. 1990. A survey of non-uniform inflow models for rotorcraft flight dynamics and control applications. *Vertica* 14:147–184.

Chen, R.T.N. and W.S. Hindson. 1987. Influence of dynamic inflow on the helicopter vertical response. *Vertica* 11:77–91.

Chopra, I. 1900. Perspectives in aeromechanical stability of helicopters. *Vertica* 14(4):457–508.

Coleman, R.P., A.M. Feingold, and C.W. Stempin. 1945. Evaluation of the induced velocity field of an idealized helicopter rotor. *NACA ARR* L5E10.

Cooper, G.E. and R.P. Harper, Jr. 1969. The use of pilot ratings in the evaluation of aircraft handling qualities. *NASA TM* D-5133.

Crespo da Silva, M.R.M. and D.H. Hodges. 1986a. Non-linear flexure and torsion of rotating beams, with application to helicopter blades: I. Formulation. *Vertica* 10:151–169.

Crespo da Silva, M.R.M. and D.H. Hodges. 1986b. Non-linear flexure and torsion of rotating beams, with application to helicopter blades: II. Results for hover. *Vertica* 10:171–186.

Drees, J.M. 1949. A theory of airflow through rotors and its application to some helicopter problems. *Journal of the Helicopter Association of Great Britain* 3:2.

* There are several research publications on various topics related to rotor blade modeling, rotor blade aeroelasticity, blade stability, flight dynamics, and aeromechanical stability. In the following, only a few sample publications are provided for reference. The reader may refer to the books (Bramwell, 1976; Johnson, 1980; Padfield,1996; and Leishman, 2000) given here for an exhaustive list of publications.

Du Val, R. 1989. A real-time blade element helicopter simulation for handling qualities analysis. *Proceedings of the 15th Annual European Rotorcraft Forum*, Amsterdam, Netherlands.

Friedmann, P. and P. Tong. 1972. Dynamic non-linear elastic stability of helicopter rotor blades in hover and forward flight (ASRL-TR-116-3, Massachusetts Institute of Technology; NAS2-6175). *NASA* CR-114485.

Friedmann, P. and P. Tong. 1973. Non-linear flap–lag dynamics of hingeless helicopter blades in hover and in forward flight. *Journal of Sound and Vibration* 30:9–31.

Friedmann, P.P. 1999. Renaissance of aeroelasticity and its future. *Journal of Aircraft* 36:105–121.

Friedmann, P.P. 2004. Rotary wing aeroelasticy: Current status and future trends. *AIAA Journal* 42:1953–1972.

Friedmann, P.P. and C. Venkatesan. 1985. Coupled helicopter rotor/body aero mechanical stability: Comparison of experimental and theoretical results. *Journal of Aircraft* 22:148–155.

Friedmann, P.P. and C. Venkatesan. 1986. Influence of unsteady aerodynamic models on aeromechanical stability in ground resonance. *Journal of the American Helicopter Society* 31:65–74.

Friedmann, P.P. and D.H. Hodges. 2003. Rotary wing aeroelasticy: A historical perspective. *Journal of Aircraft* 40:1019–1046.

Gandhi, F. and E. Hathaway. Optimized aeroelastic couplings for alleviation of helicopter ground resonance. *Journal of Aircraft* 35:582–590.

Gaonkar, G.H. and D.A. Peters. 1980. Use of multi-blade coordinates for helicopter flap–lag stability with dynamic inflow. *Journal of Aircraft* 17:112–118.

Gaonkar, G.H. and D.A. Peters. 1986. Effectiveness of current dynamic inflow models in hover and forward flight. *Journal of the American Helicopter Society* 31:47–57.

Gessow, A. and G.C. Myers. 1952. *Aerodynamics of Helicopter*. New York: Frederick Unger Publishing Co.

Glauert, H. 1926. A general theory of the autogyro. *Rep. Memo. Aeronautical Research Council* 1111.

Greenberg, J.M. 1947. Airfoil in sinusoidal motion in pulsating stream. *NACA TN* 1326.

Hansford, R.E. and I.A. Simons. 1973. Torsion–flap–lag coupling on helicopter rotor blades. *Journal of the American Helicopter Society* 18.

Heffley, R.K. and M.A. Mnich. 1987. Minimum complexity helicopter simulation math model. *NASA* CR-177476.

Hodges, D.H. 1990. Review of composite rotor blade modeling. *AIAA Journal* 28:561–564.

Hodges, D.H. and E.H. Dowell. 1974. Non-linear equations of motion for the elastic bending and torsion of twisted non-uniform rotor blades. *NASA TN* D-7818.

Hodges, D.H. and R.A. Ormiston. 1976a. Stability of elastic bending and torsion of uniform cantilever rotor blades in hover with variable structural coupling. *NASA TN* D-8192.

Hodges, D.H. and R.A. Ormiston. 1976b. Stability of elastic bending and torsion of uniform cantilever rotor blades in hover with variable elastic structural coupling. *NASA TN* D-8192.

Hohenemser, K.H. and W. Paul. 1967. Aeroelastic instability of torsionally rigid helicopter blades. *Journal of the American Helicopter Society* 12:1–13.

Hong, C.H. and I. Chopra. 1986. Aeroelastic stability analysis of a bearingless composite blade in hover. *Journal of the American Helicopter Society* 31(4):29–35.

Houbolt, J.C. and G.W. Brooks. 1958. Differential equations of motion for combined flap-wise bending, chordwise bending, and torsion of twisted nonuniform rotor blades. *NACA Report* 1346.

Huber, H. Effect of torsion–flap–lag coupling on hingeless rotor helicopter stability. *29th American Helicopter Society Forum*, Washington, D.C., AHS Preprint 371.

Huber, H.G. 1973. Effect of torsion–flap–lag coupling on hingeless rotor stability. *29th Annual National Forum of the American Helicopter Society*, Washington, D.C., AHS Preprint 731.

Johnson, W. 1977. Flap/lag/torsion dynamics of a uniform, cantilever rotor blade in hover. *NASA* TM-73248.

Johnson, W. 1980. *Helicopter Theory*. New Jersey: Princeton University Press.

Johnson, W. 1982. Influence of unsteady aerodynamics of hingeless rotor in ground resonance. *Journal of Aircraft* 19:668–673.

Johnson, W. 1988. *A Comprehensive Analytical Model of Rotorcraft Aerodynamics and Dynamics, Volume I: Theory Manual, and Volume II: User's Manual*, Palo Alto, California: Johnson Aeronautics.

Johnson, W. 1998. Rotorcraft aerodynamics models for comprehensive analysis. *Proceedings of the 54th Annual Forum of the American Helicopter Society*, Washington D.C.

Kaza, K.R. and R.G. Kvaternik. 1977. Non-linear aeroelastic equations for combined flapwise bending, chordwise bending, torsion, and extension of twisted non-uniform rotor blades in forward flight. *NASA* TM-74059.

Kaza, K.R.V. and R.G. Kvaternik. 1979. Examination of the flap–lag stability of rigid articulated rotor blades. *Journal of Aircraft* 16.

Kim, K.C. 2004. *Analytical Calculation of Helicopter Main Rotor Blade Flight Loads in Hover and Forward Flight*. Army Research Laboratory.

Kunz, D.L. 1994. Survey and comparison of engineering beam theories for helicopter rotor blades. *Journal of Aircraft* 31:473–479.

Laxman, V. 2008. Formulation of a computational aeroelastic model for the prediction of trim and response of a helicopter rotor system in forward flight. PhD thesis, Department of Aerospace Engineering, IIT Kanpur.

Leishman, J.G. 1988. Two-dimensional model for airfoil unsteady drag below stall. *Journal of Aircraft* 25:665–666.

Leishman, J.G. 2000. *Principles of Helicopter Aerodynamics*. Cambridge University Press.

Loewy, R.G. 1957. A two-dimensional approximation to unsteady aerodynamics of rotary wings. *Journal of the Aeronautical Science* 24.

Mangler, K.W. and H.B. Squire. 1950. The induced velocity field of a rotor. *Rep. Memo. Aeronautical Research Council* 2642.

McCormick, B.W. 1979. *Aerodynamics, Aeronautics, and Flight Mechanics*. John Wiley & Sons.

Nagabhushanam, J. and G.H. Gaonkar. 1984. Rotorcraft air resonance in forward flight with various dynamic inflow models and aeroelastic couplings. *Vertica* 8:373–394.

Ormiston, R.A. 1977. Aeromechanical stability of soft in-plane hingeless rotor helicopters. *3rd European Rotorcraft and Powered Lift Aircraft Forum*, Aix-en-Provence, France, p. 25.1–25.22.

Ormiston, R.A. 1991. Rotor–fuselage dynamics of helicopter air and ground resonance. *Journal of the American Helicopter Society* 36:3–20.
Ormiston, R.A. and D.H. Hodges. 1972. Linear flap–lag dynamics of hingeless helicopter rotor blades in hover. *Journal of the American Helicopter Society* 17:2–14.
Padfield, G.D. 1996. Helicopter flight dynamics: The theory and application of flying qualities and simulation modeling. *AIAA Series*.
Palika, S. 2012. *Structural Dynamic Analysis of Articulated, Hingeless, and Bearing-less Rotor Systems*. Department of Aerospace Engineering, IIT Kanpur.
Panda, B. and I. Chopra. 1985. Flap-lag-torsion stability in forward flight. *Journal of the American Helicopter Society* 30(4):30–39.
Peters, D.A. 1975. Flap–lag stability of helicopter rotor blades in forward flight. *Journal of the American Helicopter Society* 17.
Peters, D.A., D.D. Boyd, and C.J. He. 1989. Finite-state induced-flow model for rotors in hover and forward flight. *Journal of the American Helicopter Society* 34:5–17.
Pitt, D.M. and D.A. Peters. 1983. Rotor dynamic inflow derivatives and time constants from various inflow models. *Proceedings of the 9th European Rotorcraft Forum*, Stresa, Italy.
Pohit, G., C. Venkatesan, and A.K. Mallik. 2000. Influence of non-linear elastomer on isolated lag dynamics and rotor/fuselage aeromechanical stability. *Proceedings of the 41st AIAA/ASME/ASCE/AHS/ASC Structures, Structural Dynamics and Materials Conference*, Atlanta, Georgia, April 3–6, 2000, AIAA Paper 2000-1691.
Prouty, R.W. 1990. *Helicopter Performance, Stability, and Control*. Florida: R.E. Krieger Publishing Co.
Rohin Kumar, M. 2014. Comprehensive aeroelastic and flight dynamic formulation for the prediction of loads and control response of a helicopter in general maneuvering flight. PhD Thesis, Department of Aerospace Engineering, IIT-Kanpur.
Rosen, A. and P.P. Friedmann. 1978. Non-linear equations of equilibrium for elastic helicopter or wind turbine blades undergoing moderate deflection. *NASA* CR-159478.
Rosen, A. and P.P. Friedmann. 1979. The non-linear behavior of elastic slender straight beams undergoing small strains and moderate rotations. *Journal of Applied Mechanics* 46:161–168.
Seddon, J. 1990. Basic helicopter aerodynamics. *AIAA Series*.
Shamie, J. and P.P. Friedmann. 1977. Effect of moderate deflections on the aero-elastic stability of rotor blades in forward flight. *Proceedings of the 3rd European Rotorcraft and Powered Lift Aircraft Forum*, Aix-en-Provence, France, p. 24.1–24.37.
Singh, G. 2012. Helicopter flight dynamics simulation for analysis of trim, stability and control response. M. Tech. thesis, Department of Aerospace Engineering, IIT Kanpur.
Sivaneri, N.T. and I. Chopra. 1982. Dynamic stability of a rotor blade using finite element analysis. *AIAA Journal* 20:716–723.
Smith, E.C. and I. Chopra. 1991. Formulation and evaluation of an analytical model for composite box beams. *Journal of the American Helicopter Society* 36:23–35.
Stepniewski, W.Z. 1984. *Rotary Wing Aerodynamics*, Vols. 1 and 2. Dover Publications.
Theodorsen, T. 1935. General theory of aerodynamic instabilities and the mechanism of flutter. *NACA Report* 496.
Uma Maheswaraiah, M. 2000. Dynamic analysis of bearingless rotor blade with swept tips. M. Tech. thesis, Department of Aerospace Engineering, IIT Kanpur.

Vadivazhagan, G. 2012. Flap–lag and flap–torsion stability analyses of helicopter rotor blade in hover. M. Tech thesis, Department of Aerospace Engineering, IIT Kanpur.

Venkatesan, C. 1994. Treatment of axial mode in structural dynamics and aeroelastic analysis of rotor blade: A review. *Proceedings of the NASAS 94: Computational Structural Mechanics*, IIT Kharagpur, December 8–9, 1994, p. 361–372.

Venkatesan, C. 1995. Influence of aeroelastic couplings on coupled rotor/body dynamics. *6th International Workshop on the Dynamics and Aeroelastic Stability Modeling of Rotorcraft Systems*, University of California, Los Angeles, November 8–10, 1995.

Venkatesan, C. and P.P. Friedmann. 1984. Aeroelastic effects in multi-rotor vehicles with application to a hybrid heavy lift system: Part I. Formulation of equations of motion. *NASA* CR-3822.

Venkatesan, C. *Lecture Notes: Helicopter Technology*. Department of Aerospace Engineering, IIT Kanpur.

Worndle, R. 1982. Calculation of the cross-section properties and the shear stresses of composite rotor blades. *Vertica* 6:111–129.

Yeager, W.T., M.H. Hamouda, and W.R. Mantay. 1987. An experimental investigation of the aeromechanical stability of a hingeless rotor in hover and forward flight. *Journal of the American Helicopter Society* 28.

Yen, J.G., J.J. Corrigan, J.J. Schillings, and P.Y. Hsieh. Comprehensive analysis methodology at bell helicopter: COPTER. *Proceedings of the AHS Aeromechanics Specialists Conference*, San Francisco, California.

Yuan, K.A., C. Venkatesan, and P.P. Friedmann. 1992. Structural dynamic model of composite rotor blade undergoing moderate deflections. In *Recent Advances in Structural Dynamic Modeling of Composite Rotor Blades and Thick Composites, ASME AD* 30:127–155.

Zotto, M.D. and R.G. Loewy. 1992. Influence of pitch–lag coupling on damping requirements to stabilize ground/air resonance. *Journal of the American Helicopter Society* 37:68–71.

Index

Page numbers followed by f and t indicate figures and tables, respectively.

Printed in the United States
by Baker & Taylor Publisher Services